VHDL Techniques, Experiments, and Caveats

Computer Engineering Series

CHEN • *Computer Engineering Handbook,* 0-07-010924-9
CHEN • *Fuzzy Logic and Neural Network Handbook,* 0-07-073020-2
DEVADAS, DEUTZER, GHASH • *Logic Synthesis,* 0-07-016500-9
HOWLAND • *Computer Hardware Diagnostics for Engineers,* 0-07-030561-7
LEISS • *Parallel and Vector Computing,* 0-07-037692-1
PERRY • *VHDL,* Second Edition, 0-07-049434-7
ROSENSTARK • *Transmission Lines in Computer Engineering,* 0-07-053953-7
ZOMAYA • *Parallel and Distributed Computing Handbook,* 0-07-011184-8

Related Titles of Interest

KIELKOWSKI • *Inside SPICE,* 0-07-911525-X
KIELKOWSKI • *SPICE Practical Parameter Modeling,* 0-07-911524-1
MASSABRIO, ANTOGNETTI • *Semiconductor Device Modeling with SPICE,* Second Edition, 0-07-002469-3

To order or receive additional information on these or any other McGraw-Hill titles, in the United States please call 1-800-822-8158. In other countries, contact your local McGraw-Hill representative. **BC15XXA**

VHDL Techniques, Experiments, and Caveats

Joseph Pick
Synopsys Inc.

McGraw-Hill, Inc.
New York San Francisco Washington, D.C. Auckland Bogotá
Caracas Lisbon London Madrid Mexico City Milan
Montreal New Delhi San Juan Singapore
Sydney Tokyo Toronto

Library of Congress Cataloging-in-Publication Data

Pick, Joseph.
　　VHDL techniques, experiments, and caveats / Joseph Pick.
　　　　p.　　cm.
　　Includes bibliographical references and index.
　　ISBN 0-07-049906-3 (HC)
　　1. VHDL (Computer hardware description language)　I. Title.
TK7885.7.P53　　1995
621.39′2—dc20　　　　　　　　　　　　　　　　　95-22160
　　　　　　　　　　　　　　　　　　　　　　　　　　CIP

Copyright © 1996 by McGraw-Hill, Inc. All rights reserved. Printed in the United States of America. Except as permitted under the United States Copyright Act of 1976, no part of this publication may be reproduced or distributed in any form or by any means, or stored in a data base or retrieval system, without the prior written permission of the publisher.

1 2 3 4 5 6 7 8 9 0　DOC/DOC　9 0 0 9 8 7 6 5

ISBN 0-07-049906-3

The sponsoring editor for this book was Stephen S. Chapman, the editing supervisor was Stephen M. Smith, and the production supervisor was Donald R. Schmidt. It was set in Century Schoolbook by Cynthia L. Lewis of McGraw-Hill's Professional Book Group composition unit.

Printed and bound by R. R. Donnelley & Sons Company.

McGraw-Hill books are available at special quantity discounts to use as premiums and sales promotions, or for use in corporate training programs. For more information, please write to the Director of Special Sales, McGraw-Hill, Inc., 11 West 19th Street, New York, NY 10011. Or contact your local bookstore.

This book is printed on acid-free paper.

Information contained in this work has been obtained by McGraw-Hill, Inc., from sources believed to be reliable. However, neither McGraw-Hill nor its authors guarantee the accuracy or completeness of any information published herein, and neither McGraw-Hill nor its authors shall be responsible for any errors, omissions, or damages arising out of use of this information. This work is published with the understanding that McGraw-Hill and its authors are supplying information, but are not attempting to render engineering or other professional services. If such services are required, the assistance of an appropriate professional should be sought.

This book is dedicated to my sons Benji (The Origami Kid) and Daniel (The Fortuitous One), who have re-acquainted me with the wonders and magic of childhood.

Contents

Preface   xi
Acknowledgments   xv

Part 1   An Excursion into VHDL

Chapter 1. Embarkment   3

  1.1 Design Entity   3
  1.2 Architecture Body   5
  1.3 Our Traveling Companion   7

Chapter 2. The Journey   11

  2.1 File: introductory_example.vhd   11
  2.2 File: intro_static_sensitivity.vhd   29
  2.3 File: intro_concur_signal_assignment.vhd   32
  2.4 File: intro_delta_time.vhd   34
  2.5 File: cfg_introductory.vhd   44
  2.6 File: structural_decomposition.vhd   47
  2.7 File: count_3_bit_.vhd   58
  2.8 File: count_3_bit_model_1.vhd   60
  2.9 File: count_3_bit_model_2.vhd   67
  2.10 File: count_3_bit_model_2a.vhd   72
  2.11 File: count_3_bit_model_3.vhd   72
  2.12 File: count_3_bit_model_4.vhd   77
  2.13 File: count_3_bit_model_5.vhd   82
  2.14 File: count_3_bit_model_6.vhd   86
  2.15 File: count_3_bit_model_7.vhd   89
  2.16 File: clock.vhd   93
  2.17 File: count_monitor.vhd   101
  2.18 File: cfg_cnt3_m1.vhd   108
  2.19 File: count_n_bits_.vhd   113
  2.20 File: count_n_bits_model_1.vhd   116
  2.21 File: cfg_cntn_m1.vhd   118
  2.22 Logic Gates and Flip-Flops   120

2.23	File: count_3_bit_model_gates_1.vhd	127
2.24	File: cfg_cnt3_model_gates_1.vhd	131
2.25	File: cfg_cnt3_cfg.vhd	134
2.26	File: mvl4_pkg_.vhd	136
2.27	Excerpt of File: mvl4_pkg.vhd	145
2.28	File: count_3_bit_rf_.vhd	152
2.29	File: count_3_bit_rf_model_1.vhd	153
2.30	File: count_monitor_rf.vhd	157
2.31	File: encapsulate_rf_components_pkg_.vhd	160
2.32	File: encapsulate_cnt_rf.vhd	162
2.33	File: cfg_cnt_rf.vhd	164
2.34	Baggage Reduction with 1164	164
2.35	Journey's Epilogue	169

Part 2 Driving the Simulation Flow

Chapter 3. Signal-Updating Algorithms — 175

Chapter 4. Predefined Data Type Attributes — 181

Chapter 5. Predefined Signal Attributes — 189

5.1 Signal Attribute Updates — 191
5.2 Application of Signal Attributes — 197

Part 3 VHDL Techniques and Recommendations

Chapter 6. Compilation Caveats — 209

Chapter 7. Simulation Caveats — 231

Chapter 8. Model Efficiency — 247

8.1 The Efficiency War Room — 247
8.2 Examples — 248

Chapter 9. Type Conversion Tricks and Methodologies — 263

Part 4 Experimenting with VHDL

Chapter 10. Overview — 275

10.1 Background — 275
10.2 Conducting the Experiments — 276

Chapter 11. The Experiments — 277

11.1 Test_3 (wait until...) — 277
11.2 Test_5 (wait on...until...) — 278

11.3	Test_7 (Simulation cycle)	279
11.4	Test_11a (wait until compound expression)	280
11.5	Test_13 (Infinite oscillation in delta time domain)	282
11.6	Test_13a (Deadlock)	283
11.7	Test_15 (Subtype constraint checking)	283
11.8	Test_16 (Blocks and 'EVENT vs. 'STABLE)	285
11.9	Test_18b (Signal driver in inertial model)	286
11.10	Test_18c (Glitches in inertial model)	288
11.11	Test_18d (Glitches in transport model)	289
11.12	Test_113 (Transport driver preemption)	290
11.13	Test_18e (More glitches in inertial model)	292
11.14	Test_28 (Multiple array inputs)	294
11.15	Test_28a (Alias and multiple array inputs)	295
11.16	Test_27d (Single array inputs)	296
11.17	Test_63 (Procedure location option)	298
11.18	Test_110b (Procedure constraint)	299
11.19	Test_64 ('TRANSACTION in procedures)	300
11.20	Test_64a ('TRANSACTION of formal parameters)	301
11.21	Test_65 (Solution for Test_64a)	302
11.22	Test_67a (Nonefficient command decoding)	303
11.23	Test_67b (Efficient command decoding)	303
11.24	Test_68 (File declaration: Case 1)	305
11.25	Test_69 (File declaration: Case 2)	306
11.26	Test_74 ('TRANSACTION and resolved signals)	307
11.27	Test_105 (Function slices)	308
11.28	Test_120 (Circuit oscillation)	309
11.29	Test_121 (Solution 1 for Test_120)	312
11.30	Test_118 (Solution 2 for Test_120)	315
11.31	Test_89 (TIME to NATURAL)	318
11.32	Test_1 (Reference for Test_1a, 1b, 1d, and 1e)	318
11.33	Test_1a (Unexpected multisources: Case 1)	319
11.34	Test_1b (Generics and array signal drivers)	321
11.35	Test_1d (Unexpected multisources: Case 2)	322
11.36	Test_1e (Unexpected multisources: Case 3)	323
11.37	Test_124_0 (Port map actuals)	325
11.38	Test_124 (Solution for Test_124_0)	326
11.39	Test_158 (Scope and visibility)	326
11.40	Test_159 (Solution for Test_158)	328
11.41	Test_190 (Scope and visibility [cont])	328
11.42	Test_162 (& in case selector)	330
11.43	Test_172 (Reading and writing via the same buffer line)	332
11.44	Test_178 (Overriding of generic defaults)	333
11.45	Test_179 (Overriding of generic defaults [cont])	335
11.46	Test_59d (Overriding of out port initial values)	336
11.47	Test_182 (Behavior of 'DELAYED(25 ns))	340
11.48	Test_184 (Constant assignment via a function call)	341
11.49	Test_185 (Reading both array size and data from same file)	342
11.50	Test_187 (Usage of others in subprograms)	344
11.51	Test_188 (Solution for Test_187)	345

- 11.52 Test_194 (Constancy of for loop bounds) 345
- 11.53 Test_195 (Dangers of the artificial usage of inout ports) 346
- 11.54 Test_35_0 (The signal multiplexing problem) 348
- 11.55 Test_35 (Standard solution to scalar signal multiplexing) 349
- 11.56 Test_35a (Side effects of Test_35) 351
- 11.57 Test_39 (Solution for Test_35a) 352
- 11.58 Test_35b (Array signal multiplexing: error) 353
- 11.59 Test_35l (Brute force solution for Test_35b) 354
- 11.60 Test_40 (Counterexample for Test_35l's solution) 356
- 11.61 Test_35b_1 (Ideal solution for Test_35b) 357
- 11.62 Test_35j (Shadow signals and 'LAST_VALUE) 358
- 11.63 Test_35k (Array signal multiplexing: OK) 361

Appendix A Responsibilities of the In-House VHDL Guru(s)   365
Appendix B Chronology of a VHDL Modeling Project   367
Appendix C Code Walk-through Checkoff List   369
Appendix D VHDL'87 Reserved Words   373
Appendix E VHDL'93 Reserved Words   375
References   377
Index   379

Preface

The aim of this book is to enhance your understanding of the IEEE standard VHSIC Hardware Description Language (VHDL). This software medium is rapidly becoming a prominent component of the electronics industry.

VHDL was initiated by the U.S. Department of Defense to overcome the following obstacles that occurred during its procurement of digital systems:

- Lack of a standard medium for interchanging digital design information.
- Lack of a standard pathway for reimplementing electronics parts.
- Lack of a standard approach to determine which commercial off-the-shelf parts to use.
- Lack of reusable hardware description language (HDL) software components.
- Lack of an electronic communication medium in the procurement cycle.
- Lack of portable simulation models for digital systems.
- Lack of efficient techniques to replace obsolete electronic components.

The common denominator for all these deficiencies was the absence in the electronics industry of a powerful, standard communications medium. The chief objective of VHDL is to effectively fill this void.

Originally, VHDL was intended as a documentation strategy for capturing and preserving a digital device's specification in an industry-wide format that is both human and computer readable. Gradually, it became apparent that VHDL may also serve as a common language to facilitate a cohesive top-down methodology that seamlessly connects the design, simulation, synthesis, and test iteration cycles.

In 1987 VHDL was sanctioned as the official IEEE standard hardware description language. It was designated as the IEEE standard

1076-1987, and a manual was written detailing its specifictions (Ref. 1). Unfortunately this document, known as the VHDL *Language Reference Manual* (LRM), is very terse and difficult to read. It also contains numerous errors and ambiguities. According to the IEEE bylaws, any standard may be updated every five years and reapproved via a balloting procedure. So in 1992 the LRM was modified to incorporate not only clarifications but also new features and enhancements. This language update was colloquially referred to as VHDL'92 by the VHDL community. In mid-1993 Draft B of this language revision was formally approved, but it took several more months to iron out and resolve some sensitive issues such as the implementation mechanics for global variables. In September 1993 the VHDL language committee announced that a compromise had finally been reached on this matter. Since September was late into the year 1992, it no longer made sense to associate 1992 with this VHDL update. Hence this new version is now officially referred to as the IEEE standard 1076-1993. The VHDL community colloquially identifies this new language release as VHDL'93. In late 1994 the final version of the VHDL'93 LRM (Ref. 2) became available from the IEEE. However, it will probably be another year before all the VHDL tool suites will fully comply with this new standard. Until then, you and I will have to live and work with the original 1987 version of the VHDL language. As suggested by its title, this book aims to present universal VHDL coding techniques and caveats that are applicable to both 1987 and 1993 standards.

VHDL is a very robust and powerful HDL. Its mastery requires an innovative learning process that stresses practical real-world modeling techniques and caveats. Abstract VHDL concepts should be seamlessly merged with practical applications in a way that will maximize the users' productivity during the design, test, and debug iteration cycles. These pedagogical guidelines are the primary factors driving the contents of this book.

When it comes to mastering a new subject, you are probably like me. Our technical backgrounds have conditioned us to learn the most effectively via an in-depth analysis of highly descriptive examples. Consequently, this book contains a plethora of realistic VHDL source code samples. Pragmatic VHDL techniques are explicitly demonstrated so that you will not have to expend time and energy to reinvent them. A multitude of potential design and coding errors are exposed via examples, and then their creative solutions are described in full detail. By reading this text, you will benefit from the many techniques and caveats that I have accumulated over the years as a VHDL modeler, advisor, and trainer. This book will improve your VHDL productivity by strengthening your overall VHDL problem-solving skills.

The intended audience for this book spans the full spectrum of VHDL expertise and job responsibilities. Newcomers to the VHDL realm will discover in this book an innovative learning experience in which the big picture is presented right from the start via a series of tightly coupled models. All subsequent topics will therefore be viewed from a more mature and meaningful perspective. Those readers already working with VHDL will further enrich their problem-solving skills and will substantially benefit from the many advanced design and optimization techniques illustrated in this book.

The pragmatic skills gained from this text will help VHDL designers and engineers to enhance their on-the-job performance and productivity. Applications engineers and customer support personnel will be provided with a repertoire of VHDL techniques and caveats that they may reference when responding to the needs of their customers. Not only will they be able to correctly resolve a customer's query, but they also will be in a position to solidify the customer's confidence in both them and their VHDL-related product. University professors and researchers will appreciate that the fundamentals of the language are quickly covered in the first part of the book. This allots them a sufficient amount of time during the remainder of the semester to apply VHDL as a tool to complement the main thrust of their course or research activities. The rest of the book may subsequently be used as a reference source to better understand a specific topic or concept. Moreover, the real-world approach of this book will prepare university students to successfully integrate into an electronics industry that is now relying more and more heavily on VHDL.

Book Overview and Organization

This book may be viewed as a logic analyzer into VHDL that captures and vividly displays the language's characteristics and capabilities. It emphasizes what can be done, what cannot be done, and what should be done. In all three of these areas, the book is not exhaustive. How could it be? But it does strive to present those topics and issues that you will more than likely come across in your day-to-day VHDL activities.

This book is divided into four tightly coupled parts that reenforce and enhance each other.

The first part, titled "An Excursion into VHDL," will very quickly introduce you to the essential features and concepts of this language. The Excursion uses an innovative top-down approach to explain how VHDL's individual constructs fit together and interact. A multitude of complete VHDL models will systematically guide you from an abstract top-level description all the way down to a gate-level implementation. This examples-oriented introduction will give you a better appreciation for the subsequent developments in this book.

The second part, titled "Driving the Simulation Flow," will explore several of the key concepts and internal implementation algorithms that drive and control the progress of a VHDL simulation. This segment will give you a deeper insight into the underlying mechanics that implement the updating of signals.

The third part, titled "VHDL Techniques and Recommendations," consists of an in-depth examination of pragmatic coding scenarios. The main objective is to improve your VHDL problem-solving skills by presenting a series of examples all based on real-world VHDL modeling experiences and issues. The pedagogical flow from the Excursion part into the Techniques and Recommendations segment is a key element of this book. Without the Excursion, the second part would merely be a series of isolated dictionary-like snapshots of the VHDL language. However, having gone through the Excursion, you are now knowledgeable of the big picture and are well aware of the overall significance of the topics presented in this Techniques and Recommendations segment.

The final part, titled "Experimenting with VHDL," is a collection of hands-on exercises that both challenge and reenforce your understanding of how the various VHDL constructs behave and interact.

Common to all four parts is a desire to prepare you to successfully design and code in VHDL. Potential errors based on real-world coding scenarios are continually exposed and resolved.

The following three main appendexes coincide with this book's recurring theme of always touching reality:

- Responsibilities of the In-House VHDL Guru(s)
- Chronology of a VHDL Modeling Project
- Code Walk-through Checkoff List

In the final analysis I view myself first and foremost as an educator. When designing and writing this book, I was motivated by the desire that its contents should, above all, become an integral part of your VHDL education and development. I sincerely hope that you will find this book to be a valuable and entertaining journey into the real-world rhythm of VHDL.

E-Mail Connections

If you would like to discuss the content of this book or to get your own personal copy of the Excursion and Experiments models, please contact me at

jpick@synopsys.com

Joseph Pick

Acknowledgments

Many thanks to the very first VHDL team that I trained and advised at my alma mater, Westinghouse. That period is a very special part of my life, and I will always remember it fondly.

I would also like to thank the following former COMPASS/CLSI colleagues for those fascinating discussions we had regarding the many fine points of VHDL: Ken Bakalar, Daniel Barclay, Kevin Cameron, Paul Graham, Iain Finlay, Erich Marschner, Shanka Mitra, and Mitch Perilstein.

Dr. Iain Finlay deserves special praise for mastering the content and philosophy of one of my VHDL workshops. We trained together diligently for several weeks, and the results were superb. Iain presented an excellent tutorial at a West Coast conference while I, on the very same day, presented the identical tutorial at an East Coast conference.

A career enhancement thank you to Stan Mazor and Synopsys for giving me the opportunity to expand my VHDL expertise into the fascinating field of synthesis.

A very special and warm thanks to all my VHDL students who have made our classroom interactions a truly positive and enjoyable experience.

Thanks also to my two sons, Benji and Daniel, for filling my life with a deep sense of joy and satisfaction.

And of course a special thank you to my wife, Sarah, for without her I would probably still be out in the fields riding my tractor and picking grapefruit.

Part

1

An Excursion into VHDL

Chapter

1

Embarkment

1.1 Design Entity

The VHDL term *design entity* may intuitively be perceived as a model for the chip shown in Fig. 1.1. Associated with each design entity is a unique entity declaration that establishes the design entity's communication links to the outside world. These interface links are referred to as ports in VHDL and they conceptually correspond to the shaded pins of the chip in Fig. 1.1.

Though symbolized as a digital chip, the design entity may, in fact, represent any device that possesses some form of intercommunication characteristic. It could model a single gate, a *central processing unit* (CPU), a board populated by discretes, a theoretical protocol format, or even a complete system. In the latter case the entity declaration might

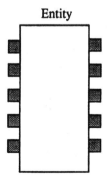

Figure 1.1 The entity declaration of a design entity establishes the external interfaces of the modeled device.

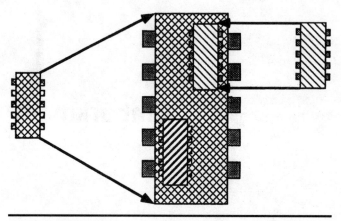

Figure 1.2 VHDL entities support both top-down and bottom-up design methodologies.

possibly not even have any interfaces at all. Instead, the design entity will merely represent a top level, self-contained system that can be simulated without any interactions to other design entity models.

As illustrated in Fig. 1.2, design entities may be decomposed into constituent lower level design entity subcomponents, or, alternatively, they may be treated as building blocks to construct even higher level design entities. These two viewpoints of the design entity show that VHDL can support both top-down and bottom-up design methodologies.

Moreover, just as a circuit board may be populated by the same identical Exclusive-OR gate, a mechanism exists in VHDL whereby multiple models of a design entity may be concurrently simulated. This option is illustrated in Fig. 1.3, where two copies of the Xor_Gate design entity are to be simultaneously simulated as part of a larger design entity called Top_Level_Entity.

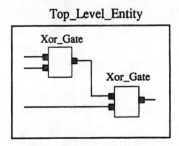

Figure 1.3 VHDL provides a mechanism by which multiple copies of the same entity may be concurrently simulated.

Recall that an entity declaration's key role is to embody the external interfaces of a modeled device. The entity declaration does not and cannot model a device's behavior. The entity declaration is merely an outer shell without any functional core. To simulate the models shown in Fig. 1.3 requires explicit information about the internal behavior of those devices represented by the various design entities. The next section introduces the VHDL construct that will add a functional substance to the design entity and allow the resulting model to be simulated.

Since the entity declaration uniquely establishes the design entity, there is a tendency in the VHDL community to reference either of these concepts via the single term *entity*. As we progress further into this book I, too, will rely on this abbreviation.

1.2 Architecture Body

Associated with each design entity is an architecture body that describes the behavior and/or the structure of a modeled device. By *behavior* I mean the simulatable functionality of the design entity, whereas by *structure* I mean the decomposition of a design entity into subcomponents as per Fig. 1.2. The option then exists for these subcomponents to be recursively partitioned into a multilayered hierarchical configuration. But somewhere along each decomposition chain a final leaf node must exist with an architecture body that is no longer structural. Otherwise, the upper level design entities will not be able to be simulated, since the simulator will not know the behavior characteristics of each design entity within the declared hierarchy.

The compound expression *architecture body* is often abbreviated within the VHDL community to the single term *architecture*. Figure 1.4 displays this book's convention that an architecture will be symbolized as a shaded circle encapsulated within a design entity icon. Assuming that the model in Fig. 1.5 does not have any further structural decompositions, it may now be simulated since the simulator is able to deduce the functionality of each of the depicted architectures.

A design entity may have multiple architecture bodies associated with it. These various architectures may represent either different modeling viewpoints or different levels of abstraction in a top-down design strategy. The key point is that the interface port names referenced by each of these architectures must be the same since these multiple architectures are all associated with the same entity declaration. Consequently, each of these architectures will have the same interfaces to the outside world, although the timings and values on the interfacing ports may be different.

Figure 1.6 shows six architecture bodies associated with the same design entity. Behave_1 and Behave_2 might reflect two different solutions to the same problem. For instance, Behave_1 might perform error

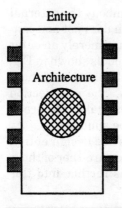

Figure 1.4 The architecture body describes the behavior and/or structural decomposition of the modeled device.

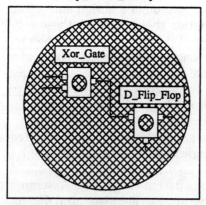

Figure 1.5 In a hierarchical decomposition, the simulator must be able to deduce the operational functionality of each architecture body within the decomposition chain.

checking via a simple parity check, whereas Behave_2 might do so via a complex check-sum algorithm. Similarly, Structure_1 and Structure_2 might reference different macrocell component decompositions in a top-down tradeoff analysis study. Before discussing the technology-based architectures, let me first emphasize that VHDL is technology independent. This means that VHDL does not imply the usage of any specific hardware implementation technology, such as ECL or *complementary metal-oxide semiconductor* (CMOS). But it is

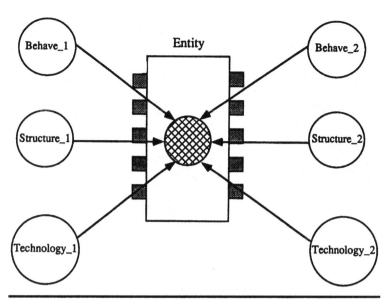

Figure 1.6 Multiple architecture bodies may be associated with the same design entity.

possible in VHDL to model certain timing and behavior characteristics that do indeed honor a specific technology. Such might be the case with the architectures Technology_1 and Technology_2.

Let us now revisit the circuit of Fig. 1.3 that contains several Exclusive-OR gates. What are our modeling options in light of the fact that multiple architecture bodies may exist for the same Xor_Gate design entity? Which of these multiple architectures is the one that is actually being simulated? Features exist within the VHDL language to explicitly pronounce which of the many available architectures is to be used during the simulation of a design entity. This entity-architecture pairing for simulation purposes is known as *binding* in the VHDL vernacular. For instance, in Fig. 1.7a the architecture Behave_1 is bound to both copies of the design entity, Xor_Gate. Hence the simulated behavior of both Xor_Gate models will be based on the VHDL code that was written for the architecture Behave_1. On the other hand, the option also exists to simulate each Xor_Gate via a different algorithm. This option is shown in Fig. 1.7b, which illustrates two different models for simulating the design entity called Xor_Gate.

1.3 Our Traveling Companion

Having introduced ourselves to the entity and architecture concepts, we are now ready to embark on our VHDL excursion. And, as with all jour-

8 An Excursion into VHDL

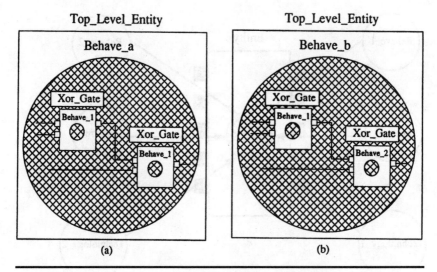

Figure 1.7 VHDL provides a mechanism by which multiple copies of the same entity may be simulated with the same (*a*) and/or different (*b*) architecture bodies.

neys in life, a friendly and multifaceted companion will certainly make the voyage more entertaining and meaningful. In our particular case, we will be traveling with the 3-bit counter shown in Fig. 1.8. This device will increment to the next value on the trailing edge of a periodic signal.

Chapter 2 will present numerous VHDL models, all of which are based on this incrementing hardware. Each of these models will highlight a key VHDL concept or design methodology. Keeping the device constant and then moving the VHDL model around it will allow you to focus better on the ensuing VHDL developments. Otherwise, each VHDL topic would have to begin by first forcing you to reorient yourself to a new and unrelated hardware environment.

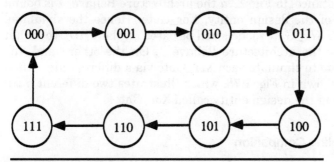

Figure 1.8 Three-bit counter increments at the trailing edge of a periodic signal.

This pedagogical approach of referencing only one device has, over the years, proven to be a successful training vehicle for rapidly accelerating the VHDL learning curve. And you should not be alarmed by the myriad of ways to represent the same device. Furthermore, these totally diverse models should not leave you bewildered as to which specific design methodology to apply. Actually, in the real world of mandated assignments your particular model will be a function of how you perceive or have been directed to perceive the targeted device. If the device is to be viewed as a finite state machine, then it shall be modeled as such. If, on the other hand, you want to model the device at the gate level, this too may be done. A key point highlighted by these diverse 3-bit counter environments is that VHDL is powerful enough to model at either the behavioral, data flow, or structural level. VHDL even allows you to mix these three design paradigms within the same model.

Moreover, the ability to generate a multitude of VHDL models for the same device is very important in the world of hardware synthesis. Optimum hardware cannot be achieved by merely optimizing a tool suite's synthesis parameters. Internal synthesis algorithms are very sensitive to the coding style of the VHDL model that is being synthesized. Consequently, just as I have done with the 3-bit counter, you too should become very adept at cranking out a diverse set of architecture bodies for the same targeted device. That way you will establish a large pool of VHDL models for your synthesis experiments and tradeoff studies.

Enough said. We are now ready to embark.

Chapter

2

The Journey

2.1 File: introductory_example.vhd

Our travels into the VHDL landscape will begin with an in-depth, line-by-line examination of the VHDL source code stored in the file introductory_example.vhd. This file, reproduced in Fig. 2.1, encapsulates a model that increments a 3-bit counter on the trailing edge of an internally generated signal.

Reading from top to bottom, we first encounter two adjacent hyphens, --, which indicate that the remainder of the line is to be interpreted as a comment. Note that successive comment lines must each be preceded by two adjacent hyphens and, as shown by the position of the line numbers, comments may begin from any column.

At line 15 of Fig. 2.1 we immediately come upon the VHDL reserved word entity. Reserved words are special code words that are anticipated by the VHDL language compiler. Appendix D tabulates all the reserved VHDL'87 words in alphabetical order. The VHDL'93 reserved words may be found in App. E. In general, the reserved word entity introduces a modeler's intention to specify an entity declaration for a VHDL design entity. The next word, Encapsulate_3_Bit_Count, is the name that I have chosen to uniquely identify this design entity. Subsequently, we have the word *is,* which is also a reserved word. Note that I have selected to write all the reserved words in lower case. This case format is not really required since VHDL is case insensitive. The compiler that processes this raw VHDL source code does not distinguish between the words entity, Entity, or ENTITY. They are all viewed as one and the same. Nonetheless, your company should mandate one of these entity patterns as a companywide standard and repeat it for all

12 An Excursion into VHDL

```
-- File Name  : introductory_example.vhd
--
-- Purpose    : Entity/Architecture pair to simulate a trailing edge
--              triggered 3 bit counter. This model contains
--              both a clock pulse generator and a 3-bit
--              counter within the same architecture body.
--
-- Author     : Joseph Pick

-- *********************************************************  LINE NUMBER
entity Encapsulate_3_Bit_Count is                             -- 15
end Encapsulate_3_Bit_Count;                                  -- 16
                                                              -- 17
architecture Introductory of Encapsulate_3_Bit_Count is       -- 18
    signal Clock : BIT := '0';                                -- 19
begin                                                         -- 20
                                                              -- 21
    process                                                   -- 22
    begin                                                     -- 23
        Clock <= not Clock after 50 ns;                       -- 24
        wait on Clock;                                        -- 25
    end process;                                              -- 26
                                                              -- 27
    Mod_Incr:                                                 -- 28
    process                                                   -- 29
        variable Current_Count : INTEGER := 0;                -- 30
    begin                                                     -- 31
                                                              -- 32
        -- Wait until the trailing edge of the clock.         -- 33
        wait until Clock = '0';                               -- 34
                                                              -- 35
        -- Increment the counter modulo 8.                    -- 36
        if Current_Count = 7 then                             -- 37
            Current_Count := 0;                               -- 38
        else                                                  -- 39
            Current_Count := Current_Count + 1;               -- 40
        end if;                                               -- 41
                                                              -- 42
    end process Mod_Incr;                                     -- 43
                                                              -- 44
end Introductory;                                             -- 45
```

Figure 2.1

---

the other reserved words. Moreover, it is highly recommended that your company endorses and drafts a VHDL style guide even before its first VHDL project gets off the ground. That way your company's VHDL models will all have the same look and feel. This conformance will make it easier for team members to read, understand, and maintain each others' VHDL models. When formulating your company's style guide, keep in mind that there is no perfect format. The best that you can strive for is some form of uniformity that will, in general, suit the purposes and integration environment of your project.

Incidentally, why would it have been inappropriate to write the following?

```
--14 entity Encapsulate_3_Bit_Count is
```

Answer: The entity statement would then be commented out since it follows two adjacent hyphens.

Line 16 begins with the reserved word *end,* which in this context serves as a delimiter announcing the termination of the entity declaration. It is optional to repeat the name of the design entity, as was done in this model, but I recommend that you do so for readability. The concluding semicolon is syntactically required, and its absence will cause an error message to be generated during the compilation of this code. This particular entity declaration does not contain any ports (pins). Hence the simulation of this design entity, Encapsulate_3_Bit_Count, is totally self-contained and does not interact with any other design entities.

By the way, VHDL is a free-form language, in that carriage returns, blank lines, and additional blank spaces may be included between words for readability without any ill side effects. For instance, lines 15 and 16 may have been combined into one long contiguous string but, for readability, were separated into two consecutive lines. Here again, the specific format to be used should be explicitly outlined in your company's VHDL style guide.

Line 18 begins with the reserved word *architecture.* This reserved word initiates a modeler's intention to write an architecture body for a VHDL design entity. Introductory is the name chosen for this architecture body. The next word, *of,* is also a reserved word, and it is followed by the name of the design entity that this architecture body is associated with. It should be well noted that in this model, the entity declaration and its associated architecture body are both resident in the same file. This file commonality is not a necessity. Later source code examples will exhibit corresponding entity declarations and architecture bodies that are stored in distinct files. Line 18 is concluded by the reserved word *is.* Here again, its absence would yield a compilation error.

Appendix A of the LRM contains a formal *Bachus-Naur Format* (BNF) description of the VHDL syntax. Though this BNF description provides a compact and concise grammatical definition of the language, newcomers to VHDL will find it to be very terse and unfriendly. It is more pedagogically sound to introduce and develop the VHDL language with a descriptive set of concrete and pragmatic examples. As such, this book will not rely on the BNF notation to formally describe new VHDL constructs.

Between the reserved words *is* (line 18) and *begin* (line 20) lies the declarative part of the architecture body. It is here that you can make declarations that may be referenced anywhere within the remainder of the architecture body. For instance, you can declare VHDL signals in this region as I have done on line 19. On a rudimentary level, the VHDL concept of a signal agrees with our intuitive understanding of this term as a communication link that transports a value from one locality to another. However, as we travel further into VHDL, we will observe that signals may, in fact, be interpreted as either a wire, a logic gate, or a register. The distinction will depend on the way that the signal has been specified to be updated in the VHDL model. These various hardware interpretations may readily be realized via a synthesis tool. In this particular example, Clock is declared to be a signal. User-defined names such as Encapsulate_3_Bit_Count, Introductory, and Clock are formally referred to in VHDL as identifiers. While on the topic of VHDL terminologies, let me also point out that VHDL has four object classes: signal, variable, constant, and file. We have just been introduced to signal. The other three object classes will also be encountered in our VHDL excursions.

The colon following Clock announces that what comes next is its data type declaration. Every declared object must be associated with a unique type that precisely defines the range of values that it can have. Data types are very important in VHDL. In fact, the compiler will not permit you to assign one object to another when they are not of the same type. This language feature, known colloquially as strong data typing, serves as your first line of defense against careless coding errors and conceptual design flaws.

In this example Clock is declared to be of type BIT. This data type comes for free with the VHDL language. By *free* I mean that this type BIT is automatically available to the VHDL user, in much the same way as the term INTEGER may be readily referenced in other programming languages. Types, such as BIT and INTEGER, that are an integral part of the VHDL language are referred to as VHDL predefined data types. Note that I have chosen to write these predefined data types in upper case. This approach is not really necessary since VHDL is case insensitive. Here again, your company's official style guide should endorse a case convention for the VHDL predefined data types.

The colon equals symbol (:=) following BIT announces that we are now going to assign a default initial simulation value to the signal Clock. It is very interesting to observe that this initial value is the character '0' and not the symbol 0. Single characters surrounded by single quotes, such as '0', are called *character literals*. In other hardware description languages, the symbols 0 and 1 are the possible options for an object of type BIT. Since VHDL also supports the data type INTEGER, it appears on the surface, as if the character literals '0' and '1' were chosen only to avoid any possible conflicts with the INTEGERs 0 and 1, respectively. Though this conflict issue may be true, the full utility of this character literal approach for BITs will become more apparent as we journey deeper into the VHDL countryside.

Line 45 concludes this architecture body with the reserved word *end*. The identifier Introductory is repeated for readability and may optionally be omitted. Here again the concluding semicolon is necessary. Between lines 20 (begin) and 45 (end Introductory;) is the statement region of this architecture body. It is here that VHDL concurrent statements may be written. Consequently, this segment is also referred to as the *architecture concurrent region*.

In this model we have two processes residing in the architecture's statement region. The first process extends from lines 22 to 26. The second one spans lines 28 to 43. Immediately, you should observe that the latter begins with an identifier followed by a colon, whereas the former does not. This difference shows that the assignment of an identifier to a process is optional. An identifier used in this context is formally known in VHDL as a *label*. The label Mod_Incr assigns a unique name to the second process. This label will henceforth be known to the VHDL simulation tool suite and may be referenced as the process's path name while interactively simulating this model. If a process does not have a user-specified label, then the VHDL tool suite will automatically generate one for it. Typically, such tool suite names are very nondescriptive and unfriendly. Imagine how impractical it would be if during the simulation debugging phase you were forced to reference a process having the name, P_Log_e_Sequence_4979_625. Not only is such a name difficult to remember, but its context is totally meaningless. By now you have probably deduced my recommendation that each process should be labeled with a highly descriptive name that captures the essence of the process's functionality.

The purpose of the first process is to generate a periodic waveform on the signal, Clock. Then, on the trailing edge of this Clock's pulse, the second process will implement the desired incrementation. It is important to realize that these two processes are communicating to each other via the signal Clock. In general, VHDL declared signals are the communication links between processes that are interfacing with each other and reside within the same architecture body in which these signals were declared.

Processes have three fundamental properties that you must always be aware of. First, when the architecture is being simulated, all the processes are theoretically running concurrently. This concurrence implies that the physical order in which processes appear in a source file is totally irrelevant to their execution order during a simulation run. This concurrent feature of processes honors the real-world environment in which hardware elements exist and operate simultaneously. But let's take a closer look at this concurrency feature. When the model is being simulated by the VHDL tool suite, do you really think that each process is being executed simultaneously? Remember that both your VHDL model and the tool suite are executed (assembly instruction) line-by-line on a sequential computer. So in this sequential medium it is impossible for all your processes to be running at exactly the same time. Rather, what you get from your external user's perspective is just the look and feel of concurrency. The VHDL tool suite managing the activities of your model contains a core engine that chronologically queues future scheduled events and removes current ones for immediate execution. In such a scheme, simulation time is incremented in discrete jumps based on the specific times at which activities are scheduled to occur. The net effect of this well-orchestrated queuing and dequeuing is that all the VHDL processes appear to be running simultaneously even though, in reality, they are not. Because of this discrete approach to the modeling of concurrency, VHDL is said to be an event-driven simulation environment.

The second fundamental property of processes concerns its sequential nature. Though every process has the look and feel of a concurrently running shell, the VHDL statements contained within it are sequentially executed in the order in which they physically appear. Sequentiality is what makes processes so useful and congenial to work with. Complex hardware devices may be modularly modeled by decomposing them into simpler concurrent subunits that implement a well-defined sequential algorithm. Then processes may be written to capture the sequential flow of these algorithms. Such a divide-and-conquer approach to VHDL modeling makes the whole endeavor very manageable.

The third fundamental property emulates hardware's nonterminating nature. Just as hardware exists and runs forever, a process will also exist for the duration of the simulation. When the final statement of a process is reached, the simulator will then loop to the top of the process and repeat another iteration of the process's code. The process is in a software infinite loop. But the process does not contain any special VHDL syntax to indicate and invoke this infinite looping behavior. From the VHDL programmer's perspective, it just happens by definition of the VHDL language. And because it just happens, you must always be alert to its occurrence and ramifications. Otherwise, your VHDL model might not behave as expected.

Let's now continue our investigation of this architecture body. To get the most out of this analysis, we will proceed as if you, the reader, would take on the role of a typical VHDL simulation engine. This internal perspective will provide you with a better appreciation of how things work in VHDL. Note that during the analysis below the terms *simulate* and *execute* will be used interchangeably.

Because the simulation is just beginning, you will set your internal time to 0 units. You are now ready to implement the VHDL initialization phase during which every process will be simulated until some form of the wait statement is sequentially reached. In our current model we have wait statements on lines 25 and 34.

Since both processes of this Introductory architecture body are running concurrently, it theoretically does not matter which of them is attacked first by you, the simulation engine. So you may as well begin the initialization phase by first executing the nonlabeled process that is generating a periodic waveform on the signal Clock.

Lines 22 (process) and 23 (begin) contain reserved words that, respectively, announce the presence of a process and the start of its simulatable code. The statement on line 24 is scheduling an assignment to the signal Clock. Because this assignment is scheduled only when line 24 is sequentially reached, it is called a sequential signal assignment. The arrow symbol (<=) signifies that the expression on its right (the right-hand side) must be scheduled to be assigned to the signal on its left (the left-hand side). The right-hand side is said to be the source of the signal assignment, whereas the left-hand side is said to be the target. On line 24 the target signal is Clock, and the source is the expression *not Clock*. The reserved word *not* is a VHDL operator that can negate objects of type BIT. So, not '1' equals '0' and not '0' equals '1'. Since Clock currently is '0' the expression *not Clock* will therefore have the value '1'. Consequently, as the simulation engine, you are now going to schedule '1' to be assigned to the signal Clock. Note well that I have now repeated the word *schedule* several times already. What is the significance of this term?

Recall that a signal may be perceived as the VHDL abstraction of a physical wire, logic gate, or register. Since these hardware elements can never be updated instantaneously, it follows that their VHDL signal counterparts should also be updated only after a certain lag time. VHDL allows you to apply an after clause as shown on line 24 to specify the propagation delay required to update a signal. This same mechanism may be also used to schedule the assignment of a value to a signal at a later desirable time. The signal's current value will be maintained until the scheduled change. In our example, Clock is currently '0', and it is now being scheduled to be updated to the value '1' 50 nanoseconds (ns) from now. The overall objective is to generate a periodic waveform with a pulse width of 50 ns. To implement this signal scheduling, you, the simulation engine, must now place this signal-scheduling request into your internally maintained time-ordered queue as shown in Fig. 2.2.

Incidentally, the time unit nanosecond is another term that comes for free with the VHDL language. Any of the following time units are available: femtosecond (fs), picosecond (ps), microsecond (μs), millisecond (ms), second (sec), minute (min), and hour (hr). Keep in mind that the 50 ns expressed on line 24 does not correspond to the progression of real time on your wristwatch. Rather, it refers to the simulation time that is generated and maintained by you, the simulation engine.

After completing this update to the time-ordered queue you, the simulation engine, will now continue to the next statement in this process which is on line 25. This statement, wait on Clock, will cause the enclosing process to be put to sleep (suspended) until an event occurs on the

18   An Excursion into VHDL

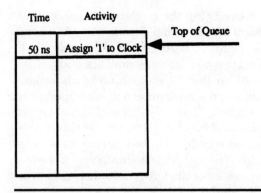

Figure 2.2 Current status of the time-ordered queue.

signal Clock. So what is an event? If a signal is updated with a value that is different from its current value, then an event is said to have occurred on the signal. In other words, this particular process will be suspended until the signal Clock will change from its current value '0' to the new value '1'. As per line 24 this update has been scheduled to occur at simulation time 50 ns. In general, the sensitivity list of a wait statement consists of those signals on which events must occur in order for the wait statement to be triggered and its host process to subsequently be awakened (reactivated). In this specific case, Clock is the only member of this wait statement's sensitivity list. You, the simulation engine, will now implement this process suspension by associating this process with the signal Clock in the signal event-monitoring queue as shown in Fig. 2.3.

Having put this process to sleep, you, the simulation engine, will now ask yourself if there are any more processes to be worked on during this

Signal	Response to Event
Clock	Reactivate waveform–generating process

Figure 2.3 Current status of the signal event-monitoring queue.

initialization phase. The answer is yes, and so you will proceed to the process labeled Mod_Incr. Specifically, you will execute its first simulatable statement that occurs on line 34. This VHDL statement is another variant of the wait construct. This particular scheme asks you, the simulator, to perform two tasks. It should come as no surprise that one of them is to confirm that the condition, Clock = '0', is satisfied. The second task, on the other hand, is not explicitly asserted by this VHDL construct. Nonetheless, prior to testing the given condition, you must first wait for an event to occur on the signal Clock. The VHDL term event has already been explained above. Hence, Clock is in the sensitivity list associated with this wait until construct. Incidentally, the statement on line 34 is the shorthand equivalent of the following statement: wait on Clock until Clock = '0'.

So what is this wait construct on line 34 really trying to achieve? Once an event occurs, only then is the wait condition to be checked. If this condition, Clock = '0', is found to be true, then you can continue simulating the process by advancing to the next executable statement following this wait construct. If, on the other hand, the condition is false, then the process will be put back to sleep on a queue and resuspended until the next event on Clock. So what does all this mean? In essence, you are going to continue into this process only after an event has occurred on Clock, and as a result of this event, Clock has been updated to '0'. So you are waiting for Clock to change from some non-'0' value to '0'. Since Clock is of type BIT, it can only be a '0' or a '1'. Hence you are waiting for Clock to transition from a '1' to a '0'. In other words, you are waiting for the trailing edge of Clock to occur. So now you know what you are waiting for. Well, are we at the trailing edge right now? What is the current value of Clock, anyway? Recall that on line 19 Clock was initialized to '0', and as yet has not been modified. (When will it actually be modified to '1'?) We are currently still progressing through the initialization phase at time 0 ns. Hence its present, initial value does, indeed, satisfy the wait condition. But the initialization of a signal is not considered to be an event. In fact, the VHDL language stipulates that a signal's initial value may be assumed to have persisted for an infinite amount of time prior to the actual simulation run. Hence, even though the wait condition is satisfied (Clock = '0'), Clock is currently not tagged as having received an event. This agrees with your intuition that a trailing edge has not yet occurred on the signal Clock since the simulation began with Clock already having the value '0'. So because an event did not occur on Clock, you are not yet permitted to test the condition Clock = '0'. Hence, even though the wait condition is currently satisfied, you, the simulator, will still not be able to proceed into this process. Instead, you must put this process to sleep until an event occurs on Clock. You, the simulator, will implement this process

An Excursion into VHDL

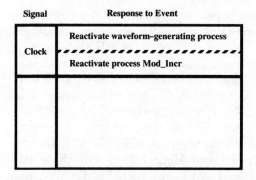

Figure 2.4 Current status of the signal event-monitoring queue.

suspension by associating this process with the signal Clock in the signal event-monitoring queue as shown in Fig. 2.4.

While we are on the topic of wait until Clock = '0', let me point out that current VHDL literature sometimes contains the statement

```
wait until Clock = '0' and Clock'EVENT;
```

Let's dissect this statement by first discussing the expression Clock'EVENT. Every signal has certain predefined VHDL attributes associated with it. We will come across several of them during our initial journeys into VHDL. The second part of this book will formally introduce and develop all of these attributes. But for now let's look into the attribute 'EVENT. First, the single quote mark is read as a *tick* in VHDL. So Clock'EVENT should be read as *Clock tick EVENT*. This attribute is a BOOLEAN data type which will be TRUE if Clock just had an event. Otherwise, it will be FALSE. Hence, we see that the above statement containing Clock'EVENT is operationally equivalent to the statement wait until Clock = '0' since in both cases the wait until clause will only be evaluated if Clock just had an event. But if Clock just had an event, then Clock'EVENT will be TRUE. Hence, the test Clock'EVENT is both redundant and inefficient in this coding scenario and may be omitted, which is what I have done on line 34. Nonetheless, you should still be aware that some synthesis tools demand the inclusion of the 'EVENT test in a wait until clause to identify a synchronous device. Perhaps as these synthesis tools mature they will accept the equivalent form wait until Clock = '0'. One final note: As per my VHDL style guide conventions, I have opted to write all VHDL attributes in upper case.

Note that both processes of this architecture contain some variant of the wait construct. Suppose one of them did not. Such a process would never be suspended. Instead, it would just loop forever, and the simulator would never be given the opportunity to work on any of the other processes. Your model would then be seriously flawed and useless, since all your processes would not be simulated. In fact, the VHDL compiler should assist you on this matter by generating a warning message if any of your processes do not contain some form of the wait construct. That way you will not inadvertently fall into this black hole.

Let's now return to our simulation activities. After suspending the process Mod_Incr, you, the simulator, must once again ask yourself if there are any more processes to be executed during this initialization phase. Recall that during the initialization phase all processes must be executed until a wait statement is reached. Since there are no more processes to be executed, you, the simulator, have formally finished with the initialization phase. You must therefore now ask yourself, "What is the first time-related activity that must be done?" On the top of the time-ordered queue is the updating of the signal Clock to '1' at simulation time 50 ns. You will therefore bump up your internal simulation time from 0 to 50 ns. Note that this time update occurred in a discrete jump, and the complete time continuum between 0 and 50 ns was, for all practical purposes, totally irrelevant. Having progressed the simulation time to 50 ns, you will next take the scheduled updating of Clock off the queue and assign '1' to it. The subsequent empty status of the time-ordered queue is shown in Fig. 2.5.

Note well that this assignment to the signal Clock was not executed while you were sequentially stepping through a process. Instead, it was done in the background outside of any existing process. Incidentally, if there would be any more signals that had been scheduled to be updated at 50 ns, they too would be taken off the queue now and modified. In our case there are no such signals, so you, the simulator, must then continue by asking the following question: "Since Clock just had an event,

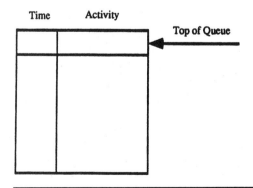

Figure 2.5 Current status of the time-ordered queue.

Processes to Execute during Current Simulation Cycle

Reactivate waveform-generating process
Reactivate process Mod_Incr

Figure 2.6 Current status of the process-pending queue.

are there any processes waiting for an event on this signal?" You will answer this question by searching through the signal event-monitoring queue. Since Clock exists in this queue, you will then remove all of its associated processes and place it into a process-pending queue as shown in Fig. 2.6. In this example there are currently two such processes. Having introduced this third queue, I must now relate to you that the existence of these queues is not specified by the VHDL language. Nowhere in the LRM will you find such names as *time-ordered queue, signal event-monitoring queue,* and *process-pending queue.* However, on the basis of my experiences with other event-driven simulation languages, I know that analogous queues must exist in order for any VHDL tool suite to implement the look and feel of concurrency.

By the way, the exact sequence of updating signals first and then reactivating the appropriate processes is explicitly spelled out in the LRM. The specific algorithm for this sequence is formally designated as the VHDL simulation cycle. For our immediate needs we may view a simulation cycle as being segmented into two well-defined halves. During the first half, signals that are scheduled to be updated at the current simulation time will be assigned their respective values. If this update is different from the signal's previous value, then the signal is immediately tagged as having received an event. Processes waiting for events on these event-tagged signals will then be awakened during the second half of this current simulation cycle. This well-defined flow of precisely when signals are updated and processes awakened establishes the deterministic mechanism via which the VHDL simulation engine coordinates the signal communications between concurrently running processes. In our particular example you, the simulation engine, updated Clock during the first half of the simulation cycle occurring at 50 ns. Then during the second half of this same simulation cycle you reactivated both processes of this architecture since they are both waiting for an event on the signal Clock. It remains now to execute these two processes that are currently resident in the process-pending queue.

Since processes run concurrently, it really should not matter what order these processes are removed from the process-pending queue for immediate execution. That being the case, you may as well remove the first one that is on top.

Having just removed the waveform-generating process from the process-pending queue you, the simulator, will now continue its execution at the first statement following the one that caused it to be suspended in the first place. Because processes are in an infinite loop, you will proceed to the top of this process and continue there at line 24. From our previous analysis we know that we must negate the current value of Clock and schedule this value to be assigned to Clock 50 ns from now. Hence, you, the simulator, must schedule '0' (= not '1') to be assigned to Clock at the simulation time of 100 ns (= 50 + 50 ns). Figure 2.7 depicts the updating of the time-ordered queue required by this scheduling. After updating this queue, you will then proceed to line 25 and, as before, declare this process to be suspended until the next event occurs on Clock. Figure 2.8 shows the resulting updated status of the signal event-monitoring queue. Note well that the scheduled negation of Clock coupled with the suspension and reactivation of this process will generate a periodic waveform on this signal, where each pulse is 50 ns wide.

Remember that you must still remove the second process Mod_Incr from the process-pending queue because it too must be awakened and executed. This time, however, you cannot proceed to the first statement after the wait statement on line 34. Instead, you must first check the wait condition to determine if it is currently TRUE. Since Clock is currently '1', the wait condition is not satisfied, and hence this process

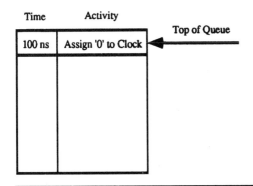

Figure 2.7 Current status of the time-ordered queue.

24 An Excursion into VHDL

Signal	Response to Event
Clock	Reactivate waveform-generating process

Figure 2.8 Current status of the signal event-monitoring queue.

must be resuspended until the next event occurs on Clock. You, the simulation engine, must therefore once again, update the signal event-monitoring queue as per Fig. 2.9.

Since the process-pending queue is now empty, there are no more processes to execute during this current simulation cycle. You, the simulator, must now go to the time-ordered queue to determine what must be done next. Since Clock must be assigned the value '0' at 100 ns, you, the simulator, must carry out the following, just as you did earlier. You must bump up the simulation time to the appropriate value, which in this case is 100 ns. Next, you will remove this scheduled updating of Clock from the time-ordered queue and update it with the value '0' during the first half of the simulation cycle occurring at exactly 100 ns. This transition from '1' to '0' means that Clock just had an event, and so you will once again remove both processes from the signal event-

Signal	Response to Event
Clock	Reactivate waveform-generating process Reactivate process Mod_Incr

Figure 2.9 Current status of the signal event-monitoring queue.

Processes to Execute during Current Simulation Cycle

Reactivate waveform-generating process
Reactivate process Mod_Incr

Figure 2.10 Current status of the process-pending queue.

monitoring queue and place them into the process-pending queue. The resulting status of the process-pending queue is illustrated in Fig. 2.10.

Once again you will first remove the waveform-generating process from the process-pending queue, loop to its beginning, schedule '0' (= not Clock) to be assigned to Clock 50 ns from now, and put this process to sleep until the next event on Clock, which should occur at time 150 ns (= 100 + 50 ns).

As for the process Mod_Incr, this time around the wait condition (Clock = '0') is finally satisfied since Clock is currently equal to '0'. Hence you, the simulator, may continue to the first executable statement following this wait condition, which will subsequently position you at line 37. The purpose of the code segment from lines 37 to 41 is to increment the internal counter called Current_Count. Where did this object Current_Count come from? Its declaration as a VHDL variable occurred on line 30 between the key words *process* and *begin*. The area between these two reserved words is, in general, known as the process declarative part. Declarations made here are applicable only to the encapsulating process. For instance, had another process also declared a variable called Current_Count, then the two objects would be distinct and unrelated. Assignments in one process to Current_Count would have absolutely no effect on the value of Current_Count in the other process. Consequently, declarations made in the process declarative part are said to be local to its host process. On the other hand, signals declared in the architecture declarative part are visible to every process contained within the architecture. We already saw this architecturewide global capability illustrated by the signal Clock. Incidentally, it is an error to declare signals in the process declarative part. Also, since global variables are not permitted in the current 1987 version of VHDL, variables cannot be declared in the architecture's global pertaining declarative part. But keep in mind that VHDL'93 will allow global variables to be declared in the architecture declarative part. The potential dangers of global variables will be discussed in more detail in a subsequent model of our VHDL journeys.

Returning our attention to line 30, we see that Current_Count is declared to be a variable of type INTEGER and is initialized to 0. INTEGER is another data type that comes for free with the VHDL language. It is important for you to be aware of the fact that this initialization of Current_Count is a one-time occurrence that will transpire just prior to the start of the VHDL initialization phase. During simulation when the process Mod_Incr loops back to the top, it will continue at line 34, which is the first executable statement following the reserved word *begin*. The variable Current_Count will never be reinitialized to 0 even when this process loops back to the top. Also, the present value of Current_Count will not be destroyed when this process is suspended. Hence, when the process is put to sleep and then reactivated, Current_Count will have the same value. These operational characteristics of Current_Count apply, in general, to all variables declared in a process declarative part.

Lines 37 through 43 will increment the INTEGER-type Current_Count as if it would be a 3-bit counter. Such an incrementer should count up from 0 to 7 and then wrap around to 0 again. The if statement beginning on line 37 will implement this incrementation algorithm. If Current_Count is currently equal to 7, then all the VHDL statements between the reserved word *if* and the reserved word *else* will be executed. In our current case, Current_Count has the value 0. Hence the if test on line 37 will fail and consequently you, the simulator, must continue to the first executable statement following the reserved word *else* (line 39) and will have to execute line 40 in which you must first add 1 to Current_Count and then immediately update Current_Count with this incremented value. Note the colon equals (:=) notation used on this line 40. This symbol can only be used if the target of an assignment is a variable. It would have been an error to use the arrow (<=) symbol here because this notation is to be used only when the target of an assignment is a signal. Note well that the assignment on line 40 is carried out immediately. Variable assignments are never scheduled to occur at a later time. Unlike signals, variables are always updated instantaneously.

The next task that you, the simulator, must now work on is to take care of the first executable statement after the *end if* reserved words on line 41. In this case you will have to loop around to the top of this process and continue at the wait until Clock = '0' statement on line 34. Hence, you, the simulator, will suspend Mod_Incr until the next event on the signal Clock. All the simulation activities outlined above will now be repeated indefinitely. Eventually, as Current_Count is incremented it will reach the value 7. In that case the if test of line 37 will be satisfied and you, the simulator, will immediately assign 0 to Current_Count as per line 38. This assignment of 0 is the decimal version of the wraparound feature of a 3-bit counter.

Now that we have seen the interactions of these two processes, let me point out that you always must be very careful where you place the various wait constructs. For instance, had the statement, wait on Clock, been at the top (just following the reserved word *begin*) then both processes would be waiting for an event on Clock. This event would never occur since the line, Clock <= not Clock after 50 ns, would never be reached and executed. Consequently, after the initialization phase the time-ordered queue would be empty. Since there would be nothing for the simulator to do, it would just terminate its own execution at time 0 ns. The desired incrementation would therefore not even occur.

The aforementioned detailed analysis of the simulation should give you an excellent feel for the fundamental principles of VHDL. Just one final remark about the colon equals (:=) notation used to assign an initial value to a signal. At first inspection there seems to be a slight disparity since signal assignments require the arrow (<=) notation, whereas variable assignments may use only the colon equals (:=) symbol. On the surface it would seem appropriate to also apply the signal assignment symbol <= when initializing signals as well. But recall that signal assignments are always scheduled to be implemented some time in the future. Signal initializations, on the other hand, occur instantaneously somewhat analogous to variable assignment updates. Hence, it does, after all, seem reasonable to use the variable update symbol to initialize signals, since signal initializations also occur instantaneously and not at some later time.

Let's conclude our analysis of this VHDL source code by pointing out that it is a behavioral model since it uses high-order language constructs such as if...then...else. As we journey deeper into the VHDL landscape, we will also come across structural and dataflow models. By the way, the terms behavioral and dataflow have many different interpretations in the VHDL literature. The only term that everyone seems to agree upon is structural. Be aware that some authors classify our current 3-bit incrementing model as a dataflow example, since it is essentially only assigning values. As you read VHDL-related conference papers and trade journal articles, do not get bogged down by these nomenclature differences. Just adapt to the authors' definitions and move on.

2.1.1 Information checkpoint

With this first, seminal example we have in effect laid a solid foundation for the rest of the book. Before you venture further along your VHDL journeys, please make certain that you are familiar with the following topics that we have established via this first model:

An Excursion into VHDL

- -- Two adjacent hyphens denote that the rest of the line is a comment.
- VHDL is not case sensitive.
- Format of the entity declaration.
- Format of the architecture body.
- Location and format of signal declarations.
- Signals declared in the architecture declarative part serve as the communication link between concurrently running processes that are interfacing with each other and reside in the same encapsulating architecture body.
- Predefined VHDL type BIT consists of the character literals '0' and '1'.
- Predefined VHDL type INTEGER.
- Processes may be optionally identified with a label.
- Processes, like hardware, run concurrently.
- Processes, like hardware, run indefinitely.
- Statements within a process are sequentially executed until a wait construct is reached.
- Location and format of variable declarations.
- Notation to initialize variables.
- Process variables maintain their respective values during the process's suspension.
- The symbol := denotes an assignment to a variable.
- The symbol <= denotes an assignment to a signal.
- The after clause of a signal assignment statement designates the future time at which the assignment is to occur.
- wait on...statement and its sensitivity list.
- wait until...statement and its sensitivity list.
- if...then...else...end if statement.
- VHDL predefined operator not.
- Definition of an event occurring on a signal.
- Activities of the VHDL initialization phase.
- Activities of the VHDL simulation cycle.
- Sequential signal assignment.
- Signals are initialized using the variable assignment notation, since the initial value is assigned immediately and not at some future delay time.

2.2 File: intro_static_sensitivity.vhd

The first model covered a tremendous amount of information. Congratulations for ploughing through and absorbing it all. We will continue our VHDL journeys by examining the model shown in Fig. 2.11.

Note that this file contains only an architecture body. Its corresponding entity declaration is in the file shown in Fig. 2.1. VHDL allows you to place these two design units into separate files. But the entity declaration must be successfully compiled before the architecture body undergoes the compilation process. Because of this compilation dependency order, the entity declaration is referred to as a primary design unit, whereas the architecture body is called a secondary design unit. Line 8 highlights one of the reasons why the entity declaration must be compiled first. The design entity name Encapsulate_3_Bit_Count is referenced on this line. If the entity declaration would not already be compiled, then the compiler would not recognize and accept this design unit's name. A compilation error would result, and the compiler would then terminate its analysis.

Incidentally, before compilation can even begin, the tool suite must be informed as to where to place the compiler's conversion of the raw

```
--  File Name  :  intro_static_sensitivity.vhd
--  Author     :  Joseph Pick
--  *********************************************** LINE NUMBER
architecture Intro_Static_Sensitivity                --  7
                        of Encapsulate_3_Bit_Count is --  8
    signal Clock : BIT := '0';                       --  9
begin                                                -- 10
    Gen_Clock:                                       -- 11
    process (Clock)                                  -- 12
    begin                                            -- 13
        Clock <= not Clock after 50 ns;              -- 14
    end process;                                     -- 15

    Mod_Incr:
    process
        variable Current_Count : INTEGER := 0;
    begin
        -- Wait until the trailing edge of the clock.
        wait until Clock = '0';

        -- Increment the counter modulo 8.
        if Current_Count = 7 then
            Current_Count := 0;
        else
            Current_Count := Current_Count + 1;
        end if;

    end process Mod_Incr;

end Intro_Static_Sensitivity;
```

Figure 2.11

VHDL source code. This specific location is known in VHDL as the current working library. The processed VHDL source code will be placed into this specified library whenever it is successfully compiled. The mechanism on how this current working library is created and managed is outside the scope of the VHDL language and is not explicitly specified in the LRM. In general, VHDL tool suites implement the current working library as a user-specified disk subdirectory. The concept of a current working library is very important in VHDL and has many ramifications. For instance, the entity declaration and all of its associated architecture bodies must be compiled into the same current working library. As per the VHDL language rules, your current working library may directly be referenced in your VHDL model via the logical name WORK. Several applications of WORK will be encountered later in our VHDL journeys.

Let's now continue with our explorations of this architecture body. Intro_Static_Sensitivity looks very similar to our previous Introductory (see Fig. 2.1) architecture. In fact, the Mod_Incr processes are identical in both these models. Consequently, we can focus our attentions on the process beginning at line 11. The true purpose of this architecture body is to introduce the VHDL syntax for a static sensitivity list. This new construct is shown on line 12. Since the signal Clock exists in the parentheses following the key word *process*, it is said to be in the process's static sensitivity list. Actually, any number of signals may have been included in this list, each separated by a comma (not a semicolon). In this particular example the process having the label Gen_Clock is theoretically equivalent to the waveform-generating process given in our previous architecture Introductory. Because of this equivalence, both architecture bodies will behave exactly the same. Consequently, there is no need to repeat the in-depth analysis that was previously done. Moreover, from this equivalence we can deduce that a process with a static sensitivity behaves as if it would have a wait on construct at its bottom just before the end process statement. This implicit wait on construct will be waiting for events on any of the signals listed in the parentheses following the reserved word *process*. In general, a sensitivity list consists of those signals that will cause a process to be reactivated. The term *static sensitivity list* implies that the members of the list are already known at compile time as opposed to being determined "on-the-fly" during run time. A process that contains at least one or more wait on or wait until constructs is said to have a dynamic sensitivity list since this list's membership does not necessarily have to be fixed during the duration of the simulation. Let me give you a realistic example that illustrates this "dynamicness." Suppose that at one point in time a process's reactivation depends only on an event on Address_Strobe. In this case the process's sensitivity list consists only of this signal. Later on, this same process might be waiting for an event only on Data_Strobe, in which case Data_Strobe is now the sole member of the process's sensitivity list. But be aware that the VHDL does not permit you to include an explicit wait statement inside a process that already has a static sensitivity list. So here is the tradeoff. It takes the simulation engine slightly more time to suspend a process having a dynamic sensitivity list than one that has a static sensitivity list. On the other hand, a process with a dynamic sensitivity list can model the behavior of complex devices more easily than processes that have a static sensitivity list. An example of a complex model is a process that must wait for events on different signals at different times during the progression of the simulation. Consider our Address_Strobe and Data_Strobe scenario from above. To begin with, it will be somewhat tedious to design a process that incorporates a static sensitivity list to model the bus interface protocol implied by these signals' names. Moreover, the additional if...then...else logic required to determine which of these signals actually had the event will more than nullify the efficiency gains earned by using a static sensitivity list. Consequently, my recommendation is to not spend your energies trying to figure out how to model complex devices with a static sensitivity list. Instead, just crank out the process and embed the necessary wait statements into it.

Here is another point regarding static sensitivity lists that you must be aware of. Since the implicit wait statement is at the bottom of a process having a static sensitivity list, such processes will always be executed once during the initialization phase. You should always be aware of the ramifications of this initial execution. Otherwise, your model might not behave as you expect it to.

2.2.1 Information checkpoint

Before you venture further along your VHDL journeys, please make certain that you are familiar with the following topics that we have established via this model:

- Entity declaration and architecture body may be in separate files.
- Entity declaration must be compiled before any of its corresponding architecture bodies.
- Purpose of the current working library.
- Sensitivity lists in general.
- Processes having a static sensitivity list.

2.3 File: intro_concur_signal_assignment.vhd

Figure 2.12 illustrates yet another architecture body for our very first design entity Encapsulate_3_Bit_Count. The main purpose of this model is to introduce you to the concept and behavior of a concurrent signal assignment.

Line 12 shows a typical concurrent signal assignment. You will immediately note that it is outside of any process construct. Recall how signal assignments work when they are embedded inside a process. In such environments, whenever the assignment statement is sequentially reached, the value currently on the right-hand side is then scheduled to be assigned at some later time to the targeted signal on the left-hand side. Because such assignments are solely a function of the process's sequential nature, they are called *sequential signal assignments*. On the other hand, concurrent signal assignments behave quite differently. For one, they are never sequentially reached, since they are not contained within a sequential region of code, such as a process. Moreover, their physical location in the file is completely irrelevant. I could have interchanged the order in which the process Mod_Incr and the concurrent signal assignment physically appear in the file, and the resulting model would still behave the same. The only thing that I

```
--    File Name  :   intro_concur_signal_assignment.vhd
--
--    Author     :   Joseph Pick
--
--    ***************************************************  LINE NUMBER

architecture Intro_Concur_Signal_Assignment           --  7
              of Encapsulate_3_Bit_Count is           --  8
    signal Clock : BIT := '0';                        --  9
begin                                                 -- 10
                                                      -- 11
    Clock <= not Clock after 50 ns;                   -- 12

    Mod_Incr:
    process
        variable Current_Count : INTEGER := 0;
    begin

        -- Wait until the trailing edge of the clock.
        wait until Clock = '0';

        -- Increment the counter modulo 8.
        if Current_Count = 7 then
            Current_Count := 0;
        else
            Current_Count := Current_Count + 1;
        end if;

    end process Mod_Incr;

end Intro_Concur_Signal_Assignment;
```

Figure 2.12

could not have done is to place line 12 inside the process. In that case the assignment would behave like a sequential signal assignment instead of a concurrent signal assignment.

So what are the key behavioral properties of a concurrent signal assignment? When will the right-hand side of the <= symbol be scheduled to go to the left-hand side? In general, the scheduling to the targeted signal will occur whenever any of the signals on the right hand side has an event. Actually, there is one exception to this rule and that occurs during the initialization phase. At this point of the simulation the initial values of the signals on the right-hand side are used when scheduling a value to the targeted signal. The very first scheduling does not wait for any events. It just will flow through analogous to a combinational circuit during its initial startup. However, all subsequent scheduling will wait for events on any of the signals named on the right-hand side. If all this sounds familiar, then my congratulations to you, for you have deduced the punch line of this model. Every concurrent signal assignment is theoretically equivalent to a process having a static sensitivity list consisting of those signals on the assignment's right-hand side. This equivalent process will contain only one sequential statement that is an exact copy of the original concurrent signal assignment. In other words, line 12 of Fig. 2.12 is theoretically equivalent to the process Gen_Clock of lines 11 through 15 of Fig. 2.11. In general, this equivalence should always be remembered, especially the fact that the implicit wait statement is at the bottom of its equivalent process. Hence, the concurrent signal assignment will always execute once during the initialization phase. You must always be aware of this initial scheduled assignment, or else your model might not behave as you expect it to.

One final note regarding concurrent signal assignments. Every concurrent signal assignment has a (static) sensitivity list whose members are those signals occurring on the right-hand side of the assignment. So in our particular example the signal Clock is the sole member of the sensitivity list associated with the concurrent signal assignment Clock <= not Clock after 50 ns.

2.3.1 Information checkpoint

Before you venture further along your VHDL journeys, please make certain that you are familiar with the following topics that we have established via this model:

- Behavior of concurrent signal assignments.
- Differences between concurrent and sequential signal assignments.

2.4 File: intro_delta_time.vhd

The file shown in Fig. 2.13 contains yet another architecture body for the design entity Encapsulate_3_Bit_Count. The main theme of this model is the VHDL term *delta time*. An understanding of this topic is very important. Otherwise, you will crank out lots and lots of VHDL code that will successfully compile, but its simulated behavior will unfortunately be contrary to your expectations. This delta time topic will be revisited and reviewed in later models of our VHDL journeys. Those subsequent models will reinforce the consequences of this concept and highlight some of the modeling pitfalls that delta time delays may unintentionally incur.

The incrementing process of this model once again remains unchanged. As before, we will focus our attentions on the process that is generating a periodic waveform on the signal Clock.

The first new VHDL construct of this file occurs on line 14. This variant of the wait statement is requesting that this process be suspended until 50 ns of simulation time have transpired. Recall from our earlier travels that the time-ordered queue contains those activities that have been scheduled to occur at some future time. Previously, we used this

```
--    File Name  :  intro_delta_time.vhd
--
--    Author     :  Joseph Pick
--
--    *************************************************** LINE NUMBER
architecture Intro_Delta_Time                              --  7
            of Encapsulate_3_Bit_Count is                  --  8
    signal Clock : BIT := '0';                             --  9
begin                                                      -- 10
                                                           -- 11
    process                                                -- 12
    begin                                                  -- 13
        wait for 50 ns;                                    -- 14
        Clock <= not Clock;                                -- 15
    end process;                                           -- 16
                                                           -- 17
    Mod_Incr:
    process
        variable Current_Count : INTEGER := 0;
    begin

        -- Wait until the trailing edge of the clock.
        wait until Clock = '0';

        -- Increment the counter modulo 8.
        if Current_Count = 7 then
            Current_Count := 0;
        else
            Current_Count := Current_Count + 1;
        end if;

    end process Mod_Incr;

end Intro_Delta_Time;
```

Figure 2.13

queue to implement the scheduling of signal updates. However, it is also appropriate to add those processes to this queue that have been suspended for a specified amount of time because of the *wait for* construct. OK, so let's do just that and add it to the queue. Next, suppose that 50 ns of simulation time have since passed. This process is now said to have timed out and must consequently be removed from the time-ordered queue. When this process is reactivated, it will continue at line 15, which contains the first executable statement following the wait construct that had caused its suspension.

To get the most out of this analysis we should once again play our previous game whereby you, the reader, will take on the role of the VHDL simulation engine. So you, the simulator, must now execute line 15 of this file, which is the statement Clock <= not Clock. Immediately, you will observe that this signal assignment is different from any of our previous signal assignments. Line 15 does not contain an after clause to explicitly stipulate when the scheduled update is to occur. But since signals are never updated instantaneously, when exactly will this signal modification be implemented? Because a time was not explicitly stated, the update to Clock must be scheduled to occur one delta time unit from now. Such delta time schedulings are, in general, implemented whenever the after clause is absent from a scheduled signal assignment, irrespective of whether the assignment is a sequential or concurrent one.

So exactly how long is a delta time delay? A delta time unit is an infinitesimal interval that never accumulates to an absolute unit such as nanoseconds or femtoseconds (the smallest absolute VHDL time unit). To better understand the delta time concept, you should think of VHDL simulation time as being represented by a two-dimensional grid. Consider the graph depicted in Fig. 2.14, which corresponds to the simulation activities of a hypothetical VHDL model. The symbol * designates those VHDL simulation times at which scheduled activities occurred.

The horizontal axis represents absolute simulation times such as 10 ns or 45 ps. A * symbol occurs at 0 ns because of the execution of the VHDL initialization phase during which each process of the model was simulated until a wait statement was reached. The * symbols at 10, 20, 40, and 50 ns imply that the model corresponding to this graph had some simulation activities at exactly these absolute times. Omissions of *'s along the horizontal axis correspond to those time intervals during which no activity was scheduled to occur. These gaps pictorially reinforce the event-driven nature of VHDL.

The vertical axis plots the advancement of delta times. From this two-dimensional perspective, it can be observed that delta times may progress indefinitely along the vertical axis while still maintaining a fixed absolute time unit on the horizontal axis. Delta time is incre-

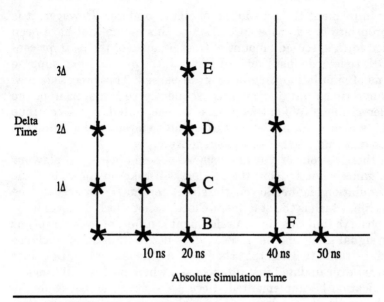

Figure 2.14 VHDL time is two-dimensional.

mented by exactly (!!) one unit whenever absolute simulation time is not increased, but there is an advancement to the next simulation cycle. We already met the term *simulation cycle* in our earlier models, but let me review it here in our current journey for quick and easy reference. A simulation cycle is segmented into two well-defined halves. During the first half, signals that are scheduled to be currently updated will be assigned their respective values. If this update is different from the signal's previous value, then the signal is immediately tagged as having received an event. Processes waiting for events on these event-tagged signals will then be awakened during the second half of this current simulation cycle. This well-defined flow of precisely when signals are updated and processes awakened establishes the deterministic mechanism via which the VHDL simulation engine coordinates the signal communications between concurrently running processes. Additionally, those processes that have timed out because of a wait for construct will also be awakened during the second half of the simulation cycle. In our previous models every simulation cycle occurred at an absolute simulation time. As per our two-dimensional graph, we now know that they really occurred at an absolute time plus 0 delta time.

In our current model the scheduling of Clock is first encountered at simulation time 50 ns. Since the simulation time is currently 50 ns, you, the simulator, must schedule Clock to get the value '1' (= not '0') at

50 ns plus 1 delta time unit. So just as before, you, the simulator, must place this scheduled assignment onto the time-ordered queue. Next you, the simulator, must loop back to the top of this process and suspend it again for another 50 ns by placing it again on the time-ordered queue. Since there are no more processes to be currently executed, you, the simulator, will now ask what must be done next as per the time-ordered queue. Since the updating of Clock is next, you will bump up the simulation time to 50 ns plus 1 delta and implement this assignment during the first half of the simulation cycle occurring at this time. Just as a reminder, recall that the updating of signals is always done in the background by the simulation engine and not while any processes are being executed. You, the simulator, will now reactivate the process Mod_Incr during the second half of this same simulation cycle occurring at 50 ns plus 1 delta, since during the first half Clock just had an event going from '0' to '1'. After this process is reactivated you must check if the wait condition is currently satisfied. But this wait condition is now false because at this point in the simulation Clock does not equal '0'. And so you, the simulator, must therefore suspend this process again. Next, you must bump up the simulation time to 100 ns, wake up the waveform-generating process, and this time schedule Clock to receive '0' (= not '1') 1 delta from now. The continuation of this model is left as an exercise for the reader. There should not be any difficulties since all the required VHDL concepts have already been fully developed. But you should be aware of a minor, though subtle, difference between this architecture body and the previous ones. In the architectures Introductory, Intro_Static_Sensitivity, and Intro_Concur_Signal_Assignment, the pulse edges for Clock's waveform occur at times 50 ns, 100 ns, 150 ns,...etc. In this current architecture Intro_Delta_Time Clock's edges occur at times 50 ns plus 1 delta, 100 ns plus 1 delta, 150 ns plus 1 delta,...etc. Hence the waveform of our current architecture is skewed by 1 delta time unit. This waveform difference will not have any severe repercussions since the incrementing processes from the various architectures will, in all cases, behave the same. They will all increment their counters immediately after the occurrence of Clock's trailing edge. And hence they will all satisfy the specification that an incrementation is to occur in response to the trailing edge of a periodic waveform.

Let us now return to the two-dimensional graph of Fig. 2.14. Consider the points B, C, D, and E along the vertical delta time axis. Since the * symbol occurs at each of these points, we know by our convention for this symbol that something must have occurred at each of these delta times. Contrast this situation to what happened along the horizontal axis. Recall that the graph had a discrete jump from 20 to 40 ns along the absolute simulation time axis. This void in absolute time occurred because there was no simulation activity during the time interval between 20 and 40 ns. Delta times, on the other hand, never skip over integral values while advancing along the vertical axis. The manner in which delta delays are scheduled guarantees that there will never be any integral gaps along the delta time axis. It is invalid to schedule a signal update to occur several delta time units into the VHDL future. The VHDL syntax provides you with only two options: You can either schedule a signal update to occur exactly 1 delta time unit later, or you can schedule a signal update to occur at any later absolute simulation time. In the former case the delta time is incremented by only 1 unit, whereas in the latter the delta time is reset to 0. In essence, signal delay times may only be specified by a method analogous to one of the following two signal assignments:

1. Clock <= not Clock after 20 ns;
2. Clock <= not Clock;

Method 1 schedules Clock to be updated during the first half of the simulation cycle that occurs at precisely 20 ns from the current absolute simulation time. Method 2 schedules Clock to be updated during the first half of the simulation cycle that occurs precisely 1 delta time from the current simulation time. By the way, Method 2 is equivalent to the following statement: Clock <= not Clock after 0 ns. So a propagation delay of 0 ns is equivalent to the delay of 1 delta time unit. And the passage of 1 delta time unit is equivalent to the advancement into the next simulation cycle without any modification to the absolute simulation time. Keep in mind that there is no VHDL syntax to update a signal after a specified number of deltas. And, anyway, there really is no practical need to do so. The last thing that you want to do is to start monitoring and counting the progression of delta delay units. You should always be aware of deltas, but you should never have to count them. The restriction that VHDL signal assignments are scheduled to occur at either an absolute time into the future (Method 1) or exactly 1 delta time unit later (Method 2) ensures that integral units along the vertical delta time axis will never be skipped over. For example, suppose that you are currently at simulation time 20 ns + 3 deltas (Point E of Fig. 2.14). Then at least one simulation activity must have been scheduled to occur at each of the following times:

20 ns + 0 delta

20 ns + 1 delta

20 ns + 2 delta

20 ns + 3 delta

In case you are wondering, let me illustrate a quick example of traveling up the delta time axis. Consider the following concurrent (not sequential!) signal assignments:

```
C <= B;              -- Concurrent assignment.

D <= C;              -- Concurrent assignment.

E <= D;              -- Concurrent assignment.

F <= E after 20 ns;  -- Concurrent assignment
```

Assume that the BIT type signals B, C, D, E, and F all have an initial value of '0'. Next, suppose that B is assigned the value '1' during the first half of the simulation cycle occurring at simulation time 20 ns. We now know that this event on B will, during the second half of this same simulation cycle, cause B's new value (= '1') to be scheduled for assignment to the signal C. Since this signal assignment statement does not contain an after clause, the update to C will occur exactly 1 delta time later. Hence C will receive the value '1' during the first half of the simulation cycle occurring at time 20 ns plus 1 delta. You can now see the (combinational) chain reaction that will next ensue. Signals D and E will successively be updated to '1' during the first half of the simulation cycle occurring at 20 ns plus 2 delta and at 20 ns plus 3 delta, respectively. The final concurrent assignment to F is scheduled using an after clause with a given delay time of 20 ns. I deliberately included this example to highlight that when a nonzero amount of time is specified in an after clause, then the current delta time units will all be absorbed into an absolute time. The scheduling of signal E's updated value to F will be done during the second half of the simulation cycle occurring at 20 ns plus 3 delta. One might therefore be tempted to state that F will be updated 20 ns later at the simulation time 40 ns plus 3 delta. However, this is not the case and, in fact, F will be updated at time 40 ns plus 0 delta. You should always be aware of this delta time absorption phenomenon whenever a nonzero absolute value is specified in a signal assignment's after clause. Otherwise, your expectations of the model's behavior will be incorrect.

By the way, many VHDL tool suites display the term *delta* in one of their simulation windows as an aid to the user. Unfortunately, the presented delta value may be somewhat confusing, since it does not correspond to the theoretical VHDL delta time that I have just described. Instead, this displayed value is a running accumulation of the VHDL simulation cycles. Consequently, its value is never reset to 0 when simulation time is advanced to an absolute time unit. The vendor intended for the customer to use this displayed delta value to identify an unintentional infinite feedback loop that causes the VHDL simulation time to advance out to infinity along either the delta or the absolute time axis. Most VHDL products have a safety net mechanism that aborts the simulation run whenever the current delta (vendor's term and not the VHDL term) exceeds either a default system or a user-specified value. So you can see that the vendor's intentions are good, but unfortunately they have incorrectly referenced the VHDL delta terminology.

While we are on the subject of vendor products, it should also be mentioned that those tool suites offering a waveform output capability do not display delta times. Otherwise, the resulting graph would have to be three-dimensional to properly display absolute time, delta time, and signal values. This three-dimensional graph would be very difficult to draw in a clear and user-comprehensible format. Instead, a two-dimensional graph is presented that exhibits only absolute times and signal values at these absolute times. So, if several delta time updates occur on a signal at the same absolute simulation time, then only the final value will be displayed on this absolute time-oriented waveform graph. Nonetheless, it should be pointed out that you do indeed have the option to monitor and record delta time updates to signals. Most tool suites provide you with the capability to generate a textual table that exhibits signal updates at both the absolute and the delta time granularity. By the way, I highly recommend that you get in the habit of routinely monitoring signal updates at the delta time level. This microscopic analysis is especially worthwhile during the preliminary stages of your debugging activities. Not only is it aesthetically appealing to see the data flow at this fine granularity level, but such a detailed analysis will also provide you with a valuable tool to isolate and identify many subtle design and coding errors.

Since assignments without an after clause look very similar to variable assignments, this is a good time to explore the topic of global variables versus signals. We could easily have discussed this subject earlier, but it is more meaningful to do so within the context of a delta delay environment. Recall that in the 1987 version of VHDL variables cannot be declared in the architecture declarative part, whereas signals can. The names of these signals henceforth become visible to all the concurrent constructs of the architecture's body. Consequently, these signal names are global with respect to the architecture body in which they are declared. On the other hand, variable names declared in a process declarative part are only visible to their encapsulating process and hence are local to it and cannot be referenced by any of the other concurrently running constructs. Let us now, for the sake of argument, assume that global variables do, indeed, exist and that they may be declared in the architecture declarative part. The following argument will expose why global variables were disallowed in the 1987 version of VHDL and why they are dangerous if misapplied in the 1993 version.

```
Update_A:
process
begin
    Global_Var_A := Global_Var_B + 1;
    ....
    wait for 55 ns;
end process;

Update_B:
process
begin
    Global_Var_B := Global_Var_A + 1;
    ....
    wait for 25 ns;
end process;
```

Figure 2.15 Global variables and nondeterminism.

The problem with global VHDL variables is that they would be (are as per VHDL'93) updated immediately within the framework of a concurrent environment. Let me show you what could go wrong based on our assumption that global variables are legal in VHDL. Consider the two processes of Fig. 2.15. As per our hypothesis, let Global_Var_A and Global_Var_B be global variables that are visible to both of the processes shown in this figure. Since these processes are running concurrently, it really should not matter which process is executed first. Suppose that during the initialization phase the VHDL tool suite purchased by your company will first attack the process Update_A. Under this choice variable Global_Var_A will be immediately updated with the incrementation of Global_Var_B's initial value. Upon reaching the wait statement, this process will then be suspended, and the tool suite will next attack the process Update_B. Here the updated value of Global_Var_A will be incremented and then immediately assigned to Global_Var_B. Now let us suppose that this same model is delivered to a customer that has a different VHDL tool suite. Suppose also that this customer's VHDL simulator will select Update_B as the first process to be executed during the initialization phase. Remember, since both processes are running concurrently, it should theoretically be irrelevant which of these two processes gets executed first. Since the process Update_B is now the first one to be handled, the variable Global_Var_B will immediately be assigned the incrementation of Global_Var_A's initial value. Already we now have a discrepancy with regards to the current value of Global_Var_B. Recall that your tool suite used the new value of Global_Var_A to update Global_Var_B. However, your customer's tool suite instead used the initial, old value of Global_Var_A to do so. Note that there now is an ambiguity in Global_Var_B's value.

Continuation of this simulation would highlight that Global_Var_A's value is also ambiguous, depending on which process was executed first. Whose simulation run is correct? Actually they both are. Unfortunately, this VHDL model is now nondeterministic: It may behave differently when executed by different VHDL simulation engines. So you can see why VHDL'87 forbids the existence of global VHDL variables. However, there is a need for them in certain modeling domains, and that is why global variables have been introduced into the VHDL'93 update. For instance, performance modeling relies heavily on globally declared objects that can be instantaneously updated and then subsequently referenced within a concurrently modeled environment. By *performance modeling* I mean the high-level simulation of a complex system to derive device utilization and bus traffic statistics. Though signals are global, it is very difficult and tedious to apply this global availability toward a performance-modeling environment. One drawback is that signals are never updated instantaneously. Another source of difficulty is that VHDL requires special handling of signals that are assigned to by more than one concurrently running construct. We will fully explore this topic of multiply driven signals later on in our travels.

And speaking of signals, I am sure that you are wondering what would have happened if the processes of Fig. 2.15 would have used signals instead of global variables. So since you asked, let's now repeat our earlier scenario, but this time within the context of signals. Consider the processes shown in Fig. 2.16. As before, suppose that your tool suite first attacks the process Update_A. Signal_A will then be scheduled to be updated 1 delta time later with the incrementation of Signal_B's initial value. Upon reaching the wait statement, this process will be suspended, and the simulator will then begin executing the process

```
Update_A:
process
begin
  Signal_A <= Signal_B + 1;
  ....
  wait for 55 ns;
end process;

Update_B:
process
begin
  Signal_B <= Signal_A + 1;
  ....
  wait for 25 ns;
end process;
```

Figure 2.16 Signals and determinism.

Update_B. Now let me ask you the following delta delay-related question. At this point has simulation time already advanced to the next delta time unit or are we still in the initialization phase? I am very confident that you unhesitatingly responded that we are currently still in the initialization phase. The scheduled delta has not yet occurred. So continuing along we are currently in the process Update_B where Signal_B is scheduled to be updated 1 delta time from now with the incremented value of Signal_A. Here is another question for you. Does this Signal_A currently have its initial, old value or its updated, new value? Once again I am confident that you know that Signal_A currently still has its initial, old value since the simulation has not yet advanced to the next delta. So after we reach the wait statement in this process it too will be put to sleep. When there are no more processes to be executed during the initialization phase, then the simulation engine will advance the time to the next delta time unit. Then both Signal_A and Signal_B will be unambiguously updated during the first half of the current simulation cycle. In both cases their respective initial values will be used in the incrementation. This discussion may be continued as in the global variable scenario, but it can readily be deduced that reversing the process execution order will still yield the same final values on Signal_A and Signal_B since both their updates will once again be scheduled to occur 1 delta time into the future. Analogous to our previous argument, their old, initial values will also be used in this reversed process execution scenario. So, in addition to mimicking the inherent delay of signal transmissions, the delta delay concept also serves as a synchronization mechanism for implementing an unambiguous communication link between concurrently running VHDL constructs.

As shown earlier the deterministic and well-defined behavior of the signal communication mechanism is due to the fact that, unlike global variables, signals can be updated at most just once during a simulation cycle. In fact, as we now know, signal updating can only be implemented during the first half of a simulation cycle. This assigned value is subsequently maintained for at least the duration of the second half of this same simulation cycle. Since process reactivation can occur only during the second half of a simulation cycle, all executed processes will see the same signal value. On the other hand, a global variable may be modified numerously during the second half and, consequently, will present different values of itself to the various concurrently running processes. Hence, the order in which processes are awakened now becomes important because different reactivation sequences may see different values for the same (VHDL'93) global variables. This nondeterminism of global variables may create havoc in your model. Hence global variables should never be used for your everyday hardware modeling activities. Rather, global variables should be reserved for those environments, such as performance modeling, where determinism is not a relevant issue. Nonetheless, irrespective of how well defined and methodical all of this sounds, caution must still be exercised when working with signals. Otherwise, the signal's old value will be used instead of the new updated value, and vice versa. This potential error in VHDL software design is a mirror image of what may go wrong when designing real hardware. A common concern that digital designers have is whether they are using the old value or the new updated value of a signal. Hence, it seems reasonable that VHDL engineers should also have similar concerns since VHDL is, in fact, modeling hardware systems and subsystems. As we travel deeper into the VHDL landscape, we will see many examples of the VHDL version of this old value versus new value timing error.

2.4.1 Information checkpoint

Before you venture further along your VHDL journeys, please make certain that you are familiar with the following topics that we have established via this model:

- Delta time.
- There are no global variables in VHDL'87.
- The dangers of using global variables in VHDL'93.

2.5 File: cfg_introductory.vhd

OK, so let's step back and see what we have accomplished so far. Currently, we have analyzed one entity declaration and four architecture bodies for the design entity Encapsulate_3_Bit_Count. Since the entity declaration has no ports, it does not need to interface with any other design entities. Hence, this design unit is totally self-contained and may be simulated as a standalone device. But which of these four architecture bodies is to be used in this simulation? As per the VHDL language, you have two options. You can either blindly rely on the VHDL default binding as specified in the LRM or you can explicitly pronounce which one to simulate via a configuration declaration. Let's first look at the default binding option. Unless otherwise specified, a simulation run will, by default, use the architecture body that was most recently compiled. So according to this default rule if the architecture Introductory was the last one compiled, then it will automatically be selected during the simulation of Encapsulate_3_Bit_Count. Clearly, using the VHDL default binding mechanism is just not the way to go. It definitely is not practical to continually have to recompile architecture bodies in order to ensure that they be included in a simulation. Moreover, for

```
-- File Name : cfg_introductory.vhd
--
-- Author    : Joseph Pick
--
-- *********************************************************** LINE NUMBER
configuration Universe_Introductory of Encapsulate_3_Bit_Count is --  7
                                                                  --  8
    for Introductory                                              --  9
    end for;                                                      -- 10
                                                                  -- 11
end Universe_Introductory;                                        -- 12
```

Figure 2.17

future reference, you would like to keep a concrete record of what constituents actually went into a specific simulation run. This is what software configuration control is all about. Hence, in general, it is recommended that you do not rely on this VHDL default option. Instead, you should use a VHDL configuration declaration to formally document the various constituents of each simulation run that you are experimenting with. Figure 2.17 shows a typical configuration declaration.

In general, the reserved word *configuration* announces a user's intention to create a configuration declaration. In this example I have decided to uniquely identify this configuration with the name Universe_Introductory. Note the following pattern very carefully. Following the reserved word *of* is the name of the design entity that we are now configuring. Since the current working library can contain only one entity declaration for this design entity, there is no ambiguity as to which one is to be used in this configuration. However, multiple architectures are possible. Hence, you must now reference the reserved word *for*, as on line 9, to explicitly indicate which of the multiple architecture bodies you are selecting for later simulation. In this configuration I have designated the architecture body called Introductory. In the VHDL vernacular I have just bound the architecture body Introductory to the design entity Encapsulate_3_Bit_Count. The reserved words *end for* on line 10 and *end* on line 12, respectively, terminate this pronounced binding and the configuration declaration. It is optional to repeat the configuration's name, and I have done so for readability. Prior to the compilation of this file, it is necessary that both the entity declaration and the identified architecture body have already been compiled into the current working library. Otherwise, a compilation error message would be generated and the compiler would discontinue its activities. The configuration Universe_Introductory is depicted in Fig. 2.18. This book will use rectangles with rounded edges to represent configurations.

Figure 2.19 shows another configuration for the design entity Encapsulate_3_Bit_Count. After compiling this second file, you will then have a choice as to which entity/architecture pair to use when simulat-

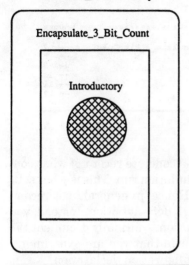

Figure 2.18 Configuration Universe_Introductory identifies Introductory as the architecture body to be used when simulating the design entity Encapsulate_3_Bit_Count.

```
-- File Name : cfg_intro_delta.vhd
--
-- Author    : Joseph Pick
--
configuration Universe_Intro_Delta of Encapsulate_3_Bit_Count is

   for Intro_Delta_Time
   end for;

end Universe_Intro_Delta;
```

Figure 2.19

ing a copy of the design entity Encapsulate_3_Bit_Count. You can request your VHDL tool suite to simulate either Universe_Introductory or Universe_Intro_Delta. So you are now in full control over which architecture body is to be simulated. Moreover, the configuration declaration provides you with the means to formally proclaim and assemble the constituents of a simulation run. This organizational capability facilitates the long-term management and control of your simulation tasks. Both the configuration declaration and the concept of model bindings will be revisited later in a more complex setting.

By the way, let me state, just for the record, that configurations are considered as VHDL primary design units. Recall that entity declarations are also primary design units, whereas architecture bodies are secondary design units.

2.5.1 Information checkpoint

Before you venture further along your VHDL journeys, please make certain that you are familiar with the following topic that we have established via this model:

- Purpose and format of a configuration declaration.

2.6 File: structural_decomposition.vhd

Figure 2.20 shows what we have accomplished so far and where we are planning to go next in our top-down exploration of both VHDL and our 3-bit incrementing companion.

On the left are those architecture bodies that we have already examined in close detail. The lone architecture on the right is where we are now going to continue. Structural_Decomposition will partition our original, abstract 3-bit counter into subcomponents that are a more authentic representation of the design entity's final hardware realization. This partitioning is the first step in our top-down refinement of the design entity Encapsulate_3_Bit_Count. Later, this decomposition will be partitioned even further until we have traveled all the way down to the gate level. Throughout our VHDL journeys it is very important that you always be aware of your bearings in terms of the big pic-

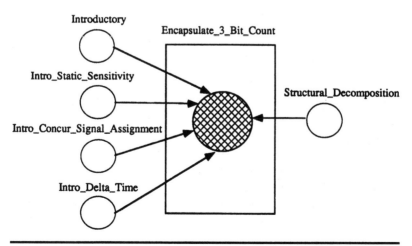

Figure 2.20 Multiple architecture bodies for the design entity Encapsulate_3_Bit_Count.

48   An Excursion into VHDL

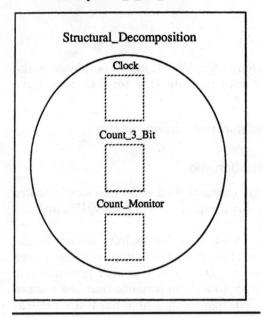

Figure 2.21 Partitioning of the design entity Encapsulate_3_Bit_Count via the architecture body Structural_Decomposition.

ture and how the various refinements fit together. If you ever get lost or become disoriented, then just review the graphic figures of this book. The main thrust of these illustrations is to serve as a road map for the progression and objectives of our top-down journeys.

Figure 2.21 illustrates a top-down decomposition of the design entity Encapsulate_3_Bit_Count into three subcomponents. The VHDL implementation of this subcomponent refinement will be achieved via the architecture body Structural_Decomposition. Note well that the boxes representing these subcomponents are all dotted instead of being solid. I am using this dotted notation to highlight the fact that, at this point in time, these components may be viewed as merely abstract concepts. Their sole purpose is to serve only as templates for their respective design entity realizations. Their corresponding entity/architecture realizations do not yet have to exist. It is as if a top-level design had been partitioned into block diagrams, and each of these blocks now has to subsequently be built by the various digital design team members. This task decomposition approach parallels the divide-and-conquer strategies of a real engineering project. In our VHDL context, a component template will be considered to be real and built only when a corresponding entity/architecture pair will have been written for it and successfully compiled into a library. In order to facilitate a top-down design strategy, VHDL allows architectures, such as Structural_Decomposition, to be compiled before their referenced components actually exist.

The overall game plan of this decomposition is to take each of the processes from the architecture body Intro_Delta_Time and encapsulate them within their own design entity. With this objective in mind, it should be apparent from the labels in Fig. 2.21 that the dotted box Clock will serve as the template for a design entity that will generate a periodic waveform. Likewise, the dotted box labeled Count_3_Bit will be a template for a 3-bit incrementer. However, the third subcomponent, Count_Monitor, does not correspond to any process in the architecture body Intro_Delta_Time. I originally added Count_Monitor to Structural_Decomposition in order to provide yet another vehicle with which to introduce further VHDL constructs. However, its inclusion immediately crystallized the observation that, as it now stands, Structural_Decomposition is in the format of an architecture body for a typical VHDL test bench. The colloquial VHDL term (i.e., not LRM term) *test bench* refers to a top-level (i.e., no ports) design entity written specifically to test and validate the functionality of a design entity. Typically, the architecture body associated with a test bench is composed of three subcomponents analogous to the exhibited partitioning for Structural_Decomposition. One of the test bench's subcomponents will correspond to a digital designer's concept of a *device under test* (DUT). The DUT is sometimes also referred to as the *unit under test* (UUT). A second subcomponent will generate the appropriate stimulus for the DUT. The injected stimulus may be created either internally, or it may be derived from reading an input file containing a set of test vectors. The third and final subcomponent will serve as a logic analyzer to monitor the key signaling activities of the DUT. In general, this observed data is recorded into an output file for later inspection and analysis. Another tactic for this logic analyzer is to compare signal values generated during a simulation run to expected values contained in an input file. The pass/fail status of these comparisons will subsequently be written to an output file. So on the basis of this analogy, you can now appreciate why the entity/architecture pair Encapsulate_3_Bit_Count and Structural_Decomposition may be interpreted as a VHDL test bench. In this specific decomposition the entity/architecture pairs for the subcomponents Clock, Count_3_Bit, and Count_Monitor will correspond to the stimulus generator, the DUT, and the logic analyzer, respectively.

The VHDL source code for the architecture body Structural_Decomposition is shown in Fig. 2.22.

50   An Excursion into VHDL

```
-- File Name : structural_decomposition.vhd
--
-- Author    : Joseph Pick
--
-- ******************************************************* LINE NUMBER
architecture Structural_Decomposition of                    --  7
               Encapsulate_3_Bit_Count is                   --  8
                                                            --  9
   component Clock                                          -- 10
          generic (PULSE_WIDTH : TIME);                     -- 11
          port    (Clock : out BIT);                        -- 12
   end component;                                           -- 13
                                                            -- 14
   component Count_3_Bit                                    -- 15
          port    (Clock : in BIT;                          -- 16
                   Dataout : out BIT_VECTOR(2 downto 0)     -- 17
                  );                                        -- 18
   end component;                                           -- 19
                                                            -- 20
   component Count_Monitor                                  -- 21
          port    (Datain : in BIT_VECTOR(2 downto 0));     -- 22
   end component;                                           -- 23
                                                            -- 24
   signal Tic_Toc   : BIT := '0';                           -- 25
   signal Data_Cnt  : BIT_VECTOR(2 downto 0) := (others => '0'); -- 26
                                                            -- 27
begin                                                       -- 28
                                                            -- 29
   Synch: Clock                                             -- 30
          generic map (PULSE_WIDTH => 50 ns)                -- 31
          port map    (Tic_Toc);                            -- 32
                                                            -- 33
   Gen_Cnt: Count_3_Bit                                     -- 34
          port map (Tic_Toc, Data_Cnt);                     -- 35
                                                            -- 36
   Logic_Analyzer: Count_Monitor                            -- 37
                   port map (Datain => Data_Cnt);           -- 38
                                                            -- 39
end Structural_Decomposition;                               -- 40
```

Figure 2.22

The architecture declarative part (lines 9 to 27) of this model contains more than just the usual signal declarations. In Structural_Decomposition we additionally have a collection of component declarations. This collection of component declarations establishes the list of design entity templates that may later be referenced and interconnected via a sequence of VHDL instantiation statements. Component declarations and instantiations are VHDL concepts that have real-world counterparts. Suppose that you are on a workstation or PC doing schematic entry. It really is a two-step process. First, you must choose from one of the many hardware symbols offered by your schematic entry tool. Next, you must (electronically) place this selected symbol onto your current schematic workspace and draw the appropriate netlist connections to the other symbols already existing in this layout. The collection of your VHDL component declarations correspond to the list of hardware symbols that are offered by your schematic entry tool. The component instantiations then correspond to your netlist configuration for these hardware symbols. Now let me ask you this question. Can you simulate your schematic if your schematic entry environment is merely an editor? Definitely not! The simulation of any schematic requires knowledge about the timing and behavioral characteristics for those symbols used in your schematics. Similarly, the architecture body Structural_Decomposition can only be compiled but it cannot be simulated, since the entity/architecture pairs corresponding to its declared components, as yet, do not exist. By the way, whenever a VHDL vendor advertises that their product can convert your schematics into VHDL, all that you will really get is an architecture body that looks very much like Structural_Decomposition. The resulting tool-generated file will consist of a collection of component declarations and their subsequent instantiations. In essence, all you will have is a netlist in the VHDL medium. You will be able to compile this file, but beware that you will not be able to simulate it. To be able to do so, you must either write the entity/architecture pairs for the listed components, or you must purchase these models from a supplier.

Having just discussed the big picture for Structural_Decomposition, let us now zero in on its fine points. Lines 10 through 13 represent the component declaration for Clock. Both *component* and *end component* are VHDL reserved words. Note the necessary placement of semicolons in this construct. Line 11 begins with the reserved word *generic* that, in general, introduces the generic parameter interface list of this component. What is the purpose of a generic parameter? Recall that the pulse-generating process in the architecture body Intro_Delta_Time had the statement wait for 50 ns. As a consequence, the periodic waveform had a pulse width of 50 ns. But what if you would want a pulse width of 20 ns? One way to achieve this is to write another architecture body with essentially the same code, but this time wait for 50 ns would be replaced by wait for 20 ns. Great, but what if your next project required a periodic clock with a pulse width of 16.33 ns? Would you then have to write yet another architecture body but this time with a wait for 16.33 ns statement? You can see that in the long run this approach is very impractical. So instead, VHDL allows you to pass constant values into your model via generic parameters that your model can reference. This way you only have to write the architecture body once using the generic parameter names. And so it is possible to write an architecture body containing the statement wait for PULSE_WIDTH, where PULSE_WIDTH is a generic parameter. Very shortly we will discuss the mechanism used by VHDL to assign a concrete value to PULSE_WIDTH. This assigned value will subsequently be used during a simulation whenever the generic parameter PULSE_WIDTH is referenced. The upper case format of PULSE_WIDTH is in conformance with my VHDL style guide convention to always write the names of constant objects in upper case. This approach is quite valid since generic parameters, such as our PULSE_WIDTH, must be treated as constants within the component's corresponding entity/architecture pair. By PULSE_WIDTH behaving as a constant, I mean that the compiler will check to guarantee that PULSE_WIDTH is never assigned a value inside the entity/architecture pair realization of the component Clock. Line 11 states that PULSE_WIDTH is of type TIME. This data type comes free with the VHDL language and is used to represent absolute simulation time. The available units, such as femtoseconds, picoseconds, and nanoseconds, have already been pointed out during one of our earlier journeys. The smallest absolute time unit is femtoseconds. Recall that delta time units are an artifact of the VHDL language and do not correspond to any concrete physical time unit. Though there are many fascinating applications for generic parameters, in general, they are used to pass into a model time-related values such as propagation delays, pulse widths, and setup and hold times. The semicolon on line 11 terminates the generic statement and is required by the compiler.

The reserved word *port* on line 12 introduces the port interface list for this component. As stated earlier, a port can be viewed intuitively as a chip's pin. The component called Clock has only one port, which is also named Clock. It is of type BIT which we already have met before. Whenever you are working with real hardware, you are always interested in the direction that the data is flowing in. Is the signal flowing into the pin, out of the pin, or is it bidirectional? In VHDL these three options are paralleled by the reserved words *in, out,* and *inout,* respectively. The directional flow of a port is formally denoted in VHDL as its mode. In our current component the port Clock has the mode out, which agrees with our intuitive understanding of how this model will eventually work. During simulation a periodic waveform will be transmitted out of the design entity that serves as the concrete realization for this template. The semicolon on line 12 terminates the port declaration and its absence would generate a compilation error. The generic declaration must always come before the port declaration.

Lines 15 through 19 illustrate the component declaration for Count_3_Bit. This component does not have any generic parameters. The port declaration announces that this component has the two ports named Clock and Dataout. Note that the declarations of these two ports are separated by a semicolon. The port Clock is of mode in, and it corresponds to the pin through which the necessary synchronizing pulse will enter. The trailing edge of this pulse will initiate the next incrementation update. The port Dataout is of mode out, and it corresponds to the group of three pins that collectively output the current incremented value. Dataout is an array of type BIT_VECTOR consisting of three elements that, as per the port declaration on line 17, may each be uniquely identified via the VHDL nomenclature Dataout(2), Dataout(1), and Dataout(0). BIT_VECTOR is yet another type that is always available for users of the VHDL language. BIT_VECTOR is defined in the LRM to be an unconstrained array type that is indexed by any NATURAL number. Let's elaborate and discuss these new terms beginning with array. An *array* is a collection of elements, each of which must be of the same type. Each element may be uniquely addressed via a value enclosed in parentheses called its *index*. The LRM definition for BIT_VECTOR allows any nonnegative integer to be used as an index. The collection of nonnegative INTEGERs is a subset (subtype) of the VHDL data type INTEGER. As per the LRM the name of this subtype is NATURAL. Furthermore, each individual element of BIT_VECTOR is of type BIT. Hence, in our model every element of the array Dataout is of type BIT. An interesting and useful feature of BIT_VECTOR is that it does not specify a precise upper or lower range bound for the indexes of this array data type. The BIT_VECTOR data type is not restricted to having a fixed number of elements. Consequently, its length is not explicitly defined and, in the VHDL vernacular, is said to be an unconstrained array type. In general, unconstrained arrays create a very flexible modeling environment. They allow you to define arrays that element-wise are fundamentally the same except that their lengths or index range bounds are different. Consequently, you do not have to keep on defining a new data type each time you want to work with an array of BITs. In fact, doing so would not even be a good modeling practice since the strong data typing VHDL environment would not permit objects of these different array data types to be assigned to each other even though, in essence, their elemental constituents are of the same type BIT. In our particular example Dataout was declared to be of BIT_VECTOR type and to have a descending index range of 2 down to 0. Alternatively, Dataout could have, just as well, been specified via the ascending range 0 to 2. Incidentally, the specific ascending or descending range that you declare does not, by itself, imply which is the *most significant bit* (MSB). In this model is Dataout(0) or Dataout(2) the MSB? In general the MSB is identified by the manner in which the array is computationally manipulated. In our subsequent VHDL journeys we will treat Dataout(2) as the MSB. By the way, the precise ascending or descending range used in your model should really be influenced by the range specification of the corresponding hardware pins. For instance, if the original hardware designer's block diagram includes Ain(31:0) as the label for a set of pins, then the corresponding VHDL port should also have the name Ain, and its indexing range should be specified as 31 *downto* 0. In the end, since VHDL is modeling hardware the hardware naming conventions should always win. While on the topic of recommended VHDL coding practices here is another useful suggestion. Whenever a port declaration contains several ports, they should all be listed according to the following categorical order. Group all the ports of mode in together and list them first. This group should then be successively followed by the ports of mode out and inout. Furthermore, within each group the ports should be arranged alphabetically. Such an ordering will greatly assist in the readability and long-term maintainability of your VHDL models.

Lines 21 through 23 show the component declaration for Count_Monitor. This component has only one port parameter named Datain that is of mode in. Its type is BIT_VECTOR and its indexing range is also 2 down to 0. Remember that this component corresponds to the logic analyzer module of a test bench environment.

Line 25 declares the signal Tic_Toc to be a BIT type object that is initialized to '0' at the startup of the simulation. Line 26 declares the signal Data_Cnt to be a BIT_VECTOR array having an index range 2 down to 0. All three of its array elements are initialized to '0' using the reserved word *others*. The statement others => '0' sets every element of the array Data_Cnt to '0'. This parenthesis notation is called an aggregate and may be used to collectively reference the elements of an array. The signals Tic_Toc and Data_Cnt will play a very important interconnecting role during our component instantiation activities that, as stated earlier, is analogous to the formation of a hardware netlist.

The architecture statement region (lines 29 to 39) of Structural_Decomposition consists only of component instantiations. This architecture body does not have any processes or concurrent signal assignments. Hence, it is an excellent representative of a pure structural VHDL model. This architecture is the VHDL version of a schematic netlist. Note well that every instantiation must be labeled. Here the specific labels are Synch, Gen_Cnt, and Logic_Analyzer. These labels are very important, and they should be well chosen and easily identifiable. Whenever you are simulating this model and you wish to travel down its hierarchical chain, then you will have to choose between these instantiation labels and not the names of their corresponding components. In this example each component is instantiated only once. There really is no restriction in doing so. For instance, it is possible to instantiate multiple copies of the component Clock. However, each of the Clock instantiations must have a different label such as Synch_50_ns or Synch_30_ns. This nomenclature requirement is analogous to drawing several identical logic gates onto a schematic and then labeling them as U1, U2, U3, etc.

Let's now take a closer look at the mechanics of an instantiation. The colon separating the name of the instantiation and the component's name is necessary. Any number of blank spaces may have been placed between Synch, the colon, and Clock with no ill effects. Line 31 contains a generic map elaboration. The reserved words *generic map* introduce the generic parameter association list. It is via such generic map statements that a generic parameter is given a concrete value. At the start of a simulation, just prior to the initialization phase, this value for the generic parameter will be passed into the entity/architecture pair associated with this Clock component template. Incidentally, this preinitialization phase is formally identified as the VHDL simulation's elaboration phase. In our particular model the generic parameter PULSE_WIDTH is associated with the value 50 ns. PULSE WIDTH is referred to as a local generic parameter since, in the final analysis, its encapsulating component is merely an intangible template. Only entity declarations and architecture bodies are considered concrete and real. The 50 ns is referred to as an actual parameter. In the VHDL vernacular the local generic parameter PULSE_WIDTH is associated with the actual parameter 50 ns. This particular correspondence is achieved using the named association method. Note well that the generic map construct is not concluded by a semicolon. Recall from our analysis of line 11 that the generic parameter list had to be terminated by a semicolon. Here, on the other hand, the generic map construct is terminated by a semicolon only if there is no port map association following it. So be very careful when you are creating the instantiation via a cut-and-paste operation of the component declaration. In any event, if you make a mistake with regards to this semicolon, then the compiler will catch it and (politely) inform you of the error.

On line 34 we have a port map construct that associates the local parameter Clock with the actual parameter Tic_Toc. In this case positional association is used instead of named. Similarly, positional association is used to associate Tic_Toc and Data_Cnt with the local parameters Clock and Dataout, respectively. Note the comma separating these two actual parameters, whereas semicolons must be used to separate the local parameter declarations (see line 16). By *positional association* I mean that the actual parameters are in the exact same linear position as their local parameter counterparts. It would have also been satisfactory to use the following named association technique:

```
port map (Clock => Tic_Toc, Dataout => Data_Cnt);
```

Named association permits you to explicitly list the desired associations in any order. The final effective result will still be the same in terms of the actual-to-local associations. Hence, it would also be acceptable to write

```
port map (Dataout => Data_Cnt, Clock => Tic_Toc);
```

Though this transposition option does exist, my recommendation, in general, is to always use named association and, even then, to still maintain the relative positional orderings of the original parameter list. This convention will greatly enhance the readability of your model.

Both named and positional may be mixed in the same association list, but once named is begun, then the remaining associations must also continue with this technique.

Note well that in our particular associations both the local and actual parameters are of the same data type. This conformance must be true in general since VHDL is a strongly typed language.

On line 37 we have the instantiation of the component Count_Monitor. This instance, labeled Logic_Analyzer, associates the local parameter Datain with the actual parameter Data_Cnt.

Observe that in this model all the actuals interconnecting the various ports are signals declared in the architecture body Structural_Decomposition. There is yet another VHDL'87 option for an actual parameter that we will come across later in our VHDL journeys. For the time being it is very instructive to return to our schematic entry analogy and observe that a signal may, in essence, be interpreted as the label for a wire that connects subcomponent pins (ports) together.

When this model is finally able to be simulated, then these instantiated components will all be running concurrently and communicating

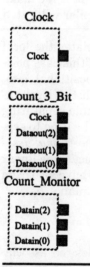

Figure 2.23 A collection of component declarations establishes the list of available design entity templates that may later be referenced and interconnected via a sequence of VHDL instantiation statements.

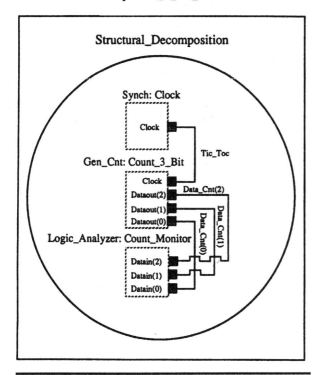

Figure 2.24 The component instantiations in the architecture Structural_Decomposition establish a schematic netlist for the design entity Encapsulate_3_Bit_Count.

with each other via the signals connecting their ports. Hence, in general, ports serve as the communication gateway between concurrently running design entities that are interfacing with each other during a simulation.

The key features of Structural_Decomposition may be summarized via Figs. 2.23 and 2.24. The first figure points out that a component declaration merely defines a symbolic template for a corresponding design entity. The dotted rectangles emphasize that these design entities do not even have to exist yet. Figure 2.24 depicts that component instantiation is merely a netlist between the declared component templates. Inspection of the wires connecting the three instantiated components yields the observation that in this particular model their pins are connected via the internal signals declared in Structural_Decomposition.

2.6.1 Information checkpoint

Before you venture further along your VHDL journeys, please make certain that you are familiar with the following topics that we have established via this model:

- Purpose and format of a VHDL test bench.
- Purpose and format of a component declaration.
- Purpose and format of a component instantiation.
- Definition and purpose of an unconstrained array type.

2.7 File: count_3_bit_.vhd

Figure 2.25 is the road map of where we are now going to travel. Recall that Structural_Decomposition is the architecture body via which the design entity Encapsulate_3_Bit_Count was partitioned into the three components: Clock, Count_3_Bit, and Count_Monitor. As per our top-down design methodology, we must now develop design entities for each of these three template components. This section will explore an entity declaration for the design entity Count_3_Bit. As shown in Fig. 2.25 this design entity will correspond to the previously declared com-

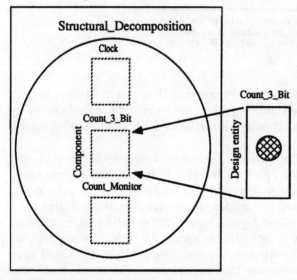

Figure 2.25 The design entity Count_3_Bit will serve as a concrete implementation for the template component Count_3_Bit.

```
--  File Name :  count_3_bit_.vhd
--
--  Author    :  Joseph Pick
--  *************************************************  LINE NUMBER
entity Count_3_Bit is                                  --  7
       port (Clock   : in BIT;                         --  8
             Dataout : out BIT_VECTOR(2 downto 0)      --  9
            );                                         -- 10
end Count_3_Bit;                                       -- 11
```

Figure 2.26

ponent having the same name. Note that the rectangle representing the entity has a solid line perimeter, whereas the component has a dotted one. As mentioned earlier this notational convention points out that design entities are in a sense real and tangible, whereas components are merely symbolic templates that are void of any concrete existence. Later sections will develop design entities for the other components Clock and Count_Monitor.

The entity declaration for the design entity Count_3_Bit is shown in Fig. 2.26. This is the first entity declaration that we have seen that has a port declaration. But, in essence, there really is nothing new here between lines 8 to 10 since we have already discussed a similar port declaration during our previous investigation of the architecture Structural_Decomposition.

This file contains only an entity declaration, thus once again pointing out that it is possible to place the entity declaration and its corresponding architecture bodies into separate files. However, remember that the entity declaration must be compiled before any of its architecture bodies.

The trailing underscore in the file's name is a technique often used in the VHDL community that you may find very convenient. When browsing through a subdirectory, this trailing underscore convention may be applied to quickly identify those files that contain only entity declarations.

Note well that the design entity name Count_3_Bit corresponds exactly to the name used in the component declaration. Moreover, the port names and declarations are identical as well. This concurrence does not have to be so. VHDL is flexible enough to allow differences in these identifiers. However, any naming discrepancy must be accounted for later during the formal binding of the design entity to the corresponding component template. A mapping must then be established between the component and entity declarations in order to account for any differences in these names. An example of this mapping technique will be shown later in our journeys. As per our current example all the names match up exactly, and so there is no need for an explicit correlation between the component and its intended corresponding design entity. According to the LRM a default binding (correlation) is always available between a component and a design entity whenever their names, generics, and ports match exactly. Whenever such an exact matchup exists and an explicit binding is not specified via a configuration declaration, then the tool suite must select this matching design entity for simulation purposes. Also, recall from our earlier journeys that if multiple architectures exist, then the most recently compiled version will be the one selected when simulating its corresponding design entity.

Another VHDL nomenclature to remember is that the port and generic parameters occurring in an entity declaration are known as formal parameters. For comparison, recall that the generic and port parameters of a component declaration are called local parameters.

2.7.1 Information checkpoint

Before you venture further along your VHDL journeys, please make certain that you are familiar with the following topic that we have established via this model:

- The hierarchical role of the design entity Count_3_Bit in the top-down decomposition of the design entity Encapsulate_3_Bit_Count.

2.8 File: count_3_bit_model_1.vhd

The road map in Fig. 2.27 reveals the scope of our upcoming VHDL journeys and their relation to our previous travels. Recall that our current game plan is to develop a design entity that will be used during simulation as a concrete realization for the component template Count_3_Bit. Its entity declaration has already been discussed in the previous subsection. As per Fig. 2.27's panorama, we are now going to closely analyze and explore numerous architecture bodies for the design entity Count_3_Bit. Previously, we had done the desired incrementation in the INTEGER domain. Henceforth this activity will be implemented in the BIT domain. That way the '0' and '1' outputs of our model will have the same look and feel as real hardware. But more importantly our true objective is to exploit these various architectures as a vehicle to introduce either new VHDL features, caveats, or yet a further top-down design path. In either case the various models will reflect a designer's viewpoint of how he/she wishes to perceive the 3-bit counter in a modeling project. It will become very apparent that, in general, VHDL's robustness allows the designer to build models at any appropriate level of abstraction. In our particular case the 3-bit incrementation may be modeled via either a bit-wise algorithm (Model_1), a state machine (Model_2 through Model_4), a data flow

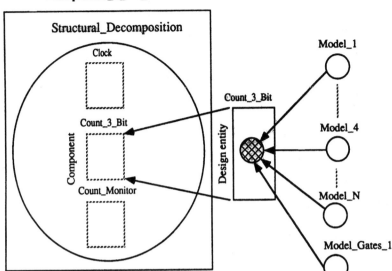

Figure 2.27 Multiple architecture bodies will be presented and explored for the design entity Count_3_Bit.

description (Model_5 through Model_7), or even a gate level decomposition (Model_Gates_1).

So let's begin with Model_1. Its VHDL source code is shown in Fig. 2.28.

Line 7 states that this architecture body called Model_1 is associated with the design entity Count_3_Bit. Recall that the entity declaration for Count_3_Bit must already have been successfully compiled into our current working library. Otherwise, the design entity name Count_3_Bit would not be known to the compiler. This lack of name visibility would then cause the compiler to generate an error message and to terminate its activities. Very shortly we will see that there are other visibility reasons for compiling the entity declaration before any of its associated architecture bodies.

This architecture Model_1 contains only one process, and there really is no need for any interprocess communications. Thus it is technically not necessary to work with signals, and, consequently, there are no signal declarations in the architecture declarative part. However, the process Bit_Incr does have a variable declaration on line 13. Current_Count is declared to be a BIT_VECTOR type having range 2 down to 0. Hence, Current_Count will be three elements wide where each element is a BIT. As per one of our earlier discussions, the MSB of Current_Count is still undeterminable at this point. On line 14 an initial

An Excursion into VHDL

```
-- File Name : count_3_bit_model_1.vhd
--
-- Author    : Joseph Pick
--
-- ******************************************************* LINE NUMBER
architecture Model_1 of Count_3_Bit is                      --  7
                                                            --  8
begin                                                       --  9
                                                            -- 10
   Bit_Incr:                                                -- 11
   process                                                  -- 12
      variable Current_Count : BIT_VECTOR(2 downto 0)       -- 13
                             := ('0','0','0');              -- 14
                          -- := "000";                      -- 15
                          -- := (others => '0');            -- 16
   begin                                                    -- 17
                                                            -- 18
      -- Wait until the trailing edge of the clock.         -- 19
      wait until Clock = '0';                               -- 20
                                                            -- 21
      -- Increment the counter.                             -- 22
      Incr_Loop:                                            -- 23
      for K in 0 to 2 loop                                  -- 24
         if Current_Count(K) = '0' then                     -- 25
            Current_Count(K) := '1';                        -- 26
            exit Incr_Loop;                                 -- 27
         else                                               -- 28
            Current_Count(K) := '0';                        -- 29
         end if;                                            -- 30
      end loop Incr_Loop;                                   -- 31
                                                            -- 32
      -- Transfer out incremented value via signal          -- 33
      -- assignment to port parameter.                      -- 34
      Dataout <= Current_Count;                             -- 35
                                                            -- 36
   end process Bit_Incr;                                    -- 37
                                                            -- 38
end Model_1;                                                -- 39
```

Figure 2.28

value is assigned to this variable. This initialization expression is in the format of an aggregate which is the VHDL term for any parenthesized list of one or more elements separated by a comma. The aggregate on line 14 agrees with our intuitive understanding of arrays as a collection of positionally important individual elements. It follows very easily that this notation infers that each element of Current_Count be assigned an initial value of '0'. Lines 15 and 16 are other notational styles that we could have used. Note well that, because of the leading two hyphens, both of these lines are commented out. Otherwise, we would have a compile error. The double quotes delimiting the expression on line 15 introduces the interpretation that a BIT_VECTOR object may also be viewed as a string of individual BITs. This viewpoint makes a lot of sense because a BIT is defined in terms of a character literal and a collection of character literals is a string. Formally speaking, "000" is referred to as a *bit string literal*. Traversing this string from left to right each element '0' will be, respectively, assigned to Current_Count(2), Current_Count(1), and Current_Count(0). The (others

=> '0') notation shown on line 16 has already been discussed, and it is shorthand for the assignment of '0' to every element of Current_Count.

Line 20 contains our old buddy, wait until Clock = '0' which, just as a friendly reminder, will cause this process to be suspended until the occurrence of Clock's trailing edge.

Lines 24 through 31 is a VHDL for loop statement that will implement an incrementation algorithm for a string of '0's and '1's. To understand the mechanics of this algorithm it is best to first increment an arbitrary bit stream by hand using pen and paper. Note the ensuing patterns that develop and what action must be taken, depending on whether the *least significant bit* (LSB) is a '0' or a '1'. The VHDL code on lines 24 through 31 implements an algorithm that can readily be derived by manually executing a bit stream incrementation. Line 24 begins with the VHDL reserve word *for*. The purpose of a for statement is to repeatedly loop through a sequence of code. In this example we will repeat those statements between lines 25 and 30. Continuing on line 24 we have the loop iterator, which in this case is named K. Note that the loop iterator was not declared in the process declarative part. It is implicitly declared via this for statement, and its data type is determined by the range of its allowable values. In this example K's range is given to be 0 to 2. This range implies that K is an INTEGER. Note well that the implemented algorithm will force Current_Count(2) to serve as the MSB. On each iteration through lines 25 to 30 the loop iterator K will successively take on the values 0, 1, and 2. Unlike some other programming languages, VHDL does not permit the loop iterator to be incremented by any arbitrary number. Each new value assigned to K must be the immediate successor of its previous value. Also the loop iterator is only visible within its encapsulating for loop construct. It would be illegal to reference K outside of lines 25 through 30. Incidentally, had a variable been declared in this process with the same name K then certainly this variable would be visible outside of the for loop. But inside this for loop construct only the loop iterator would be visible. Actually, there is a way to make such a variable visible inside the for loop, but my recommendation is to avoid the problem completely by making sure that your declared objects (variables, signals, constants) never have the same name as any of your loop iterators. Returning to this for loop we have inside of it an if statement that spans lines 25 through 30. This if statement uses the loop iterator K as an index into the array Current_Count. Recall that since Current_Count is a BIT_VECTOR, it can only be indexed by a NATURAL data type. Since NATURAL is a subtype of INTEGER, it follows that it is valid to use the INTEGER K as an index. On the first pass through K has the value 0, so at this point we are concerned with the element Current_Count(0). If this value is currently '0', then those statements between the reserved words *then* and *else* would be executed. This being the case line 26 would be executed next and Current_Count(0) would be changed to '1'. Since Current_Count(0) is a constituent of the variable Current_Count it too is, in essence, a variable and hence its update is immediate. Having changed Current_Count(0) to '1' we are now done and the incrementation is complete. Hence it is no longer required to do another iteration of the for loop and we should jump out of it immediately. This jumping out is achieved via the reserved word *exit* on line 27, which causes the simulation to continue at the first executable statement following the end loop delimiter. In our model this would correspond to line 35. By the way, it is optional to label for loops but, for readability, I always recommend that you do so when using the *exit* statement. Now let's step through this algorithm again, but this time assume that on the first iteration of this for loop Current_Count(0) is '1'. In this case the if test will fail, and, consequently, only those statements between the *else* and *end if* statements will be executed. Current_Count(0) will immediately be changed to '0' and we will then have to repeat another iteration of this for loop but with the loop iterator having the updated value 1. Computation-wise, this addition of '1' to Current_Count(0)'s value has propagated a '1' to be added to the element Current_Count(1). These algorithmic steps will be repeated until we have either exhausted all the elements of the array Current_Count or there is no longer a need to propagate a '1'. In any event, after the for loop is completed, the next statement to be executed is on line 35. At this point the variable Current_Count contains the incrementation of the value it had just prior to the current trailing edge of Clock.

The assignment Dataout <= Current_Count opens up several important issues. First of all, where did the name Dataout come from anyway? And why was it accepted by the compiler? The following visibility discussion could also have been conducted during our analysis of line 20 where we came across the name Clock. Let's look back at the entity declaration shown in Fig. 2.26 in which Dataout (Clock) is declared to be a port of mode out (in). Here now is another justification for why the entity declaration must be compiled before any of its corresponding architecture bodies. Not only must the design entity's name be visible to all its children architectures but so must the names of its ports and generic parameters. And as illustrated here, these name visibilities must occur even if the entity and its architectures reside in physically different files. Hence, when the compiler reaches line 35, it then checks to see if the parent entity declared an object called Dataout since the architecture encapsulating this name did not. Having found Dataout in the entity declaration, the compiler will then confirm that it has the correct mode.

Since Dataout is the target of an assignment, it must be valid for the VHDL model to write to this port. Hence this port cannot be of mode in since such ports can only be read and cannot be written to. In actuality, it turns out that Dataout's mode is OK since it is declared as an out port. Had it been of mode in then Dataout could only have appeared on the right-hand side of an assignment. By the way, the arrow (<=) notation tells us that the target of the assignment on line 35 must be a signal. Hence, this confirms what we intuitively already suspected. VHDL ports are, in fact, signals and, as we saw in the architecture Structural_Decomposition, ports are the communication gateways between concurrently running design entities. Another important feature highlighted by line 35 is that VHDL allows a variable to be assigned to a signal as long as their data types are the same. Though the objects are referenced as a whole the implication is that this signal assignment is accomplished element-wise going systematically from left to right (or right to left). Since Dataout and Current_Count were both declared to be indexed by the range 2 down to 0, this signal assignment is equivalent to the following set of individual signal assignments: Dataout(2) <= Current_Count(2), Dataout(1) <= Current_Count(1), and Dataout(0) <= Current_Count(0). After these respective assignments are scheduled to occur 1 delta time from now, the simulation will then return to the top of the process, and the next statement to be executed is line 20. Consequently, this process will be suspended again until the next trailing edge on Clock. On the basis of these developments it is apparent that the value driven by the signal Dataout will be incremented on each trailing edge of the signal Clock. Now here is a question to think about. What is the value on Dataout before the occurrence of the very first trailing edge? Let's ease into this matter. First of all, let's backtrack a little and provide a rigorous and formal definition for the data type BIT. The LRM defines this data type as follows: type BIT is ('0', '1'). In the VHDL vernacular BIT is an enumeration type consisting of two elements. In general, an enumeration type is a parenthesized list that explicitly itemizes all of the members of a data type. The left to right sequence in which the members of this list physically appear defines an ascending order relationship. VHDL defines several useful attributes for enumeration data types. One of them is 'LEFT which is read as "tick left." For example, BIT'LEFT is that element in the enumeration list that is visually farthest to the left in the list. Hence BIT'LEFT is the value '0'. The significance of all this is that if a VHDL engineer does not assign an initial value to a signal (or to a variable), then the VHDL tool suite will automatically assign the signal's (variable's) 'LEFT value to it. So, for instance, if I declare a signal or a variable to be of type BIT but do not assign an initial value to it, then the simulation tool will initialize it to '0'. Incidentally, the term type mark is used in VHDL to refer to and identify an object's declared type or subtype. So the type mark of Clock and Dataout is BIT and BIT_VECTOR, respectively. Furthermore, in the case of composite data types, such as arrays, a similar initialization is done element-wise. Hence, any signal or variable of BIT_VECTOR type that is not formally initialized by the VHDL modeler will be set to all '0's by the simulation engine.

Having said all this, let us now return to the question regarding Dataout's value before the first trailing edge. I always had the option to assign a default value to Dataout in its port declaration. This technique will be shown later in our VHDL journeys. For the time being I declined this option and instead relied on the simulation tool suite to set each of Dataout's elements to BIT'LEFT which, as we just developed, is '0'. As a result Dataout will be all '0's and at simulation startup will drive the binary equivalent for the decimal value zero. We also now see why Current_Count must have an initial value of all '0's. Had Current_Count been initialized to "010", then 1 delta time unit after the first trailing edge Dataout would be updated to "011" This modification would then correspond to a jump from decimal zero ("000") to decimal 3 ("011"), which would be incorrect. Consequently, it is very important to coordinate the initial values of Dataout and Current_Count so that they will be the same.

One final note about this architecture body. Because of the high-order language (HOL) constructs used here, such as *for loops* and *if...then...else* statements, this architecture body may be considered as a behavioral model.

2.8.1 Information checkpoint

Before you venture further along your VHDL journeys, please make certain that you are familiar with the following topics that we have established via this model:

- Explicit initialization techniques by VHDL engineer for an array data type object.
- The mechanics of the for loop statement.
- Port (and generic) names declared in an entity declaration are visible to any of the entity's corresponding architecture bodies even if the entity declaration and the architecture bodies reside in physically different files.
- Ports are signals.
- Ports of mode out may only appear on the left side of an assignment statement.
- The role of 'LEFT in the automatic assignment of initial values to signals and variables when the VHDL engineer has not explicitly assigned one.

2.9 File: count_3_bit_model_2.vhd

We are now going to develop a state machine version of our 3-bit incrementing device. Recall the bubble notation used in Fig. 1.8 to visually capture the data flow specifications of this counter. Each of the depicted bubbles will now represent a unique machine state, and the bit pattern contained inside of it will correspond to the machine's output while in that state. For those with a background in finite state automata theory, our 3-bit counter will be implemented as a Moore finite state machine since the device's output will be solely a function of its current state.

Line 7 of Fig. 2.29 tells us that we are now dealing with yet another architecture body for the design entity Count_3_Bit. In order to label each state of our machine with a meaningful name, we will invoke the VHDL capability that allows users to define their own enumeration data types. We have already come across enumeration types during our description of VHDL's definition for the type BIT. Recall that its definition is: type BIT is ('0', '1'). Lines 13 and 14 show an analogous type declaration that itemizes the various states of our intended incrementing machine. This new user-defined data type is called States_Enum_Type. Each member of its parenthesized list represents a machine state that uniquely corresponds to one of the bubbles of Fig. 1.8. For instance, STATE0 is the machine state name for the bubble that contains the binary pattern "000". This bit stream "000" will be the output of the state machine when it is in machine state STATE0. In general, STATEN, is the state in which the machine's output is equivalent to the decimal number N (N = 0...7).

Each element of an enumeration list is called an enumeration literal. Since VHDL is case insensitive, the compiler will view STATE0 and StaTE0 as being the same. However, I prefer to use all upper case letters for enumeration literals since in many ways they are essentially constant values. Earlier I had stated that my VHDL style guide is to write constants in upper case. This constant viewpoint is quite valid for enumeration literals. To see why, let me first discuss how a similar environment can artificially be created in those (ancient) programming languages that do not support enumeration types. Back then the programmer had no other option but to declare constants having names such as STATE0 and STATE1. Consequently, whenever he or she referenced these constant objects in his or her program, the compiler would convert these names into their previously equated numerical values. In a sense, down at the CPU machine code level, something analogous happens with VHDL enumeration types, but the actual implementation is transparent and hidden from the user. Somewhere in the translation path of VHDL into the targeted host machine language a unique constant value is associated with each enumeration lit-

```
-- File Name  :  count_3_bit_model_2.vhd
--
-- Author     :  Joseph Pick
--
-- *************************************************************** LINE NUMBER
architecture Model_2 of Count_3_Bit is                            -- 7
                                                                  -- 8
begin                                                             -- 9
                                                                  -- 10
   State_Machine:                                                 -- 11
   process                                                        -- 12
      type States_Enum_Type is (STATE0, STATE1, STATE2, STATE3,   -- 13
                                STATE4, STATE5, STATE6, STATE7);  -- 14
                                                                  -- 15
      variable Current_State : States_Enum_Type := STATE0;        -- 16
                                                                  -- 17
   begin                                                          -- 18
                                                                  -- 19
      case Current_State is                                       -- 20
                                                                  -- 21
         when STATE0 =>                                           -- 22
            Dataout <= (others => '0');                           -- 23
            Current_State := STATE1;                              -- 24
                                                                  -- 25
         when STATE1 =>                                           -- 26
            Dataout <= ('0', '0', '1');                           -- 27
            Current_State := STATE2;                              -- 28
                                                                  -- 29
         when STATE2 =>                                           -- 30
            Dataout <= (1 => '1', others => '0');                 -- 31
            Current_State := STATE3;                              -- 32
                                                                  -- 33
         when STATE3 =>                                           -- 34
            Dataout <= ('0', '1', '1');                           -- 35
            Current_State := STATE4;                              -- 36
                                                                  -- 37
         when STATE4 =>                                           -- 38
            Dataout <= ('1', others => '0');                      -- 39
            Current_State := STATE5;                              -- 40
                                                                  -- 41
         when STATE5 =>                                           -- 42
            Dataout <= (2 | 0 => '1', others => '0');             -- 43
            Current_State := STATE6;                              -- 44
                                                                  -- 45
         when STATE6 =>                                           -- 46
            Dataout <= "110";                                     -- 47
            Current_State := STATE7;                              -- 48
                                                                  -- 49
         when STATE7 =>                                           -- 50
            Dataout <= (others => '1');                           -- 51
            Current_State := STATE0;                              -- 52
                                                                  -- 53
      end case;                                                   -- 54
                                                                  -- 55
      wait until Clock = '0';                                     -- 56
                                                                  -- 57
   end process State_Machine;                                     -- 58
                                                                  -- 59
end Model_2;                                                      -- 60
```

Figure 2.29

eral. Though nothing is specified in the LRM on this implementation matter, most translations will typically associate an enumeration literal with its physical position in the enumeration list. The leftmost entry in the enumeration list is said to be in position 0. So at the CPU machine code level our STATE0 will be equivalent to the constant number 0. Similarly, STATE5 will correspond to the constant number 5.

This implementation method justifies my nomenclature viewpoint that enumeration literals are essentially constant values.

Returning to our model we have, on line 16, a declaration for Current_State. This variable is of type States_Enum_Type and is given the initial value STATE0.

In general, VHDL engineers use case statements to model state machines. In Model_1 the case statement spans lines 20 through 54. The reserved word *case* must be followed by the case selector expression and the reserved word *is*. In this case statement Current_State serves as the case selector. VHDL requires that the case selector must be either a discrete type or a one-dimensional array. The term *discrete* refers to any of the following data types: INTEGER, subtype of INTEGER, or an enumeration type. In other words, VHDL does not permit the case selector to be a real number. The reason for this disallowance will become apparent once the mechanics of the case statement are explained.

Inside each case statement is a collection of when statements. Each when statement is called a when arm of the case statement. Lines 22 through 24 is an example of a when arm. VHDL requires that the reserved word *when* be followed by a value that is known at compile time. This value is referred to as a *case* choice. The execution of the case statement will begin with the comparison of the case selector to the first case choice. If there is a match, then the statements within the first when arm are executed, and upon its completion, the next statement to be executed will be the first statement after the reserved words *end case*. If this comparison fails, then the case selector will be compared with the next case choice that belongs to the immediately following when arm. These comparisons will continue until either a match is found or all the comparisons to the case choices have been exhausted. Clearly, the case selector and the various case choices must be of the same type. Otherwise, the comparisons would be invalid in VHDL's strong data-typing environment. It should now make sense why the case selector cannot be a real number. Computers can only approximate real numbers and so, depending on the CPU's internal bit representation for them, different real numbers might incorrectly be interpreted as being equal. Hence, equality checks in the real-number domain are neither reliable nor portable. Therefore VHDL is quite correct to disallow real numbers as case selectors. As for the requirement that the case choices be known at compile time we have to once again think of the actual machine code implementation options. Case statements would be very inefficient and difficult to implement if the case choices would be allowed to take on values that are dynamically derived during a simulation run. In other words, case choices can be neither variables nor signals. The best way to implement case statements at the host CPU's machine code level is to use a jump table where the various case choices serve as offsets into this table. Since each offset must be a fixed value, it follows that the case choices must also be fixed. Our earlier discussion showed how a vendor's tool suite might associate an enumeration literal with a fixed integer, namely, its fixed position within its enumeration list. Hence, the enumeration literal's corresponding numerical value is, in essence, fixed and known at compile time. This fixed property explains why enumeration literals may be used as case choices.

Having laid all this groundwork for the case statement, let us now focus on the activities of each when arm. In our current model the purpose of each when arm is to schedule a signal assignment to the port Dataout and to update the machine's internal state in anticipation for the next time that this case statement is entered. The net effect will be that on each trailing edge of Clock the previous value of Dataout will be incremented.

This VHDL program also introduces several additional techniques for assigning values to arrays. We are already familiar with the methods used in lines 23, 27, 35, 47, and 51. Let's now investigate some of the other available options. Line 31 uses a named aggregate notation. It implies that the element of Dataout indexed by 1 should be assigned the value '1', and all the other remaining elements should be assigned '0'. So, in other words, Dataout will collectively be assigned the bit string "010", which is the binary equivalent of the decimal number 2. Line 39 displays a positional aggregate notation combined with the reserved word *others*. Since Dataout was defined as an array 2 down to 0, this positional notation implies that the element Dataout(2) will receive '1'. The expression *others* => '0' means that the remaining elements will receive the value '0'. The net result is that Dataout will collectively receive the bit string "100", which is the binary equivalent of the decimal number 4. The vertical bar notation (|) used on line 43 should be read as the word *and*. So 2 | 0 => '1' states that both the elements Dataout(2) and Dataout(0) will receive the value '1'. Several remarks are now in order. Named associations may appear in any physical order in the aggregate list. So the aggregate (0 => '0', 2 => '1', 1 => '0') is equivalent to (1 => '0', 0 => '0', 2 => '1'). However, aggregate notations cannot combine both named and positional associations in the same list. (This topic will be discussed in Part 3 of this book.) Another restriction to keep in mind is that whenever the reserved word *others* is used in an aggregate, it must be the last member of the list.

Upon exiting the case statement, we will execute line 56 that will suspend this process until the next trailing edge of Clock.

By the way, you may have noticed a slight disparity between the successive values of Current_State and Dataout. Let me elaborate on this discrepancy. Suppose that on a particular trailing edge of Clock, Current_Count has the value STATE3. Then the when arm having the case choice STATE3 will be executed. Consequently, Dataout will be scheduled to receive, 1 delta time later, the binary equivalent for the decimal number 3. So far so good! This assignment is exactly what we want. However, line 36 will cause the variable Current_State to immediately be updated to STATE4. Hence, even after Dataout is updated 1 delta time later, there will still be a disagreement in the machine state's output and its current internally maintained state. So this is not a pure implementation of a Moore state machine. The problem is that in this model the wait condition is at the bottom of the process instead of at the top. This location combined with the fact that Current_State is immediately updated forces the when arms to be written such that the above discrepancy came about. But, in all honesty, there really is no serious damage done. The values on Dataout will be correctly incremented on each successive trailing edge of Clock and the current machine state does not have to be communicated to any other process. Hence the disparity, though it exists, has no dire consequences.

Another feature of this model that you must be aware of is that since the wait statement is at the bottom, the case statement will be executed once during the initialization phase of the simulation. Since Current_State is initially STATE0, the first when arm will be executed during this initialization phase. Dataout will subsequently be scheduled to receive all '0's at simulation time 0 ns plus 1 delta time unit. Hence Dataout will be updated even before the first trailing edge of Clock. On the surface this might look like a problem but, here again, there really is none. Recall from our previous discussions that at simulation startup the port Dataout will already be all '0's. Consequently, it will not matter that 1 delta time after the initialization phase Dataout will be assigned its current value of all '0's. Hence from the user's perspective there really is no change in Dataout's value, and everything appears just fine. Here again the root of the superficial discrepancy is that the wait condition is at the bottom of the process. Herein lies the hidden agenda for this architecture Model_1. I wanted to draw your attention to the fact that the placement of the wait statement within a process will definitely influence the coding scenario and output of your VHDL model. Subsequent models will provide further insight into this theme.

2.9.1 Information checkpoint

Before you venture further along your VHDL journeys, please make certain that you are familiar with the following topics that we have established via this model:

- User-defined enumeration data types.
- Mechanics of the case statement.

72   An Excursion into VHDL

- Options available to assign a fixed value to an object of type BIT_VECTOR.
- VHDL engineer must always be aware of the ramifications of placing a wait statement at the bottom of a process.

2.10 File: count_3_bit_model_2a.vhd

Figure 2.30 displays a state machine model for our 3-bit counter that is, in spirit, very similar to the model we have just completed analyzing. The main difference is that the wait condition is now at the top of the process. The purpose of this model is to highlight how this new location of the wait statement forces us to carefully rethink the relationship between Dataout and Current_State and their respective updating values. As it turns out, this approach is much more satisfying since the Model_2 discrepancy between the current machine state and the current output is completely done away with. One delta time unit after the trailing edge of Clock the values of Dataout and Current_State will be in synch. Thus we have here a true Moore state machine where the output is in line with the machine's current state. Also, we do not have to worry about Dataout being updated before the first trailing edge. In short, this model's approach is much cleaner.

2.10.1 Information checkpoint

Before you venture further along your VHDL journeys, please make certain that you are familiar with the following topics that we have established via this model:

- VHDL engineer must always be aware of the ramifications of placing a wait statement at the top of a process.

2.11 File: count_3_bit_model_3.vhd

Though the previous state machine models are quite valid there is a more general method that has many practical advantages. Figure 2.31 shows that two processes will now be used instead of just one. The first process, labeled Sequential_Circuit, will synchronously update the signal Current_State that models the register holding the current state of this state machine. This state update will then propagate to the second process of this model. This second process, labeled Combinational_Circuit, will subsequently respond to the machine state's update by outputting the appropriate incremented value and by deriving the next state. This next state will only be assigned to (registered into) Cur-

```
-- File Name  :  count_3_bit_model_2a.vhd
--
-- Author     :  Joseph Pick
--
-- *********************************************************** LINE NUMBER
architecture Model_2a of Count_3_Bit is                         -- 7
                                                                -- 8
begin                                                           -- 9
                                                                -- 10
   State_Machine:                                               -- 11
   process                                                      -- 12
      type States_Enum_Type is (STATE0, STATE1, STATE2, STATE3, -- 13
                                STATE4, STATE5, STATE6, STATE7); -- 14
                                                                -- 15
      variable Current_State : States_Enum_Type := STATE0;      -- 16
                                                                -- 17
   begin                                                        -- 18
                                                                -- 19
      wait until Clock = '0';                                   -- 20
                                                                -- 21
      case Current_State is                                     -- 22
                                                                -- 23
         when STATE0 =>                                         -- 24
            Dataout <= ('0', '0', '1');                         -- 25
            Current_State := STATE1;                            -- 26
                                                                -- 27
         when STATE1 =>                                         -- 28
            Dataout <= (1 => '1', others => '0');               -- 29
            Current_State := STATE2;                            -- 30
                                                                -- 31
         when STATE2 =>                                         -- 32
            Dataout <= ('0', '1', '1');                         -- 33
            Current_State := STATE3;                            -- 34
                                                                -- 35
         when STATE3 =>                                         -- 36
            Dataout <= ('1', others => '0');                    -- 37
            Current_State := STATE4;                            -- 38
                                                                -- 39
         when STATE4 =>                                         -- 40
            Dataout <= (2 | 0 => '1', others => '0');           -- 41
            Current_State := STATE5;                            -- 42
                                                                -- 43
         when STATE5 =>                                         -- 44
            Dataout <= "110";                                   -- 45
            Current_State := STATE6;                            -- 46
                                                                -- 47
         when STATE6 =>                                         -- 48
            Dataout <= (others => '1');                         -- 49
            Current_State := STATE7;                            -- 50
                                                                -- 51
         when STATE7 =>                                         -- 52
            Dataout <= (others => '0');                         -- 53
            Current_State := STATE0;                            -- 54
                                                                -- 55
      end case;                                                 -- 56
                                                                -- 57
   end process State_Machine;                                   -- 58
                                                                -- 59
end Model_2a;                                                   -- 60
```

Figure 2.30

rent_State at the next trailing edge of Clock. Such a two-process method is indicative of a real hardware implementation, which is based on the interplay between a sequential and a combinational circuit. Typically, the sequential circuit synchronously updates a register holding the current machine state. This update then triggers the combinational

74 An Excursion into VHDL

```
--  File Name  :  count_3_bit_model_3.vhd
--
--  Author     :  Joseph Pick
--
-- ************************************************************ LINE NUMBER
architecture Model_3 of Count_3_Bit is                          --  6
                                                                --  7
    type States_Enum_Type is (STATE0, STATE1, STATE2, STATE3,   --  8
                              STATE4, STATE5, STATE6, STATE7);  --  9
                                                                -- 10
    signal Current_State : States_Enum_Type;                    -- 11
    signal Next_State    : States_Enum_Type;                    -- 12
                                                                -- 13
begin                                                           -- 14
                                                                -- 15
    Sequential_Circuit:                                         -- 17
    process                                                     -- 18
    begin                                                       -- 19
                                                                -- 20
        wait until Clock = '0';                                 -- 21
                                                                -- 22
        Current_State <= Next_State;                            -- 23
                                                                -- 24
    end process Sequential_Circuit;                             -- 25
                                                                -- 26
    Combinational_Circuit:                                      -- 27
    process (Current_State)                                     -- 28
    begin                                                       -- 29
                                                                -- 30
        case Current_State is                                   -- 31
                                                                -- 32
            when STATE0 =>                                      -- 33
                Dataout    <= (others => '0');                  -- 34
                Next_State <= STATE1;                           -- 35
                                                                -- 36
            when STATE1 =>                                      -- 37
                Dataout    <= ('0', '0', '1');                  -- 38
                Next_State <= STATE2;                           -- 39
                                                                -- 40
            when STATE2 =>                                      -- 41
                Dataout <= "010";                               -- 42
                Next_State <= STATE3;                           -- 43
                                                                -- 44
            when STATE3 =>                                      -- 45
                Dataout <= ('0', '1', '1');                     -- 46
                Next_State <= STATE4;                           -- 47
                                                                -- 48
            when STATE4 =>                                      -- 49
                Dataout <= ('1', others => '0');                -- 50
                Next_State <= STATE5;                           -- 51
                                                                -- 52
            when STATE5 =>                                      -- 53
                Dataout <= ('1', '0', '1');                     -- 54
                Next_State <= STATE6;                           -- 55
                                                                -- 56
            when STATE6 =>                                      -- 57
                Dataout <= "110";                               -- 58
                Next_State <= STATE7;                           -- 59
                                                                -- 60
            when STATE7 =>                                      -- 61
                Dataout <= (others => '1');                     -- 62
                Next_State <= STATE0;                           -- 63
                                                                -- 64
        end case;                                               -- 65
                                                                -- 66
    end process Combinational_Circuit;                          -- 67
                                                                -- 68
end Model_3;                                                    -- 69
```

Figure 2.31

circuit to respond by generating the appropriate output and decoding (deriving) the machine's next state. Such a close resemblance to the real hardware configuration explains why synthesis tools prefer VHDL state machines to be modeled via such a two-process technique. By the way, the communication of the current state from one process to another forces Current_State to now be implemented as a signal instead of as a local process variable. This signaling requirement levies two conditions upon us. First of all, the signal declaration for Current_State must be placed in the architecture declarative part instead of in the process declarative part, as per our previous state machine model. The name Current_State subsequently becomes accessible (visible) throughout the architecture body, thus allowing both processes to reference it. The second requirement forced upon us by Current_State's declaration is that the name of its data type must already be known to the compiler at the time of this signal's declaration. This type visibility is achieved by moving the type declaration for States_Enum_Type from the process declarative part into the architecture declarative part. Lines 8 through 11 of Fig. 2.31 show both these type and signal declarations. On line 11 I deliberately did not specify an initial value for the signal Current_State. Do you recall what its initial value should be? Previously, I had mentioned that if a VHDL user does not initialize a signal or a variable, then the tool suite will automatically assign its 'LEFT value. Hence, Current_State will be initialized to the enumeration literal STATE0, which is the leftmost member in the list for States_Enum_Type.

The process Sequential_Circuit will first wait for the trailing edge of Clock (line 21). When this edge occurs, then the current value of Next_State will be scheduled to be assigned to Current_State 1 delta time later. This update will cause Current_State to be identified as having received a VHDL event. Consequently, the process Combinational_Circuit will be reactivated since it was waiting for an event to occur on Current_State. Upon reactivation this process will continue with the case statement beginning on line 31. Recall that since this process has a static sensitivity list, the waiting for events occurs at the process' bottom. The only exception to this rule occurs during the simulation initialization phase during which the process is executed until the implicit wait statement is reached at its bottom. There is fundamentally nothing new in the various when arms of this case statement. Dataout is scheduled to receive a value that will honor the current machine state, and the signal Next_State is updated with the value that is to be assigned to Current_State during the next trailing edge of Clock. So 1 delta time later Dataout will have a value that reflects the incrementation of its previous value. Note that this update to Dataout will occur 2 deltas after the trailing edge of Clock. Please be aware that I only counted these delta time units for pedagogical purposes. In general, you should never have to count the number of delta delays. But, nonetheless, you should always be aware of them!

Observe that the signal Next_State will also be initialized to the enumeration literal STATE0. Is this good or bad? Since I want this state machine to transition from STATE0 to STATE1 on the first trailing edge of Clock (see lines 21 and 23) should I have initialized Next_State to STATE1? Actually, it really does not matter what Next_State is initialized to. Recall that the process Combinational_Circuit has a static sensitivity list. Consequently, this process has an implicit wait statement at its bottom. Hence, it will be executed once during the initialization phase. At that point in time Current_State is equal to STATE0, and so during the initialization phase the first when arm will be entered and Next_State will be scheduled to receive STATE1 1 delta time later at time 0 ns plus 1 delta. So by the time the first trailing edge occurs, Next_State will already have the appropriate value STATE1. Hence, it really does not matter what initial value is assigned to Next_State since whatever value is assigned will be overwritten with STATE1.

One final note regarding the format of this VHDL state machine model. It unfortunately is not general enough. A typical Moore state machine environment must also allow for input data to influence its state transitions. Actually, the general Moore state machine may be modeled analogous to this Model_3 architecture. It will also consist of two processes. One of these processes will update the current state and the other process will represent the combinational circuit responding to this update. The first process will be identical to our current Sequential_Circuit process. The second process must additionally honor the input data stream. Consequently, the combinational modeling process will consist of a static sensitivity list containing the current state signal and all of the input signals. Each when arm of the case statement will additionally contain if...then...else...logic to derive the next machine state based on the input data values.

For those familiar with Mealy state machines, you can readily see how the aforementioned description may be adapted to create a general VHDL Mealy state machine model. It too will be segmented into a Sequential_Circuit process and a Combinational_Circuit process. The Moore and Mealy Sequential_Circuit processes will be identical. The general Mealy Combinational_Circuit will also contain if...then...else...logic, but unlike its Moore counterpart, the Mealy machine state's output will also be conditionally updated. Such an approach concurs with the Mealy state machine's requirement that its output must be a function of both the machine's current state and its input(s).

2.11.1 Information checkpoint

Before you venture further along your VHDL journeys, please make certain that you are familiar with the following topics that we have established via this model:

- Type declarations must be made prior to their use so that their names will become visible to the compiler.
- Optimum technique for writing a Moore VHDL state machine.

2.12 File: count_3_bit_model_4.vhd

The name of the architecture shown in Fig. 2.32 suggests that it will not function correctly. BAD_Model_4 has many sentimental memories for me since it is a variation of my very first VHDL model. The year was 1989 and I had already read two books on VHDL, had attended a 2-day VHDL training seminar, and had read an IEEE VHDL tutorial as well. You would think that I was adequately prepared to write a simple VHDL model. Wrong!

On the surface BAD_Model_4 looks fine. Individually, each when arm seems quite reasonable. First, Dataout is scheduled to be updated 1 delta time later with a value based on the machine's current state. The process is then suspended until the next trailing edge of the signal Clock. When this edge occurs, then Current_State is scheduled to be updated 1 delta later to the next appropriate state. Sounds like each when arm is doing the right thing. But let's take a closer look at the situation. Suppose that the case statement is entered with Current_State equal to STATE2. Dataout is therefore scheduled to receive "010" (line 25). This signal scheduling is then followed by the process being suspended (line 26) until the next trailing edge. Upon process reactivation, Current_State will subsequently be scheduled to be updated to STATE3 1 delta time later (line 27).

Having completed this when arm the model's execution then continues at the first statement following the reserved words *end case*. This next executable statement should be line 49 but it is commented out. This omission is deliberate since my very first VHDL model did not contain it either. Since line 49 is merely a comment, the process's execution must continue by looping around to the top and reentering the case statement. But herein lies the problem. And I now have some interesting questions for you. Has the next delta time unit already occurred? Does Current_State have its new updated value of STATE3 or does it still have its old value of STATE2? These questions are really intended to quiz you on your knowledge of the delta delay concept. If you answered that we did not yet advance to the next delta and that Cur-

```
--    File Name  :  count_3_bit_model_4.vhd
--
--    Author     :  Joseph Pick
--
--    ************************************************** LINE NUMBER
architecture BAD_Model_4 of Count_3_Bit is              --  7
    type States_Enum_Type is (STATE0, STATE1, STATE2, STATE3,  --  8
                              STATE4, STATE5, STATE6, STATE7); --  9
    signal Current_State : States_Enum_Type := STATE0;  -- 10
begin                                                   -- 11
    State_Machine:                                      -- 12
    process                                             -- 13
    begin                                               -- 14
        case Current_State is                           -- 15
            when STATE0 =>                              -- 16
                Dataout <= (others => '0');             -- 17
                wait until Clock = '0';                 -- 18
                Current_State <= STATE1;                -- 19
            when STATE1 =>                              -- 20
                Dataout <= ('0', '0', '1');             -- 21
                wait until Clock = '0';                 -- 22
                Current_State <= STATE2;                -- 23
            when STATE2 =>                              -- 24
                Dataout <= "010";                       -- 25
                wait until Clock = '0';                 -- 26
                Current_State <= STATE3;                -- 27
            when STATE3 =>                              -- 28
                Dataout <= ('0', '1', '1');             -- 29
                wait until Clock = '0';                 -- 30
                Current_State <= STATE4;                -- 31
            when STATE4 =>                              -- 32
                Dataout <= ('1', others => '0');        -- 33
                wait until Clock = '0';                 -- 34
                Current_State <= STATE5;                -- 35
            when STATE5 =>                              -- 36
                Dataout <= ('1', '0', '1');             -- 37
                wait until Clock = '0';                 -- 38
                Current_State <= STATE6;                -- 39
            when STATE6 =>                              -- 40
                Dataout <= "110";                       -- 41
                wait until Clock = '0';                 -- 42
                Current_State <= STATE7;                -- 43
            when STATE7 =>                              -- 44
                Dataout <= (others => '1');             -- 45
                wait until Clock = '0';                 -- 46
                Current_State <= STATE0;                -- 47
        end case;                                       -- 48
        -- ***** wait on Current_State; *****           -- 49
    end process State_Machine;                          -- 50
end BAD_Model_4;                                        -- 51
```

Figure 2.32

rent_State still has its old value STATE2 then you have comprehended the essence of our earlier delta time discussions. If you feel somewhat uneasy about the subject, then please return to the in-depth analysis of the file intro_delta_time.vhd.

So if the case selector Current_State is still equal to STATE2, then the same when arm will now be repeated as was just previously done. Dataout, which is currently equal to "010", will be scheduled to receive this very same bit string again 1 delta time from now. Line 26 will once again cause this process to be suspended until the next trailing edge of Clock. One delta time after this process is suspended, Dataout and Current_State will be updated to "010" and STATE3, respectively. As you can see there is total disarray in this model's behavior, and it is not providing the output when we want it. The problem here is that Current_State still had its old value when the simulation reentered the case statement. Uncommenting line 49 will produce the desired effect. That way the process will only sequence back to the top after Current_State has an event. This event will occur when the signal will be updated to the next state. Since our 3-bit counter always advances to a new machine state, it is quite all right to wait for an event on Current_State. However, it is conceivable that a more general state machine might remain in the same state for several clock periods. Under this circumstance it would be incorrect to wait for an event on Current_State since this signal might not be updated to a different machine state. Consequently, in the more general case the following statement should be used instead: wait on Current_State'TRANSACTION. Let's dissect and explore this wait condition. First consider Current_State'TRANSACTION. Recall that every signal has certain VHDL predefined attributes associated with it. We have already come across the BOOLEAN-valued attribute 'EVENT in one of our earlier journeys. 'TRANSACTION is a BIT-valued attribute that toggles every time its prefix signal receives an update, irrespective of whether the update is the same or different from its preupdated value. Any type of update to a signal is called a *transaction* or an *activity*. Recall that an update that is different from the signal's previous value is called an *event*. So an event is a subset of a transaction but not the other way around. A very important feature of the attribute 'TRANSACTION is that its application to a signal results in a signal as well. For instance, Current_State'TRANSACTION is a signal, and so it is valid to talk about an event occurring on it. An event will occur on Current_State'TRANSACTION whenever this signal toggles from one BIT value to another. Since Current_State'TRANSACTION toggles whenever Current_State is updated, the statement wait on Current_State'TRANSACTION is, in effect, waiting for Current_State to receive any value, even its preupdated one. Hence, a more general state machine should use wait on Current_State'TRANSACTION. Another alternative that we could have applied in this model is to use the statement wait for 0 ns, which will cause the process to be suspended for 1 delta time unit. Since Current_State is scheduled to be updated 1 delta time later, it follows that by the time this process is reactivated the signal Current_State will already have its updated value.

The astute reader will probably realize that all my problems with this model would evaporate completely had Current_State been a variable instead of a signal. A variable Current_State would update immediately. Consequently, there would not be a need to wait for it to be updated. The process would not have to be suspended with a wait statement as per the signal version. In all honesty, Current_State does not have to be a signal. Since this architecture body contains only one process, there is no need for any signals to facilitate interprocess communications. Nonetheless, I decided back in 1989 to model Current_State as a signal object so that it would exhibit real signaling qualities such as lag times in its updating. Well, I got what I deserved. But then, on the other hand, this signaling approach had the positive effect of pointing out to me that it is very easy to write a VHDL model that on the surface looks correct but in actuality may be totally wrong. The old value of a signal may incorrectly be used instead of the intended updated value or vice versa. Hence, the VHDL designer must always ask whether it is the old value or the new value that is currently being applied. Since VHDL is concerned with the modeling of real hardware, it should come as no surprise that this old versus new value has a counterpart in the design of hardware. It is very typical for a hardware designer to ask whether his/her circuit is using the old value or the new updated value of a signal.

Returning to my initial VHDL education, or lack of it, you can see that the root of my problem was that I had not completely understood the essence of the delta delay concept. The books and training classes had failed to properly present this very important topic. In order to not propagate this pedagogical flaw, I vowed that any VHDL course (book) of mine would introduce delta delays very early and then repeatedly reinforce their modeling consequences. That way the chances would be very good that my students (readers) would not duplicate my own initial misunderstandings of this very delicate subject. And if on those rare occasions that they would accidentally commit a delta delay error, then my hope was that the course would make them alert to the symptoms of the problem. This familiarity would substantially assist them in quickly isolating the particular old versus new signal usage that was creating havoc with their model. One of the niceties of this old versus new problem is that once it is tracked down, it is very easy to remedy. The many positive experiences of my students on this matter proved that I was right to stress the mechanics and consequences of the delta delay. Unfortunately, such a deep emphasis on delta delays did backfire once, which leads me to yet another anecdote that I would like to share with you.

Back in 1990 I served as a trainer, modeler, and advisor on a key VHDL project. About three weeks after I had trained one of my colleagues, he approached me with a 3000-line VHDL model that at one time had functioned correctly. However, a recent one-line addition totally destroyed his model, and the outputs were now in total disarray. He did not know where to even begin his debugging activities. By that time I had already successfully completed a state machine for a bus interface model. Out of curiosity I went straight to his VHDL state machine to see what he had done. What I found there was both alarming and, in hindsight, quite humorous. Buried in one of his state machines was the following code sequence:

```
wait for 0 ns;
wait for 0 ns;
wait for 0 ns;
```

Clearly, my colleague wanted to suspend his state machine process until the passage of 3 delta time units. When questioned about this matter my colleague responded with, "But, Joe, you always stressed that we should be aware of delta delays!" Yes, this emphasis is true. But I never told anyone to count delta time units, which is what my colleague had done in his model. It now became obvious what had happened. His extra line of code required him to include yet another wait for 0 ns since this addition forced his process to be suspended for 4 delta time units before it could correctly proceed. So, as per this VHDL modeling methodology, every time my colleague would modify his source code he would then be forced to laboriously step through his simulation line by line and recount the number of delta delays that must transpire before the process could correctly resume. This technique is very tedious and highly error prone. Even worse is the problem that the VHDL model is now no longer scalable. Code cannot be surgically included or deleted. Major, in-depth analysis is required every time the model is updated. But actually the real culprit of this hazardous model is that my colleague had written a state machine that was doing too much. To understand what I mean by doing too much, think back to the architecture BAD_Model_4 of this subsection. Yes, we had an old signal versus new signal error. But the real source of the problem was that each when arm was concerned with too many independent activities. Each when arm had to update Dataout, identify the trailing edge of Clock, and update Current_State as well. If our state machine had been simple like Model_3, then we would not have had any problems. Herein lies the moral to this anecdote. When designing synchronous state machines, it is recommended that the process managing the state transitions should synchronously schedule only those signals that will trigger other processes to begin their respective executions. Of course the signal corresponding to the current state should be one of these synchronously updated signals. These updated signals will then create a ripple effect by triggering the reactivation of other processes. These subsequently triggered processes may be viewed as representing combinational circuits. Such a state machine modeling approach not only reflects the real hardware configuration, but it also simplifies the overall VHDL model. The VHDL designer will then no longer have to constantly adjust and adapt to delta delay effects with superficial wait for 0 ns statements.

2.12.1 Information checkpoint

Before you venture further along your VHDL journeys, please make certain that you are familiar with the following topics that we have established via this model:

- Definition of a transaction occurring on a signal.
- The statement wait for 0 ns causes a 1 delta time suspension.
- Consecutive wait for 0 ns statements indicates that the model's correct behavior demands that delta delays be manually counted and that this sum subsequently be incorporated into the overall design. This poor modeling methodology should be reevaluated, and the model should be redesigned.
- Processes updating a machine's state should minimize their other signal scheduling activities. Otherwise, delta delay effects will constantly have to be accounted for via the appropriate number of consecutive wait for 0 ns statements.

2.13 File: count_3_bit_model_5.vhd

Consider the gate level schematic shown in Fig. 2.33. It was derived via the application of Karnaugh map minimization techniques and excitation tables to the state machine diagram of Fig. 1.8. Our excursion into VHDL will now continue with a model that captures the functionality of this schematic. Note well that I deliberately used the word *functionality*. This terminology implies that we are still relying on abstract concepts instead of working with concrete gate level component models. The predefined VHDL logical operators will be applied to capture the essence of the logic gates shown in Fig. 2.33. Pay close attention to the wires labeled Q2, Q1, and Q0. They represent the outputs of the trailing edge-triggered flip-flops and will be referenced in our subsequent models.

Model_5 of Fig. 2.34 shows an architecture body based on the schematic drawn in Fig. 2.33. It has only one process, and so there is no need to declare any signals. The declared variables Q2, Q1, and Q0 (lines 12 through 14) will correspond to the flip-flop outputs shown in the schematic because they will be synchronously updated. On each trailing edge these variables will be updated as per the Boolean equations given on lines 18 through 20. Let's take a closer look at these Boolean equations. They were derived by manually following the

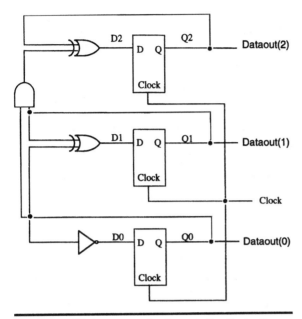

Figure 2.33 Minimum gate implementation for trailing edge-triggered 3-bit counter.

```
--   File Name  :  count_3_bit_model_5.vhd
--
--   Author     :  Joseph Pick
--
--   ************************************ LINE NUMBER
architecture Model_5 of Count_3_Bit is   --  6
                                         --  7
begin                                    --  8
                                         --  9
   Gate_Functions_Implied:                --  10
   process                               --  11
      variable Q0 : BIT := '0';          --  12
      variable Q1 : BIT := '0';          --  13
      variable Q2 : BIT := '0';          --  14
   begin                                 --  15
      wait until Clock = '0';            --  16
                                         --  17
      Q2 := Q2 xor (Q1 and Q0);          --  18
      Q1 := Q0 xor Q1;                   --  19
      Q0 := not Q0;                      --  20
                                         --  21
      Dataout <= Q2 & Q1 & Q0;           --  22
                                         --  23
   end process Gate_Functions_Implied;   --  24
                                         --  25
end Model_5;                             --  26
```

Figure 2.34

dataflow of the various wires shown in the schematic. For instance, line 18 originated with the following dataflow analysis for the updating of Q2. On each trailing edge of Clock the new value of Q2 will be set to the value of D2, which is the flip-flop's input. Tracing backward we find that D2 is connected to the output of an Exclusive-OR gate that has as its inputs the old, pretrailing edge value of Q2 and the output of an AND gate. The inputs to this AND gate are the old, pretrailing edge values of Q1 and Q0. Translation of this manual-tracing effort leads directly to the Boolean equation shown on line 18. Similar analysis will provide the equations on lines 19 and 20. Notice how the functionality of the logic gates are being duplicated by the VHDL predefined operators *and, xor,* and *not.*

Line 22 introduces another available VHDL predefined operator. The ampersand symbol (&) is used to denote concatenation. Individually each of Q2, Q1, and Q0 are of type BIT and hence their concatenation may be intuitively interpreted as a BIT_VECTOR consisting of three elements. In fact, the formal VHDL definition for concatenation supports this BIT_VECTOR viewpoint. Now recall that Dataout was declared as a BIT_VECTOR having the range 2 down to 0. Hence, as per VHDL's strong data-typing environment, the assignment on line 22 is valid since both the target and the destination are BIT_VECTORs. This scheduled assignment of one array to another will be done positionally going from left to right. Consequently, Q2 will be assigned to Dataout(2), Q1 to Dataout(1), and Q0 to Dataout(0). Based on our Boolean equations Q2 serves as the MSB, and so here again, as in our previous models, Dataout(2) will take on the role as the MSB. The net result of this assignment is that each trailing edge of Clock will cause the previous outputed value of Dataout to be incremented.

Let's now use Model_5 as a vehicle to introduce some more VHDL modeling issues. First and foremost is the observation that since the variables Q2, Q1, and Q0 are immediately updated, their physical order within the process must comply with the overall dataflow sequence. Otherwise, the resulting boolean equations would inappropriately mix new variable values with old ones. For instance, the following order of boolean equations would not correctly increment the previous value of the concatenation Q2 & Q1 & Q0.

```
Q0 := not Q0;

Q1 := Q0 xor Q1;

Q2 := Q2 xor (Q1 and Q0);
```

As per this given order, Q2 will be updated using the old pretrailing edge value of Q2 and the new posttrailing edge values of Q0 and Q1.

Such a mixture of old and new values do not honor the intent of the hardware configuration that this model is attempting to capture. Hence when using variables to model gate level schematics, it is very important to remember that the variable updates take effect immediately. If your sequence of Boolean equations do not accurately follow the flow of the signal updates, then your model will be in error. This modeling requirement sounds easy enough, but it sometimes is difficult to honor, especially when the modeled circuit is complex and contains many feedback loops. In this case you might have to introduce temporary variables that do not have any corresponding wire counterparts on the schematic.

Another worthwhile point to mention is that since Dataout is a port having mode out, it can only appear on the left-hand side of a signal assignment. In other words, it would have been illegal to have used the previous values of Dataout instead of Q2, Q1, and Q0. The following statement would have been rejected by the compiler:

```
Dataout(2) <= Dataout(2) xor (Dataout(1) and Dataout(0));
```

By the way, just a reminder, based on our previous examples. Objects declared as signals or variables do not possess this restriction. They may appear on both sides of an assignment statement as long as they are of the same data type. The key distinction is that in this model Dataout is a port of mode out.

Another point to be made has to do with synthesis. In general, we tend to view variables as merely software artifacts devoid of any hardware interpretations. However, in many cases the pattern used to update variables can influence a synthesis tool to map them onto real hardware components such as wires, logic gates, or flip-flops. For instance, the synchronous updating of Model_5's Q2, Q1, and Q0 variables will induce the synthesis tool to map them onto trailing edge-triggered flip-flops.

One final remark about this architecture. It is said to be a dataflow model since it contains only assignment statements.

2.13.1 Information checkpoint

Before you venture further along your VHDL journeys, please make certain that you are familiar with the following topics that we have established via this model:

- Existence of VHDL operators such as and, not, and xor.
- Existence of the BIT concatenation operator &.
- Order of a sequence of variable assignments is very important. Otherwise, the VHDL description would not match the modeled hardware.
- Ports of mode out may only appear on the left side of an assignment.

2.14 File: count_3_bit_model_6.vhd

The purpose of Model_6 (see Fig. 2.35) is to strengthen and reinforce our understanding of two fundamental VHDL themes. Earlier in our journeys we explored the differences between variable and signal assignments and between sequential and concurrent signal assignments. Model_6 will review these differences but do so in a context that will give you a deeper insight and appreciation for these topics.

At first glance, the VHDL source code of Fig. 2.35 looks very much like our previous Model_5. But closer inspection reveals a crucial difference. Model_6 uses signals, whereas Model_5 uses variables. Actually, Model_6's usage of signals is really not necessary since this architecture contains only one process, and there is no need for any interprocess communications via signals. By the way, you should always strive to use variables instead of signals. This preference is due to the execution overhead required by signals. The simulation engine must schedule their future assignments and monitor their updates. These signal-managing activities are achieved via adding to, deleting from, or searching through linked list data structures. Such overhead operations take time and hence slow down the overall simulation; therefore, it is more efficient to work with variables than signals. Nonetheless, for pedagogical reasons signals were deliberately used in Model_6.

```
--    File Name  :  count_3_bit_model_6.vhd
--
--    Author     :  Joseph Pick
--
-- ****************************************    LINE NUMBER
architecture Model_6 of Count_3_Bit is        --  6
                                              --  7
    signal Q0 : BIT := '0';                   --  8
    signal Q1 : BIT := '0';                   --  9
    signal Q2 : BIT := '0';                   -- 10
begin                                         -- 11
                                              -- 12
    Gate_Functions_Implied:                   -- 13
    process                                   -- 14
    begin                                     -- 15
        wait until Clock = '0';               -- 16
                                              -- 17
        Q0 <= not Q0;                         -- 18
        Q1 <= Q0 xor Q1;                      -- 19
        Q2 <= Q2 xor (Q1 and Q0);             -- 20
                                              -- 21
        --*** Dataout <= Q2 & Q1 & Q0; ***    -- 22
                                              -- 23
    end process Gate_Functions_Implied;       -- 24
                                              -- 25
    Dataout <= Q2 & Q1 & Q0;                  -- 26
                                              -- 27
end Model_6;                                  -- 28
```

Figure 2.35

Lines 18 through 20 contain the Boolean equations that we saw in Model_5, but this time they are in a jumbled order. Recall that in Model_5 their order was crucial. Otherwise, the model would behave incorrectly. So why is it possible in Model_6 to mix up the sequential ordering of these same equations? The answer to this question lies at the heart of the difference between variables and signals. Variables are updated immediately and hence the equations of Model_5 had to be written in a specific order or else a new updated value would be used instead of the required old value. Signals, on the other hand, are always updated at a later time. Since a delay time was not specified in Model_6 via an after clause, we know, from our earlier travels, that the update lag time will be equal to 1 delta time unit. Let's now walk through this simulation to confirm that the jumbled boolean equations will not cause any problems. On the trailing edge of Clock we will first execute line 18 in which the signal Q0 is scheduled to be updated 1 delta time from now. The value of this update will be equal to the Boolean equation on the right-hand side of the assignment. Hence this scheduled update will use the old pretrailing edge value of Q0. After scheduling this signal update, the simulation engine will then proceed to line 19. Here Q1 is scheduled to be updated 1 delta time later. The Boolean equation on the right-hand side of this assignment references the signals Q0 and Q1. At this point in time does Q0 still have its old pretrailing edge value? Or has it already been updated as per the Boolean equation on line 18? Stated differently, has a delta time unit transpired during our progression from line 18 to line 19? This last question is the most important one, and its answer is a definite *no!* A delta time unit has not advanced yet. Consequently, the signal Q0 still has its old, pretrailing edge value. Similarly, the Boolean equation on line 20 will use the old, pretrailing edge values of Q1, Q0, and Q2. The upshot of this in-depth analysis is that because signal assignments are used, the three Boolean equations may be given in any sequential order.

Let's now continue to the second review topic. Recall that assignments such as those on lines 18 through 20 are called sequential signal assignments. Suppose that line 26 would be moved to replace the comment on line 22. The former line 26 would then function as a sequential signal assignment. Here is another question for you. When the simulation engine reaches this new, uncommented line 22, will the concatenation operation use the old pretrailing edge values of Q2, Q1, and Q0 or will it use their posttrailing edge values instead? Note that essentially this question is the same one that I asked earlier. Here again, the response should be the same. We would be using the old pretrailing edge values of Q2, Q1, and Q0. But is this what we want to do? No it is not! The trailing edge should result in the incrementation of

Dataout's previous pretrailing edge value. Hence Dataout should be updated with the posttrailing edge values of Q2, Q1, and Q0. Actually, if line 22 would be used instead of line 26, then our model's output would be one clock period behind the real targeted hardware. It would be as if a buffering flip-flop would be placed between the incrementing circuit and each of the device's three output lines. If you really insist on using line 22, then we would have had to precede it with our infamous friend wait for 0 ns. Recall that the process would then be suspended for exactly 1 delta time unit. During this time interval the signals Q2, Q1, and Q0 would be updated so that when line 22 is executed 1 delta time later, their new updated values would already be available. But the usage of wait for 0 ns seems very artificial and looks more like a cute software trick than a true reflection of the actual hardware. Now I will admit that I like cute tricks, but here in this case it really is not necessary. Moreover, if your overall objectives are to write synthesizable models, be aware that the statement wait for 0 ns does not synthesize. So for all these reasons it is best to omit line 22. Instead, Dataout should be updated via a concurrent signal assignment. Line 26 contains this construct and recall what it means. During the initialization phase the initial values of Q2, Q1, and Q0 will be concatenated, and this result will be scheduled to be assigned to Dataout 1 delta time later. Henceforth, the signals Q2, Q1, and Q0 are in the sensitivity list associated with this signal assignment. So whenever any of these signals has an event during the first half of a simulation cycle, then the concatenation expression will be reevaluated during the second half of this same cycle, and the result will be scheduled to go to Dataout 1 delta time later. So as per these developments, Dataout will be updated to the incrementation of its previous value exactly 2 delta time units after the trailing edge of Clock. Incidentally, please note that this counting of delta time units is for instructive purposes only. As per our earlier discussions, you should never have to resort to such counting practices. Remember my motto that you should always be aware of delta delays but you should never have to count them.

The architecture Model_6 is also a dataflow model since it consists solely of assignment statements.

2.14.1 Information checkpoint

Before you venture further along your VHDL journeys, please make certain that you are familiar with the following topics that we have reviewed via this model:

- Differences between variable and signal assignments.
- Differences between sequential and concurrent signal assignments.

2.15 File: count_3_bit_model_7.vhd

The purpose of Model_7 (Fig. 2.36) is to introduce VHDL's block construct. As always, let's begin with the big picture. Every block must have a label. There is no labeling option as is the case with the process construct. Line 14 shows that in this example Block_Implementation will serve as the required label. Line 15 begins with the reserved word *block* and then continues with a parenthesized Boolean expression that is known as the *block's guard expression*. Very shortly a lot more will be said about this expression's important role within the block. Lines 14 and 15 could have been combined into one single line, but for readability I recommend that you use the coding style illustrated in Fig. 2.36. The reserved word *begin* on line 16 denotes the start of the block's statement region. The reserved words *end block* on line 24 terminate the block construct. It is optional to repeat the block's label as done on this line and, in general, I always do so for readability.

Each VHDL construct within a block statement region is theoretically running concurrently. For comparative purposes, recall that the encapsulated code within a process is executed sequentially line by line in the order of their physical occurrence within the process. Since each statement within a block statement region is running concurrently, we know that the physical order in which they are listed is completely irrelevant. It follows that the assignments on lines 18, 19, 20, and 22 are all concurrent signal assignments and hence may be written in any physical order. From our past journeys we are already familiar with the opera-

```
--   File Name  :  count_3_bit_model_7.vhd
--
--   Author     :  Joseph Pick
--
--   *******************************************  LINE NUMBER
architecture Model_7 of Count_3_Bit is           --  6
                                                 --  7
    signal Q0 : BIT := '0';                      --  8
    signal Q1 : BIT := '0';                      --  9
    signal Q2 : BIT := '0';                      -- 10
                                                 -- 11
begin                                            -- 12
                                                 -- 13
    Block_Implementation:                        -- 14
    block (Clock = '0' and (not Clock'STABLE))   -- 15
    begin                                        -- 16
                                                 -- 17
        Q0 <= guarded not Q0;                    -- 18
        Q1 <= guarded Q0 xor Q1;                 -- 19
        Q2 <= guarded Q2 xor (Q1 and Q0);        -- 20
                                                 -- 21
        Dataout <= Q2 & Q1 & Q0;                 -- 22
                                                 -- 23
    end block Block_Implementation;              -- 24
                                                 -- 25
end Model_7;                                     -- 26
```

Figure 2.36

tional characteristics of concurrent signal assignments. However, in a block construct the story becomes just a little bit more complicated. Note that the signal assignments on lines 18 through 20 contain the reserved word *guarded,* whereas the assignment on line 22 does not. Let's first discuss line 22 since that is the easier one and, besides, there really is nothing new here anyway. The assignment to Dataout is simply a concurrent signal assignment, similar operationally to any that we have already seen before. Aside from the forced scheduling due to the initialization phase, this signal assignment on line 22 will henceforth be sensitive only to events occurring on any of the signals Q2, Q1, and Q0. Whenever such an event occurs during the first half of a simulation cycle, during the second half of this same cycle these three signals will be concatenated, and the result will be assigned to Dataout 1 delta time unit later. Since this signal assignment to Dataout excludes the reserved word *guarded*, the statement itself could well have been written outside of the block construct. Functionally, there would have been no difference. However, I deliberately placed it inside the block statement part to emphasize that the usage of the reserved word *guarded* is always an option. But as line 22 shows I decided to decline this option for reasons that will be explained in the upcoming paragraphs.

Let's now delve deeper into the finer points of the block construct. If a block has a guard expression then a unique BOOLEAN signal called *GUARD* is implicitly defined for this block. Since GUARD is a signal, it has a sensitivity list. The members of a GUARD's sensitivity list are those signals appearing in the block's guard expression. If an event occurs on any of these signals, then GUARD will be updated to the current value of the guard expression. Now comes the tricky part. Since GUARD is a signal, you would expect some lag time in its update, perhaps the passage of at least 1 delta time unit. But it turns out that there is no delay at all. Suppose that a signal in GUARD's sensitivity list has an event during the first half of a particular simulation cycle. Then GUARD will be updated during that fuzzy interval between the first half and second half of this same simulation cycle. Consequently, GUARD will be updated during the same simulation cycle in which an event occurred on any of the signals in its sensitivity list. Since VHDL simulation time does not progress during the span of a simulation cycle, GUARD may be viewed as being updated immediately. So in this respect GUARD behaves like a variable. But since it has a sensitivity list, it also behaves like a signal. So what is GUARD? Is it a signal or is it a variable? Though these described characteristics seem somewhat incongruent, VHDL dogma (i.e., the LRM) insists that GUARD is a signal.

Let's now take a closer look at the guard expression (Clock = '0' and (not Clock'STABLE)). In our previous journeys we have already come across the VHDL attributes 'EVENT and 'TRANSACTION. 'STABLE (read as tick STABLE) is yet another BOOLEAN attribute that is associated with signals. Suppose that the signal Clock has an event during the first half of a particular simulation cycle. Since Clock just changed and is currently not stable, it should come as no surprise that Clock'STABLE will be set to FALSE for the duration of this simulation cycle. Remember that this same scenario will cause Clock'EVENT to be set to TRUE. Suppose further that Clock does not change during the next simulation cycle. This stability in Clock's value will cause Clock'STABLE and Clock'EVENT to be set to TRUE and FALSE, respectively. So it follows that value-wise Clock'EVENT will always be equal to the logical negation of Clock'STABLE. In other words, Clock'EVENT will always be equal to the expression (not Clock'STABLE). So the attributes 'EVENT and 'STABLE are very closely related to each other. BUT there is one major difference. As per the LRM Clock'STABLE is a signal, whereas Clock'EVENT is not. Since Clock'STABLE is a signal, it is eligible to be a member of a sensitivity list. Consequently, based on Model_7's guard expression, we now know that GUARD's sensitivity list will contain not only the signal Clock, but it will also contain the signal Clock'STABLE. Hence, GUARD will be reevaluated whenever Clock or Clock'STABLE has an event.

Let's now focus our attention on the significance of the reserved word *guarded*. Signal assignments qualified by the reserved word *guarded* will be scheduled to occur if at least one of the following two rules are satisfied:

1. GUARD just had an event and is currently TRUE.
2. Any of the signals on the right-hand side just had an event and the GUARD is currently TRUE.

These constraints can best be understood via a concrete scenario. Suppose that GUARD is currently FALSE and that Clock just experienced a trailing edge. Hence Clock just transitioned from '1' to '0'. Hence the relational expression Clock = '0' has the value TRUE. Since Clock just had an event, Clock'STABLE is now FALSE and hence the expression (not Clock'STABLE) is therefore TRUE. Since both Clock and Clock'STABLE are in the GUARD's sensitivity list and at least one of them experienced an event, the guard expression is consequently reevaluated. Since the composite expression (Clock and (not Clock'STABLE)) is TRUE, it follows that GUARD's previous FALSE value will be replaced with the value TRUE. Recall from our earlier discussions that GUARD will be updated to TRUE during that fuzzy interval between the first and second half of the simulation cycle in which Clock had the event. GUARD's modification means that it just experienced an event and is currently TRUE. Therefore as per the aforementioned rule 1 the Boolean equations on lines 18 through 20 will be scheduled to be assigned to the targeted signals Q2, Q1, and Q0. These Boolean equations are identical to those used in the architectures Model_5 and Model_6.

Putting all these facts together we see that the desired incrementation will occur on each trailing edge of Clock. Sounds great! But there is one more operational detail to point out. Let's continue into the next simulation cycle. Since Clock did not have an event during this cycle, Clock'STABLE will change from FALSE to TRUE. This change means that Clock'STABLE just had an event. But remember that Clock'STABLE is in GUARD's sensitivity list. Consequently, the guard expression must be reevaluated and then assigned to GUARD. Though Clock = '0' is TRUE, the expression (not Clock'STABLE) is now FALSE. Since TRUE and'ed together with FALSE is FALSE, it follows that both the guard expression and GUARD now have the value FALSE. Since GUARD is FALSE rules 1 and 2 guarantee that events on the signals Q2, Q1, and Q0 will not trigger the guarded signal assignments on lines 18 through 20. Incidentally, had GUARD remained TRUE then we would be in a lot of trouble. Consider the guarded signal assignment on line 18. If GUARD would continue to be TRUE after the trailing edge of Clock then, as per rule 2, events on Q0 would induce the guarded signal assignment on line 18. Since each negation of Q0 causes a new event on this signal, line 18 would be reexecuted over and over again. Model_7 would then behave incorrectly. Even worse, Model_7 would go into an infinite oscillation. Simulation time would advance only in delta time units, and it would do so out to infinity.

Let's now return to the signal assignment on line 22 that is not under the control of the signal GUARD. What would happen if this signal assignment would be guarded as well? As per the earlier analysis, Dataout would consequently be updated with the concatenation of the old pretrailing edge values of Q2, Q1, and Q0. The concatenation of the new trailing edge-triggered values of Q2, Q1, and Q0 would only be assigned to Dataout after the arrival of Clock's next trailing edge. The resulting model would erroneously be one period behind the actual hardware. Hence to avoid this modeling error we must exclude the reserved word *guarded* from the signal assignment to Dataout. This argument shows that you must be very careful when using guarded signal assignments, since their existence controls the overall behavior of your VHDL model.

A major disadvantage of a block statement is that in terms of simulation speed it is not as efficient as a process-based model. For instance, during the simulation of Model_7, the simulation engine has to monitor and maintain the following signals: Clock, Clock'STABLE, Q2, Q1, and Q0. Our previous journeys have shown that there exist process-oriented models that also simulate a 3-bit incrementer but do not have to monitor as many signals as Model_7 does. This situation is true in general. One can always design a process-based model that will do the same job as a block can, and yet it will simulate much more efficiently. Nevertheless, block statements are still used by some members of the VHDL community, since blocks very compactly capture the specifications of a hardware device. By the way, in case you are wondering, my personal preference is to use processes instead of blocks.

One final remark about blocks. Since Clock'EVENT is always equal to (not Clock'STABLE), you might be tempted to replace Model_7's guard expression with (Clock = '0' and Clock'EVENT). In our particular example such a substitution would be a grave mistake. The problem is that Clock'EVENT is not a signal, and, consequently, GUARD's sensitivity list will consist of only the signal Clock. In this case GUARD would still be TRUE 1 delta after Clock's trailing edge. The signal assignment statement on line 18 would then oscillate indefinitely, and the desired trailing edge-triggered incrementation would not occur. Hence, we cannot substitute Clock'EVENT for (not Clock'STABLE) in Model_7's guard expression. The topic of guard expressions determining edge-triggered or latchlike behavior will be further investigated by Test_16 in the Experiments segment of this book.

2.15.1 Information checkpoint

Before you venture further along your VHDL journeys, please make certain that you are familiar with the following topics that we have established via this model:

- Format of a block statement containing a guarded expression.
- Concurrency of statements within a block's statement region.
- Definition and consequences of the guard expression.
- Definition and signaling characteristics of GUARD.
- Definition and behavior of guarded signal assignments.
- Similarities and differences between the attributes 'EVENT and 'STABLE.
- Edge-triggered devices modeled by blocks should use the attribute 'STABLE instead of 'EVENT.

2.16 File: clock.vhd

Figure 2.37 illustrates where we are going to travel next and how this part of the journey fits into a slice of our top-down design flow. Just as a quick refresher of our past journeys, you should now take a few

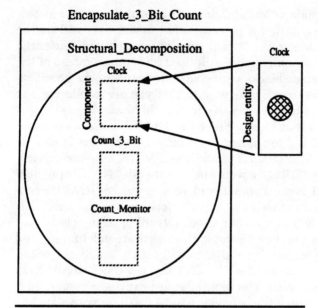

Figure 2.37 The design entity Clock will serve as a concrete implementation for the template component Clock.

moments to reflect on how our previous models all fit together to form the big picture. While doing so you should appreciate that, in essence, these models additionally serve as a vehicle to facilitate your understanding of how the various VHDL language constructs also fit together and interact. Figures 2.20, 2.21, 2.25, and 2.27, viewed in the order in which they are listed, should greatly assist you in your understanding of what we have done so far and the intent of the next leg of our journey.

Analogous to our past activities with the design entity Count_3_Bit, we are now going to write an entity declaration and an architecture body for the design entity called Clock. As a friendly reminder, the component Clock is dotted in Fig. 2.37 because it is only a template and, in essence, does not really exist. In fact, the design entity Clock does not even have to exist at the time that the component Clock is declared. Recall that design entities come into existence only after their corresponding entity declaration and at least one architecture body have been written and both have been compiled into a library. The arrows of Fig. 2.37 emphasize my intent to write a design entity called Clock that during simulation will correspond to the template component Clock.

Let's now examine the source code for this design entity Clock. In this particular case, both the entity declaration and its corresponding architecture body have been included into the same file. Lines 6 through 9 of Fig. 2.38 represent the entity declaration for the design

```
-- File Name : clock.vhd
--
-- Author    : Joseph Pick
--
-- ************************************ LINE NUMBER
entity Clock is                         --  6
    generic (PULSE_WIDTH : TIME := 20 ns); --  7
    port    (Clock : out BIT);          --  8
end Clock;                              --  9
                                        -- 10
architecture Model_1 of Clock is        -- 11
                                        -- 12
begin                                   -- 13
                                        -- 14
    Gen_Clock:                          -- 15
    process                             -- 16
        variable Clock_Var : BIT := '0'; -- 17
    begin                               -- 18
                                        -- 19
        wait for PULSE_WIDTH;           -- 20
        Clock_Var := not Clock_Var;     -- 21
        Clock <= Clock_Var;             -- 22
                                        -- 23
    end process Gen_Clock;              -- 24
                                        -- 25
end Model_1;                            -- 26
```

Figure 2.38

entity Clock. There is only one port, called Clock, and it is of mode out and of type BIT. Preceding this port declaration is the declaration for the generic parameter PULSE_WIDTH. Note well that the VHDL language explicitly stipulates that generic declarations must always come before the port declarations. Any other ordering would result in a compilation error.

The generic parameter PULSE_WIDTH is of type TIME and on line 7 is assigned a default value of 20 ns. Let's investigate this default assignment a bit more closely. First of all, look back at line 31 of Fig. 2.22. During the instantiation of the component Clock, an actual parameter of 50 ns was associated with this component's local generic parameter PULSE_WIDTH. In VHDL's hierarchical scheme this 50 ns has higher priority and will, during simulation, override the 20 ns default value. If an actual value had not been assigned using the generic map technique, then the default 20 ns would have been used during the simulation of the design entity Clock. However, you should be very aware that if PULSE_WIDTH is neither given a default value nor associated with an actual parameter via the generic map construct, then an error condition will be noted by your VHDL tool suite. Contrast this situation with the way that signals and variables are initialized. If you, the user, do not assign a default value to a declared signal or variable then, as per the LRM (Language Reference Manual), the simulation engine will subsequently initialize the object to its (type mark)'LEFT value. Recall that the term type mark is used in VHDL to refer to and identify an object's type or subtype. So, for instance, a non-initialized signal of type BIT will be initialized by the tool suite to '0' since, as you recall, BIT'LEFT is equal to this value. Generic parameters, on the other hand, are handled quite differently by VHDL. If a generic parameter is not given a value via either a default assignment or a generic map association, then the VHDL tool suite will not provide a (type mark)'LEFT default value. Instead, an erroneous modeling environment will be identified, and the model will not be simulated. So when exactly will the VHDL tool suite actually identify this error condition and come to a screeching halt? Recall that during the initialization phase the simulation will execute the source code in each process of the overall model until an explicit or implicit wait statement is reached. As stated earlier, an implicit wait statement will occur at the bottom of those processes having a static sensitivity list. Now just prior to the initialization phase of a simulation is the model's elaboration phase. As per the LRM, it is during this phase that real computer memory is allocated for the various signals and variables declared in your VHDL model. Furthermore, the subcomponents of your model are linked together during this elaboration phase, and the correctness of these interconnections is investigated. It is during these investigations that a check is made to ensure that every generic parameter has somehow been assigned a user-specified value. This assigned value may take the form of either an explicitly proclaimed default value or an actual parameter associated via a generic map construct. If neither of these options is carried out, as mentioned earlier, an elaboration error will be noted, and the tool suite will terminate its simulation activities. In many tool suites the elaboration phase is totally transparent to the VHDL modeler since it automatically comes up when the simulation command is given. If all goes well then the modeler will not even be notified about the occurrence of the elaboration phase. In fact, the only time the user is ever aware of the elaboration phase's existence is when it identifies an error condition and does not permit the simulation effort to proceed into the initialization phase. The second and third parts of this book will illustrate a myriad of subtle errors that are identified during the elaboration phase.

One final note about the entity declaration shown in Fig. 2.38. The design entity's name matches exactly with the Clock component that was declared in the architecture Structural_Decomposition (Fig. 2.22). Moreover, the entity's formal generic and port parameters also match exactly in name and position to the local parameters of the component Clock. These exact one-to-one matches do not have to occur, but any such discrepancies must later be accounted for. This naming topic has already been touched upon in Sec. 2.7 during our development of the entity declaration for Count_3_Bit. The exact details and mechanics of overcoming these naming differences will be explored later in our VHDL travels.

The architecture Model_1 amplifies two important VHDL topics. First, the manner in which PULSE_WIDTH is used is very significant. The formal generic parameter PULSE_WIDTH cannot be on the left-hand side of any assignment statement. Aside from the allowed initial default allocation, in general, formal generic parameters can only be read but cannot be written to. So for all practical purposes they are essentially constant objects for the duration of any simulation run. This constant characteristic of PULSE_WIDTH explains why I have written it in upper case. As per my VHDL-coding conventions, I always write the names of constants in upper case.

The second main feature of this architecture is the usage of the temporary variable Clock_Var. It would be incorrect to write

```
Clock <= not Clock.
```

Recall that in our previous models it was quite valid to write such an assignment. In those models Clock was an explicitly defined signal, and signals may occur on both sides of an assignment. Although Clock is also a signal in our current model, the key point to remember is that now Clock is a port. Being a port means that the model must always honor the direction of the port's dataflow. In our current model Clock is of mode out so it can only appear on the left-hand side of a signal assignment. Consequently, we are forced to introduce the variable Clock_Var, negate it, and then schedule this computed result to be assigned to the out port Clock. This temporary variable trick is used quite often by the VHDL community whenever an analogous out port scenario is confronted. However, its usage is both inefficient and an obstacle to the documentation flow of the model. Alternatively, you may circumvent this temporary variable requirement by applying any of the following design strategies.

The first alternative method is depicted in Fig. 2.39 and it solves the problem by declaring Clock to be an inout port. This way Clock may appear on both sides of a signal assignment as it does on line 20 of Fig. 2.39. Unfortunately, such an approach has many shortcomings and pitfalls. On the surface there seems to be an immediate documentation problem. Anyone reading the entity declaration might incorrectly deduce that Clock is a bidirectional port. An even more dangerous flaw is that Clock's inout port status will allow other external ports and signals to influence its effective value. These multiple sources on Clock might modify it so that it will no longer generate the intended periodic waveform. An example of this unwanted predicament is vividly illustrated by Test_195 in the Experiments segment of this book. In short, my recommendation is not to artificially use the mode inout so that a port's name may appear on both sides of a signal assignment statement.

```
-- File Name : clock_inout.vhd
--
-- Author    : Joseph Pick
--
-- *****************************************  LINE NUMBER
entity Clock is                               --  6
   generic (PULSE_WIDTH : TIME);              --  7
   port    (Clock : inout BIT);               --  8
end Clock;                                    --  9
                                              -- 10
architecture Model_1 of Clock is              -- 11
                                              -- 12
begin                                         -- 13
                                              -- 14
   Gen_Pulse:                                 -- 15
   process                                    -- 16
   begin                                      -- 17
                                              -- 18
      wait for PULSE_WIDTH;                   -- 19
      Clock <= not Clock;                     -- 20
                                              -- 21
   end process Gen_Pulse;                     -- 22
                                              -- 23
end Model_1;                                  -- 24
```

Figure 2.39

```
-- File Name : clock_buffer.vhd
--
-- Author    : Joseph Pick
--
-- *****************************************  LINE NUMBER
entity Clock is                               --  6
   generic (PULSE_WIDTH : TIME);              --  7
   port    (Clock : buffer BIT);              --  8
end Clock;                                    --  9
                                              -- 10
architecture Model_1 of Clock is              -- 11
                                              -- 12
begin                                         -- 13
                                              -- 14
   Gen_Pulse:                                 -- 15
   process                                    -- 16
   begin                                      -- 17
                                              -- 18
      wait for PULSE_WIDTH;                   -- 19
      Clock <= not Clock;                     -- 20
                                              -- 21
   end process Gen_Pulse;                     -- 22
                                              -- 23
end Model_1;                                  -- 24
```

Figure 2.40

Another solution to the temporary variable requirement is to use the mode buffer as shown on line 8 of Fig. 2.40.

The mode buffer was introduced into the VHDL language to allow a port's value to be read even though the data flowing through it is in the out direction. Recall that port of mode out may only be written to but cannot be read in a VHDL model. Hence line 20 of Fig. 2.40 is now valid. Sounds like our problem is solved but there are two major drawbacks to using buffer ports. First, a port of mode buffer can have only one source that is driving it. In other words, a buffer port cannot be used in a multisourced tristated environment. But there is an even worse restriction. As you hierarchically build (bottom-up) or decompose (top-down) your model, you will, unfortunately, find that a buffer port used as an actual (term has already been used) in a port map association can be mapped only onto another buffer port. Note well that this restriction is quite valid in light of the intended information flow of the various port modes. So if you are working with a multilevel hierarchical model, my recommendation is that you should not use this buffer technique. The topic of buffer restrictions will be revisited in the Techniques and Recommendations segment of this book.

The VHDL'93 update committee recognized all of the above weaknesses and so came up with the following solution. A new signal attribute called 'DRIVING_VALUE is now part of the VHDL'93 update, and, as its name suggests, it will contain the local process-driven value (not effective value, as per a multisourced environment) of a signal. Hence VHDL'93 allows you to declare an out port, such as Clock, and then later write the following signal assignment:

```
Clock <= not Clock'DRIVING_VALUE;
```

A key advantage of applying 'DRIVING_VALUE is that synthesis tools will not have to cope with any of the previously described weak solutions. The synthesis engineer's intentions will be immediately known.

However, an even better modeling scheme is shown in Fig. 2.41, and it avoids all of the deficiencies mentioned earlier. In this approach we are, on the one hand, back to an out port description, but we now have two significant advantages in terms of overall simulation efficiency.

```
-- File Name : clock_model_2.vhd
--
-- Author    : Joseph Pick
--
-- ************************************** LINE NUMBER
architecture Model_2 of Clock is           -- 6
                                           -- 7
begin                                      -- 8
                                           -- 9
    Gen_Clock:                             -- 10
    process                                -- 11
        constant PERIOD : TIME := 2 * PULSE_WIDTH;  -- 12
    begin                                  -- 13
                                           -- 14
        Clock <= '1' after PULSE_WIDTH,    -- 15
                 '0' after PERIOD;         -- 16
        wait for PERIOD;                   -- 17
                                           -- 18
    end process Gen_Clock;                 -- 19
                                           -- 20
end Model_2;                               -- 21
```

Figure 2.41

First, there is no need for a temporary variable. Second, the process Gen_Clock will be reactivated and suspended only once per Clock period. In all our previous variations the waveform-generating process had to be reactivated and suspended on both Clock's leading and trailing edges. Hence this new method is at least a twofold improvement over any of the previous coding strategies. Let's now zero in and examine the key features of the architecture body contained in Fig. 2.41. Line 12 shows a standard software optimization technique that you can apply to many analogous coding scenarios. Suppose that line 17 would instead have been written as

```
wait for 2 * PULSE_WIDTH;
```

Then every time line 17 would be sequentially reached, a multiplication operation would be executed. But PULSE_WIDTH is a generic parameter, and so it has a fixed value for the duration of the simulation. Consequently, the product (2 * PULSE_WIDTH) is also fixed, and so it really does not make any sense to continually recompute this constant valued product, especially in light of the fact that multiplication requires anywhere from 20 to 30 CPU cycles. CPU resources should not be wasted by such meaningless recomputations. Instead, a constant declaration is made, as per line 12, that computes the product once and for all during the elaboration phase. Then this constant name PERIOD is referenced as in line 17. The product is no longer being recomputed over and over again. Line 12 is similar to our earlier signal and variable declarations. *Constant* is a reserved word and is followed by the constant's designated name. A colon precedes the constant's data type and a colon equal (:=) symbol introduces the value that the constant object is to have for the duration of the simulation. It would have been illegal to place an expression following the := notation that contains either a variable or a signal object. However, generic parameters are permitted, as shown on line 12, since at the conclusion of the elaboration phase all generic parameters are fixed and regarded as *constant* valued objects.

The signal assignment on lines 15 and 16 displays a technique in VHDL whereby a projected waveform may be compactly written and assigned to a signal. Several points must be made regarding this construct. There is no limit to the number of future assignments that can be included, but each projected assignment must be separated by a comma (not a semicolon), and the listed times must be in ascending order. The listed waveform times are all referenced to the simulation time at which the signal assignment is sequentially reached. So, for instance, suppose that the current simulation time is 100 ns when lines 15 and 16 are reached. Then Clock will be scheduled to receive the values '1' and '0' at (100 ns + PULSE_WIDTH) and (100 ns + PERIOD), respectively. Incidentally, in our current model this waveform technique is used to propagate a signal with a 50 percent duty cycle. It readily follows that this waveform-generating construct may easily be modified to produce a non-50 percent duty cycle.

One final remark regarding this architecture body. Actually, what follows is just a friendly reminder of a topic that we have already seen before in one of our previous VHDL models. During the initialization phase Clock will be scheduled to receive the values '1' at time (0 ns + PULSE_WIDTH) and '0' at time (0 ns + PERIOD). But what will Clock's value be until the first scheduled assignment of '1' is actually carried out? If your answer is '0' then congratulations! And if you are not quite certain about how this value '0' was derived, then let me refresh your memory with the following key phrases: out port and BIT'LEFT. If you are still not sure about the '0' value, then please review Sec. 2.8 in which the file count_3_bit_model_1.vhd is explored and discussed.

2.16.1 Information checkpoint

Before you venture further along your VHDL journeys, please make certain that you are familiar with the following topics that we have established via this model:

- The potential overriding of a formal generic parameter's default value.
- The VHDL tool suite will not automatically assign a default initial value to a generic parameter.
- Ports of mode out cannot be on the right-hand side of an assignment.
- The VHDL'87 and VHDL'93 strategies to circumvent the inability to have an out port on the right-hand side of an assignment.
- The VHDL waveform-generating construct that allows the scheduling of a projected list of values to a signal.

2.17 File: count_monitor.vhd

Figure 2.42 illustrates where we are going to travel next and how this part of the journey fits into a slice of our top-down design flow. Note well the similarities between this figure and Fig. 2.37. Analogously, we are now going to develop an entity/architecture pair that, during simulation, will serve as a concrete realization for the component Count_Monitor. The VHDL source code for this design entity is shown in Fig. 2.43. The purpose of this model is to monitor the output signal of the 3-bit counter and to write its NATURAL number equivalent to an output file for later inspection and analysis. As per the architecture

102 An Excursion into VHDL

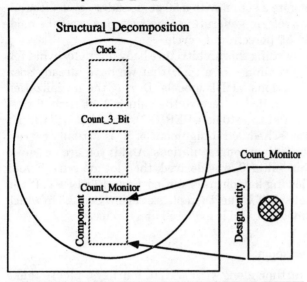

Figure 2.42 The design entity Count_Monitor will serve as a concrete implementation for the template component Count_Monitor.

Structural_Decomposition (see Sec. 2.6) the incrementing signal will, during simulation, be connected to the Datain port of this design entity. Before we explore how data may be written into a file, let's first look into how a 3-bit value may be converted into a NATURAL number. Recall that a NATURAL number is any nonnegative integer. The conversion algorithm shown in lines 35 to 39 is the standard technique that is often presented in the VHDL literature; it is very straightforward and easy to comprehend. Unfortunately, depending on its implementation, the expression 2**I could potentially be very inefficient during a simulation run. The real objective of the expression 2**I is to shift the binary representation of 2 by I places to the left. If the compiler is smart enough to do so and the host CPU has a barrel shifter, then there is no inefficiency problem. However if the compiler cannot transform the power of two operations into a shift left, then the expression 2**I will be computed using either exponentiation or repeated multiplications. Both of these calculations will be very costly in terms of CPU time. In the Techniques and Recommendations segment of this book (Part 3) I will apply VHDL data type attributes to achieve an efficient BIT_VECTOR to NATURAL conversion algorithm.

Another interesting point about this algorithm is the requirement on line 32 to reset Current_Count to 0. Suppose that this line would be

```
-- File Name : count_monitor.vhd                              LINE NUMBER
-- Author    : Joseph Pick
--
-- *********************************************************
entity Count_Monitor is                                       --  6
   port (Datain : in BIT_VECTOR(2 downto 0));                 --  7
end Count_Monitor;                                            --  8
                                                              --  9
use STD.TEXTIO.all;                                           -- 10
architecture Model_1 of Count_Monitor is                      -- 11
                                                              -- 12
begin                                                         -- 13
                                                              -- 14
   Display:                                                   -- 15
   process                                                    -- 16
      variable Current_Count    : NATURAL;                    -- 17
                                                              -- 18
      file Outfile : TEXT is out "counter_values";            -- 19
--    In VHDL'93 the above line MUST be replaced by           -- 20
--    file Outfile : TEXT open WRITE_MODE is "counter_values";-- 21
                                                              -- 22
      variable Outline : LINE;                                -- 23
                                                              -- 24
   begin                                                      -- 25
                                                              -- 26
      write (Outline, STRING'(" TIME      COUNTER VALUE"));   -- 27
      writeline (Outfile, Outline);                           -- 28
                                                              -- 29
      Main_Loop:                                              -- 30
      loop                                                    -- 31
         Current_Count := 0;                                  -- 32
                                                              -- 33
         Convert_Loop:                                        -- 34
         for I in 0 to 2 loop                                 -- 35
            if Datain(I) = '1' then                           -- 36
               Current_Count := Current_Count + (2**I);       -- 37
            end if;                                           -- 38
         end loop Convert_Loop;                               -- 39
                                                              -- 40
         write (L      => Outline,                            -- 41
                Value  => Now,                                -- 42
                FIELD  => 6);                                 -- 43
         write (L      => Outline,                            -- 44
                Value  => Current_Count,                      -- 45
                FIELD  => 8);                                 -- 46
         writeline (Outfile, Outline);                        -- 47
                                                              -- 48
         wait on Datain;                                      -- 49
      end loop Main_loop;                                     -- 50
                                                              -- 51
   end process;                                               -- 52
                                                              -- 53
end Model_1;                                                  -- 54
```

Figure 2.43

omitted. Then the algorithm on line 37 would be incorrect because subsequent BIT_VECTOR to NATURAL conversions would begin by using the final value for Current_Count as computed during the previous conversion activity. This analysis highlights a very important fact regarding the usage of processes to implement algorithms. Variables essential to the algorithm must be reset appropriately at the start of the algorithm's execution. Otherwise, the previous unrelated value will be used which will result in incorrect computations.

Let's now focus our attention on how to write data into an output file. To begin with, you must first have a use clause such as the one on line 10. The reserved word *use* is followed by what VHDL refers to as a "selected name." In this example the selected name is STD.TEXTIO.all. Let's dissect this selected name piece by piece. First we have STD, which is the name of a VHDL predefined library. We have already encountered the predefined library WORK. The STD is the logical name of another library that is specified by the VHDL language to contain two highly specialized packages. A package in VHDL may be thought of as a container of declarations that may be used by any model that references the package's contents via a use clause, as was done in this current example. The library STD may, as per VHDL'93, contain only the two packages STANDARD and TEXTIO. Among the declarations contained in STANDARD are those for BIT, BIT_VECTOR, TIME, BOOLEAN, INTEGER, and NATURAL. Based on our previous journeys, you can see that VHDL models, in general, will at one point or another have need for some of these declarations. The VHDL language acknowledges this need and will automatically include the contents of the package STANDARD. In other words, a VHDL modeler will never have to write use STD.STANDARD.all, where the reserved word *all* implies the potential usage of all the contents of this package. TEXTIO is the other package contained in the library STD, and its usage is handled slightly differently. This package contains those declarations that are required for the reading and writing of ASCII data. Since every model does not require this read/write capability, it follows that the package TEXTIO should not automatically be linked into each and every model. Consequently, application of the package TEXTIO requires that the VHDL engineer explicitly request its inclusion via a use clause as was done on line 10 of our current model.

Writing to a file requires several steps, the first of which has already been taken care of when we wrote use STD.TEXTIO.all. Appropriate object declarations must be made next. This is done on lines 19 and 23. Line 19 declares the object Outfile to be the logical name of a file that contains data of type TEXT. *File* is a VHDL reserved word and the type TEXT is declared in the package TEXTIO to be a file that can contain VHDL STRINGs. STRING is declared in the STANDARD package to be an array of CHARACTERs. Consequently, TEXTIO files contain ASCII-related data. The logical name Outfile will be used within the body of this model instead of the actual physical name of the output file. The concrete physical name in double quotes on line 19 is the name of the file as it will appear in your subdirectory filing system. The reserved word *out* on line 19 announces that the file counter_values can only be used as an output file by this model. If the reserved word *in* had been used instead, then this declared file could only be read from and not written to. In case you are wondering, VHDL does not allow a file to be declared as being accessible for both read and write operations. But actually we have an even more serious problem on our hands. Suppose that the physical file that you declare for a write operation already exists in your subdirectory due to a previous simulation run. Will a subsequent simulation run append to this file or will it simply overwrite it? In VHDL'87 there is no clear-cut answer since the 1987 LRM is very vague on this subject. Consequently, different vendors took different interpretations and went down opposite paths. Some will append while others will overwrite. Consequently, it is possible in VHDL'87 to write a VHDL model that is not portable between several different tool suites since the final output files may not be the same. To amend this potential ambiguity, VHDL'93 now has a new enumeration data type called FILE_OPEN_KIND that can be used to specify how a newly opened file is to be used. Line 21 shows the correct format anticipated by a VHDL'93 compliant compiler. In this particular case the enumeration literal WRITE_MODE implies that the file counter_values is to be opened for writing only and furthermore any preexisting files with the same name will be overwritten. If I had used APPEND_MODE, then the preexisting file with the same name counter_values would then be appended to during subsequent simulation runs.

Let's now return to our necessary file I/O declarations. Line 23 declares the variable Outline to be of type LINE where LINE is declared in the TEXTIO package as a pointer to an array of STRINGs. In essence, Outline will be used as (a pointer to) a buffer line into which the data must first temporarily be written to. This requirement is in accordance with the VHDL language, whereby the actual writing to a file is a two-step process. First the data must be written into a temporary buffer line, and then this buffer line must be written into the file. These two steps are illustrated in lines 27 and 28. The overall purpose of this write operation is to place an appropriate header into the file counter_values. Both the procedures write and writeline are contained in the package TEXTIO. A procedure is a special form of a subprogram. A subprogram is a sequence of declarations and statements that collectively may be invoked repeatedly from different locations within a subprogram model. There are two kinds of subprograms: functions and procedures. For those not familiar with the concept of a subprogram, you may think of it as a macrocell that may be repeatedly used in a schematic. A function is a VHDL expression that algorithmically generates and returns only one value. Hence a function may appear on the right side of an assignment. A procedure, on the other hand, is a VHDL statement that optionally can update the parameters that it is called with. Since a procedure is a statement and not an expression, a procedure cannot exist on the right-hand side of an assignment. Instead, a procedure exists by itself as a stand-alone statement. Line 27 will write the header into the temporary buffer line, and line 28 will write this buffer line to the output file having the logical name Outfile. The term STRING' (read as STRING tick) can be explained as follows. The package TEXTIO has overloaded the procedure write for the various standard data types available to us via the STANDARD package. Overloading means that the same name *write* may be used when writing objects of type BIT, BIT_VECTOR, STRING, INTEGER, etc. into a buffer line. The actual data type being written is used by the compiler to distinguish between these various procedures having the same name. Now suppose that I wished to write "010101" into a buffer line. How is this expression to be interpreted? Is it a BIT_VECTOR or is it a STRING? Since it can be both, we must help the compiler to resolve this ambiguous situation. To do so, we can rely on the VHDL capability to explicitly identify an object's data type. This action is known as the qualification of an expression. For instance, suppose that I wanted the compiler to interpret this string as a BIT_VECTOR, then I would have written BIT_VECTOR'("010101"). On the other hand, suppose that I had intended this expression to be viewed as a string. Then I would have written STRING'("010101"). So far so good! But what about the qualification STRING'(" TIME   COUNTER   VALUE") that exists on line 27. Why is it necessary to qualify this quoted expression? What is the ambiguity here? It seems quite obvious that this expression is of type STRING and not of type BIT_VECTOR. What is the problem? Actually, what is obvious to you and me is not so readily obvious to the compiler. It turns out that the LRM does *not* require a compiler to enter a double-quoted expression in order to determine the expression's data type. Hence though it is obvious to you and me that the double-quoted expression on line 27 is not a BIT_VECTOR, the compiler is *not* obligated to research this matter. Consequently, without a type qualification such as STRING' or BIT_VECTOR', the compiler will not be able to resolve this data typing ambiguity. Instead, the compiler will simply generate an error message and terminate its compilation activities. And so that is why you must use STRING' on line 27.

At this point, it is very important that I discuss the topology (layout) of this model. The process Display begins by writing key headers into the file and then enters the infinite loop Main_Loop in which the input Datain is first converted into a nonnegative number and then is subsequently written into the output file via the two-step technique described earlier. As per line 49, the encapsulating process Display will then be suspended until an event occurs on the input port Datain. Based on our overall game plan, as expressed in the architecture Structural_Decomposition, we know that such an event will occur every time our 3-bit counter is incremented. Note well the significance of the infinite loop labeled Main_Loop. What would happen if we would not have it? Recall that every process is in an infinite loop. Consequently, omitting this embedded infinite loop will cause lines 27 and 28 to be executed every time Datain has an event. In this case the file headers would be repeatedly written instead of just the intended one time at the top of the file. You must always be aware of the infinite looping nature of processes. Otherwise, your models might not behave as expected.

One final observation about this model concerns lines 41 through 47. The first two write procedures use named association techniques, whereas the aforementioned write procedure on line 27 used positional associations. Recall that we had a similar option during our earlier generic map and port map associations. Note that lines 43 and 46 both reference a formal parameter called FIELD. This parameter specifies the number of columns to be used for the outputed data. The write subprogram on line 27 used the default value for FIELD that was declared in this procedure's specification. Consequently, in that usage a value did not have to be supplied for the parameter FIELD since the default was relied upon. Note also the identifier Now on line 42. *Now* is the name of a function that is declared in the STANDARD package. The output of this function is the current absolute simulation time. Recall that by absolute simulation time I mean that delta times are discarded (truncated). By including the current simulation time I have time tagged each incrementation event with the absolute time of its occurrence.

To see how easily things can go wrong consider the model shown in Fig. 2.44. On the surface all looks well. The variable Current_Count is reset on line 24, and an explicit infinite loop is included as well. Take a few minutes to step through this model as if you would be a VHDL simulation engine. In this model the process Display has a static sensitivity list, and so there is an implicit wait on Datain at the bottom of this process. Unfortunately, this bottom location is outside of the loop Main_Loop. This means that once we enter Main_Loop during the initialization phase we will never be able to suspend the encapsulating process, since the suspending wait condition is outside of Main_Loop. This model is yet another example highlighting the importance of code walkthroughs during which you are emulating the VHDL simulator's behavior.

2.17.1 Information checkpoint

Before you venture further along your VHDL journeys, please make certain that you are familiar with the following topics that we have established via this model:

- The VHDL predefined library having the name STD.
- The VHDL predefined packaged STANDARD and TEXTIO.
- The use clause may be applied to make the contents of a package visible.

```
-- File Name : count_monitor_bad.vhd
--
-- Author    : Joseph Pick
--
-- ****************************************************** LINE NUMBER
use STD.TEXTIO.all;                                        -- 6
architecture Model_Bad_1 of Count_Monitor is               -- 7
                                                           -- 8
begin                                                      -- 9
                                                           -- 10
  Display:                                                 -- 11
  process (Datain)                                         -- 12
    variable Current_Count    : NATURAL;                   -- 13
    file Outfile : TEXT open WRITE_MODE is "counter_values";-- 14
    variable Outline : LINE;                               -- 15
                                                           -- 16
  begin                                                    -- 17
                                                           -- 18
    write (Outline, STRING'("TIME       COUNTER VALUE"));  -- 19
    writeline (Outfile, Outline);                          -- 20
                                                           -- 21
    Main_Loop:                                             -- 22
    loop                                                   -- 23
      Current_Count := 0;                                  -- 24
                                                           -- 25
      Convert_Loop:                                        -- 26
      for I in 0 to 2 loop                                 -- 27
        if Datain(I) = '1' then                            -- 28
          Current_Count := Current_Count + (2**I);         -- 29
        end if;                                            -- 30
      end loop Convert_Loop;                               -- 31
                                                           -- 32
      write (L     => Outline,                             -- 33
             Value => Now,                                 -- 34
             FIELD => 6);                                  -- 35
      write (L     => Outline,                             -- 36
             Value => Current_Count,                       -- 37
             FIELD => 8);                                  -- 38
      writeline (Outfile, Outline);                        -- 39
    end loop Main_loop;                                    -- 40
                                                           -- 41
  end process;                                             -- 42
                                                           -- 43
end Model_Bad_1;                                           -- 44
```

Figure 2.44

- The VHDL predefined data type NATURAL.
- The predefined exponentiation operator **.
- Algorithm to convert from a BIT_VECTOR to a NATURAL number.
  - A more efficient algorithm will be shown in the Techniques and Recommendations segment of this book.
- VHDL engineers should always be cautious when using processes to implement an algorithm.
  - Variables not reset might produce incorrect results.

2.18 File: cfg_cnt3_m1.vhd

After successfully compiling the design entities Count_3_Bit, Clock, and Count_Monitor, we are now ready to simulate the architecture

Structural_Decomposition. There are several simulation options available. We can apply default binding or we can use a configuration declaration as was previously done in Sec. 2.5. Default binding will have some operational restrictions. For instance, Structural_Decomposition must be the most recently compiled architecture body associated with the design entity Encapsulate_3_Bit_Count (introduced in Sec. 2.1). In this case you can command your simulator to merely simulate Encapsulate_3_Bit_Count. Another option that you have is to pronounce that the entity/architecture pair Encapsulate_3_Bit_Count and Structural_Decomposition are to be simulated. If your tool suite allows this technique, then it is no longer necessary that Structural_Decomposition be the most recently compiled architecture body associated with Encapsulate_3_Bit_Count. In both these cases the simulator tool suite will search the current working library for entity declarations that exactly match those components declared and instantiated in Structural_Decomposition. If this search fails, then an error message will be generated and the simulation will immediately terminate. If the search is successful, then the matching design entities' most recently compiled architecture bodies will be used in the simulation. No choices are to be made regarding Clock and Count_Monitor since there is only one architecture body associated with each of these design entities. On the other hand, Count_3_Bit has numerous architecture bodies, and so in this case it might not be such a good idea to rely on the default-binding option. It would be very impractical to always have to remember to recompile that specific architecture that you wish to use in your simulation. Moreover, the reliance on the default binding does not provide a self-documenting record of what you have just simulated. Consequently, it is recommended that you use a configuration declaration whenever component instantiations are made. Incidentally, an interesting by-product of using a configuration declaration is that it can be used to textually capture and display the complete hierarchy of a simulated design entity. Figure 2.45 contains a configuration declaration called Universe_Model_1. It is a variation of the configuration declaration that we have already explored in Sec. 2.5. Please take a few minutes now and return to that section to recall our first contact with this construct.

Line 7 of Fig. 2.45 states that Universe_Model_1 is to serve as a configuration for the design entity Encapsulate_3_Bit_Count. Recall from our previous journeys that this design entity has five architecture bodies associated with it (see Fig. 2.20). Line 9 states that of all these available architecture bodies we are now going to exclusively work with the architecture Structural_Decomposition. Recall that Structural_Decomposition had three component instantiations, labeled Synch, Logic_Analyzer, and Gen_Cnt, respectively. We are now going to account for each of these instantiated components. In effect, in this con-

110 An Excursion into VHDL

```
-- File Name : cfg_cnt3_m1.vhd
--
-- Author    : Joseph Pick
--
-- *********************************************************** LINE NUMBER
configuration Universe_Model_1 of Encapsulate_3_Bit_Count is  --  7
                                                              --  8
    for Structural_Decomposition                              --  9
                                                              -- 10
        for Synch: Clock                                      -- 11
            use entity WORK.Clock(Model_1);                   -- 12
        end for;                                              -- 13
                                                              -- 14
        for Logic_Analyzer: Count_Monitor                     -- 15
            use entity WORK.Count_Monitor(Model_1);           -- 16
        end for;                                              -- 17
                                                              -- 18
        for Gen_Cnt: Count_3_Bit                              -- 19
            use entity WORK.Count_3_Bit(Model_1);             -- 20
        end for;                                              -- 21
                                                              -- 22
    end for;                                                  -- 23
                                                              -- 24
end Universe_Model_1;                                         -- 25
```

Figure 2.45

---

figuration we will explicitly specify which design entity should be used to represent each of these component instantiations during a simulation run. This accountability is achieved on lines 11 through 21, and it is a two-stage operation. First, each instantiated component must be associated with a design entity, and second, a unique architecture body must be associated with this designated design entity. The formal VHDL term for these respective associations is the word binding. Note how important it is to properly indent the VHDL source code. Lines 11 through 21 are indented relative to the for construct on line 9. Hence it is now visually apparent that lines 11 through 21 refer to the hierarchical decomposition defined by the architecture Structural_Decomposition. Let's take a closer look at the VHDL mechanics behind these respective bindings.

Lines 11 through 13 show the application of the VHDL for...end for construct to account for the instantiation of the component Clock that is labeled Synch. If Structural_Decomposition would not have an instantiation for Clock that is labeled Synch, then a compilation error would be reported, and the compiler would subsequently terminate its activities. Incidentally, what I have just described has the following implicit by-products. First, both the entity Encapsulate_3_Bit_Count and its architecture Structural_Decomposition must be compiled before this configuration Universe_Model_1 is compiled. And second, all three must be compiled into the same current working library. If any of these conditions are not satisfied, then the compilation of Universe_Model_1 will fail since the compiler will not recognize the name of the entity and architecture pair that is being configured. Line 12 explicitly announces those bindings that are to be used for the instantiation described on line 11. Note well that the indentation of line 12 relative to line 11 also adds greatly to the readability of this VHDL model. Let's now look at this line 12 in more detail. It begins with the VHDL reserved word *use,* and this construct explicitly announces that we are now going to formally carry out a binding activity. The next word *entity* fine-tunes this binding activity by stating that we are going to be referencing an entity. Later on in our journeys we will come across another option that I could have used instead of an entity. (Sneak preview: alternatively, I could have applied the reserved word *configuration.*) Following this reserved word *entity* is the expanded name (a VHDL term) for a design entity. Note the strategic positioning of the period in this entity's full name. Actually, this expanded name very lucidly captures the essence of our binding intentions. It explicitly states that we are going to use from the library WORK a design entity called Clock. Moreover, of all (in Clock's case there is only one) the architecture bodies associated with this design entity Clock, we are going to select the architecture called Model_1. Note well the parentheses surrounding the architecture's name. Omission of these parentheses will result in a compilation error. If either Clock or Model_1 are not resident in the current working library, then a compilation error will be reported. Thus, here again, we have a specific well-defined dependency tree for our compilation units. Continuing along with our analysis, recall that the generic and port parameters of the component Clock are identical to those of the entity declaration for Clock. Hence there is no need to mention these parameters in our binding activities. Very shortly in our journeys we will meet a situation in which the respective parameters do not match.

Lines 16 and 20 will analogously take care of the bindings for the instantiations labeled Logic_Analyzer and Gen_Count. Figure 2.46 graphically captures all the binding activities conducted in this configuration Universe_Model_1. Note well that this is the first time that the rectangles internal to Structural_Decomposition are drawn with solid lines instead of dotted ones. This feature vividly emphasizes the fact that we now finally have real concrete design entities instead of just abstract component templates. To simulate this configuration you must merely inform your simulator that you wish to simulate Universe_Model_1. The simulator will then take care of the necessary bookwork that must be done to incorporate the various models of your declared hierarchy and to manage all the port connections and dataflows that you have created.

At this point I would like to present an analogy that will solidify the physical realities of what we have just done in our abstract world of

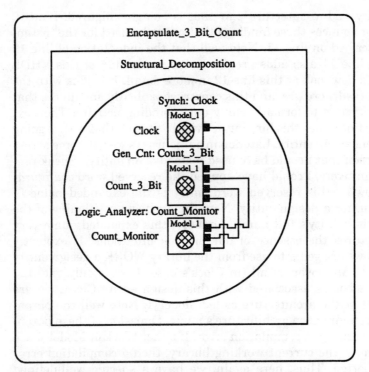

Figure 2.46 Entity/architecture bindings specified in the configuration Universe_Model_1.

VHDL software design. Recall how the concept of component declarations and instantiations were made more familiar via an analogy to your schematic entry activities (see Sec. 2.6). Let's now take a slightly different viewpoint of these VHDL topics. Component declarations may alternatively be perceived as the creation of sockets on a printed circuit board. Component instantiations are then equivalent to connecting the appropriate sockets with wires. The binding of a design entity to a component instance via a configuration declaration is analogous to the populating of a printed circuit board with discrete chips. Each binding activity permits us to make a choice with respect to which pin-compatible chip we wish to place into the prewired sockets. It is at this point in time that I am able to make decisions such as, "Do I want to use INTEL's 486 or AMD's 486?".

Figure 2.47 depicts yet another configuration for the design entity Encapsulate_3_Bit_Count. The only difference between this configuration and the aforementioned one (besides their names, of course) occurs on line 20.

```
-- File Name : cfg_cnt3_m2.vhd
--
-- Author    : Joseph Pick
--
-- ********************************************************* LINE NUMBER
configuration Universe_Model_2 of Encapsulate_3_Bit_Count is  --  7
                                                              --  8
   for Structural_Decomposition                               --  9
                                                              -- 10
      for Synch: Clock                                        -- 11
         use entity WORK.Clock(Model_1);                      -- 12
      end for;                                                -- 13
                                                              -- 14
      for Logic_Analyzer: Count_Monitor                       -- 15
         use entity WORK.Count_Monitor(Model_1);              -- 16
      end for;                                                -- 17
                                                              -- 18
      for Gen_Cnt: Count_3_Bit                                -- 19
         use entity WORK.Count_3_Bit(Model_2);                -- 20
      end for;                                                -- 21
                                                              -- 22
   end for;                                                   -- 23
end Universe_Model_2;                                         -- 24
```

Figure 2.47

This configuration specifies that the architecture Model_2 will be bound to the design entity Count_3_Bit. Consequently, when Universe_Model_2 is being simulated, then the architecture Model_2 will be used as the concrete realization for the instantiation Gen_Cnt of the component Count_3_Bit. For comparison, recall that a simulation of the previous configuration Universe_Model_1 will use the architecture Model_1 instead of this Model_2. Figure 2.48 illustrates the configuration Universe_Model_2.

2.18.1 Information checkpoint

Before you venture further along your VHDL journeys, please make certain that you are familiar with the following topics that we have established via this model:

- Format of the configuration declaration to create the following bindings:
  - Design entity to a specific component instantiation.
  - Architecture Body to a design entity that has already been bounded to an instantiation of a component.

2.19 File: count_n_bits_.vhd

One of the key objectives of VHDL is to write reusable models and subprograms. In general, design for reuse is a very important methodology

114   An Excursion into VHDL

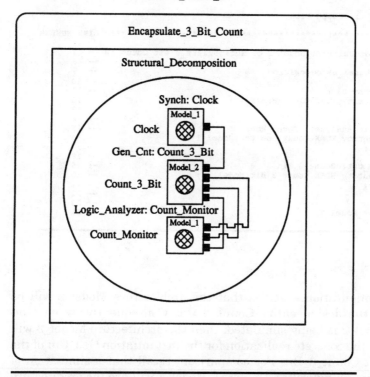

Figure 2.48 Entity/architecture bindings specified in the configuration Universe_Model_2.

that has many positive ramifications. Design reuse can greatly improve your overall productivity by leveraging previously completed work toward the needs of current projects. Later in our VHDL journeys I will show you how subprograms can be made elastic so that they can adjust to any input array length. The purpose of this current section is to present a technique that parameterizes design entities so that they will be able to model similar devices irrespective of the device's data path size. Design entity parameterization can easily be achieved via the use of generic parameters. Recall that when I first introduced this construct (Sec. 2.6) I mentioned that typically generic parameters are used as a means to pass in timing-related information such as propagation delays, pulse width, setup and hold times, and high-to-low and low-to-high delay values. Additionally, generic parameters may also quantify the width of a device's data path. For instance, suppose that we wanted to write a VHDL model for an N-bit incrementer instead of for just a 3-bit counting device. This N-bit model may then subsequently be used as a counter for any value of N. Hence for all practical

```
--  File Name :  count_n_bits_.vhd
--
--  Author    :  Joseph Pick
--
--  ****************************************************** LINE NUMBER
entity Count_N_Bits is                                          --  7
        generic (NUMBER_OF_BITS : NATURAL);                     --  8
        port    (Clock_In  : in BIT;                            --  9
                 Dataout_N : out BIT_VECTOR(NUMBER_OF_BITS downto 1) -- 10
                );                                              -- 11
end Count_N_Bits;                                               -- 12
```

Figure 2.49

purposes the model has to be designed, coded up, and tested only once. An entity/architecture pair will no longer have to be written anew every time a different-sized incrementer is to be modeled in VHDL. Let's first parameterize the entity declaration using the technique shown in Fig. 2.49.

The key points about this entity declaration for Count_N_Bits occur on lines 8 and 10. On line 8 a generic parameter called NUMBER_OF_BITS is declared and it is of type NATURAL. This generic parameter is then used on line 10 to define the index bounds for the out port Dataout_N. Note well that different values of NUMBER_OF_BITS will, in essence, produce a family of models each having a different array size for its out port. Hence this entity declaration may be used to model numerous hardware devices that have one input pin and one or more output pins. But it is important for you to realize that once simulation actually starts, then the array size for Dataout_N will be fixed and cannot be dynamically modified. The application of NUMBER_OF_BITS to determine Dataout_N's array size is done during the VHDL elaboration phase. Recall that the elaboration phase occurs just prior to the initialization phase. During the elaboration phase your VHDL tool suite links all your various VHDL models together, allocates memory for declared signals and process variables, passes generic parameter values into the model, and conducts numerous VHDL language-related checks that could not be executed by the compiler.

By the way, in case you are wondering how this parameterized N-bit incrementer will fit into our current modeling environment, let me give you the following sneak preview. This N-bit incrementer will serve as a concrete realization for the component Count_3_Bit declared in the architecture Structural_Decomposition. In VHDL terminology we are going to bind Count_N_Bits to Count_3_Bit via a configuration declaration. An excellent educational by-product of this binding activity will be the illustration of how one can (and must) account for any naming differences between a component declaration and the design entity that will serve as its concrete realization during a simulation run.

2.19.1 Information checkpoint

Before you venture further along your VHDL journeys, please make certain that you are familiar with the following topic that we have established via this model:

- The application of generic parameters to parameterize the array size of an entity's ports and by so doing to create generic reusable design entities.

2.20 File: count_n_bits_model_1.vhd

As per our sneak preview above we know that the design entity Count_N_Bits will be a concrete realization for the component Count_3_Bit. Since the design entity Count_3_Bit is also a realization for this very same component we should look at its various architecture bodies to determine which one of these can most easily be adapted to model a generic N-bit counter. After a quick glance at the VHDL listings in Secs. 2.8 through 2.15 we see that architecture Model_1 is the most promising. The other architectures simply do not contain any potential parameterization hooks. An example of a potential parameterization hook is Model_1's for loop. The beauty of this for loop is that its upper and lower bounds may very easily be adjusted to honor any array length. In one of our later journeys I will show you the exact details on how to make a for loop adjust to any input length. To make the N-bit model even more oriented to the real world, I decided to encapsulate the resulting N-bit incrementation algorithm into a function called Inc. That way the resulting VHDL model becomes highly modular. Design modularity is another important divide-and-conquer strategy that you should always aim for. Continuing with this code modularization theme, I placed the function Inc into the package Math_Bit_Vectors_Pkg so that it would be available (visible) to every model that references this package. I compiled this package Math_Bit_Vectors_Pkg into the library having the logical name MVL_Packages_Lib. Note well that during the compilation of Math_Bit_Vectors_Pkg the library MVL_Packages_Lib was also announced to the tool suite as the current working library. But during the compilation of this section's N-bit counter the library MVL_Packages_Lib is no longer considered as the current working library. Instead, it is referred to as a resource library. Once again let me reiterate that the mechanics for the creation and management of libraries (either resource or current working) is the responsibility of your VHDL tool suite. The VHDL language does, however, give you the capability to make the logical name of any library available (visible) to your VHDL model. Line 6 of Fig. 2.50 shows how the logical name of a resource library can be made visible via a library clause. The use clause on line 7 makes the contents of this package available (visible) throughout this architecture body. I

```
-- File Name :  count_n_bits_model_1.vhdl
--
-- Author    :  Joseph Pick
--
-- *********************************************************** LINE NUMBER
library MVL_Packages_Lib;                                       --  6
use MVL_Packages_Lib.Math_Bit_Vector_Pkg.all;                   --  7
architecture Model_1 of Count_N_Bits is                         --  8
                                                                --  9
begin                                                           -- 10
                                                                -- 11
    Bit_Incr:                                                   -- 12
    process                                                     -- 13
        variable Current_Count : BIT_VECTOR(NUMBER_OF_BITS downto 1) -- 14
                               := (others => '0');              -- 15
                                                                -- 16
        begin                                                   -- 17
        wait until Clock_In = '0';                              -- 18
                                                                -- 19
        Current_Count := Inc(Current_Count);                    -- 20
                                                                -- 21
        Dataout_N <= Current_Count;                             -- 22
                                                                -- 23
    end process Bit_Incr;                                       -- 24
                                                                -- 25
end Model_1;                                                    -- 26
```

Figure 2.50

could have written use MVL_Packages_Lib.Math_Bit_Vector_Pkg.Inc to isolate the specific subprogram's name that I intend to call in this model. However, it is much more convenient to simply apply the reserved word *all,* as was done on line 7. Continuing along we see that line 14 applies the generic parameter Number_Of_Bits to define the array bounds for the variable Current_Count. As a reminder, observe that any generic parameter declared in an entity declaration will automatically be visible to any of the entity's architecture bodies. The variable Current_Count will be used as the design entity's internal counter. Line 18 is our old friend again. It will suspend this process until the occurrence of a trailing edge on the input port called Clock_In. When this trailing edge transpires, then the process will be reactivated, and line 20 will call the function Inc to instantaneously increment the present value of Current_Count. Observe that a function call may be perceived as an expression on the right-hand side of an assignment. The output of this function will be the incremented value of Current_Count. Continuing our examination of line 20 we have that this incremented value will then be scheduled for assignment to the port Dataout_N 1 delta time later. After the bookkeeping (internal queue updating) for this signal scheduling is completed, then execution of this process will loop back to its top since, as you recall, processes are in an infinite loop. As per line 18 this process will then once again be suspended until Clock_In's next trailing edge. As a friendly reminder, recall that the Current_Count, like all other variables, will maintain its incremented value during the suspension of its encapsulating process.

2.20.1 Information checkpoint

Before you venture further along your VHDL journeys, please make certain that you are familiar with the following topics that we have established via this model:

- Application of the reserved word *library* to make the name of a resource library available (visible).
- Application of the use clause to make the contents of a package available (visible).
- The placement of subprograms into a package so that they may be available (visible) to any user of the package.
- The usage of a function as an expression on the right-hand side of an assignment.

2.21 File: cfg_cntn_m1.vhd

As stated earlier the design entity Count_N_Bits will serve as a concrete realization for the component Count_3_Bit that was declared and instantiated in the architecture Structural_Decomposition. Based on our previous journeys we know that a configuration declaration can be used to bind Count_N_Bits to Count_3_Bit. But this binding by itself is insufficient since their *port* parameter names are different. Moreover, Count_N_Bits has a *generic* parameter declaration whereas Count_3_Bit does not. Consequently, this *generic* parameter must be accounted for via a *generic map* construct. This associating of an actual value with the *generic* parameter NUMBER_OF_BITS is shown on line 20 of Fig. 2.51.

Though this *generic map* construct uses named association, I could just as well have applied a positional association scheme. But for readability my recommendation is to always use named associations along with maintaining each parameter's original position. On lines 21 and 22 the VHDL construct *port map* is used to account for the differences in the *port* parameter names. Theoretically, this is what is going on in terms of the parameters. Recall that the *generic* and *port* parameters occurring in a component declaration are known, in VHDL, as local parameters. During component instantiation actual parameters are associated with these local parameters. Furthermore, recall that the *generic* and *port* parameters occurring in an *entity* declaration, are referred to as formal parameters. And so it follows that during the *configuration* declaration, the formal parameters of the design entity are associated with their respective component local parameters. If these names all agree, then VHDL's default binding may be implicitly used and nothing more needs to be done. However, if any of these names differ, then they must be accounted for via a *generic* and/or *port* map construct. Another possibil-

```
-- File Name : cfg_cntn_m1.vhd
--
-- Author   : Joseph Pick
--
-- ********************************************************** LINE NUMBER
configuration Universe_N_Model_1 of Encapsulate_3_Bit_Count is --   6
                                                               --   7
    for Structural_Decomposition                               --   8
                                                               --   9
        for Synch: Clock                                       --  10
            use entity WORK.Clock(Model_1);                    --  11
        end for;                                               --  12
                                                               --  13
        for Logic_Analyzer: Count_Monitor                      --  14
            use entity WORK.Count_Monitor(Model_1);            --  15
        end for;                                               --  16
                                                               --  17
        for Gen_Cnt: Count_3_Bit                               --  18
            use entity WORK.Count_N_Bits(Model_1)              --  19
                generic map (NUMBER_OF_BITS => 3)              --  20
                port map    (Clock_In => Clock,                --  21
                             Dataout_N => Dataout);            --  22
        end for;                                               --  23
                                                               --  24
    end for;                                                   --  25
end Universe_N_Model_1;                                        --  26
```

Figure 2.51

ity is that the corresponding design entity may have some formal parameters that do not correspond to any of the component's local parameters. An example of this situation occurs in our present N-bit counter model in which there is no local parameter to correspond to the formal parameter NUMBER_OF_BITS. Lines 20 through 22 display how in our particular example all these differences were accounted for. Incidentally, the order of these two mapping constructs is well defined by the VHDL language. If present, the *generic map* construct must always come before the *port map* construct and there is no semicolon separating them as there is in the *component* and *entity* declarations. Furthermore, suppose that the *generic* parameter NUMBER_OF_BITS had not been assigned a value of 3. There would then have been a length mismatch between Dataout_N and Dataout, and an error condition would have been identified during the elaboration phase.

By the way, another practical application of the above *generic map* technique is to use this input path to back annotate postplacement and routing timing information. These derived timing numbers may then be used in a VHDL simulation in order to obtain more accurate gate level simulation results.

2.21.1 Information checkpoint

Before you venture further along your VHDL journeys, please make certain that you are familiar with the following topic that we have established via this model:

- The method to account for any differences in a component's and design entity's *generic* and *port* parameter lists when the component and design entity are bound together in a *configuration* declaration.

2.22 Logic Gates and Flip-Flops

Figure 2.52 shows that we are now going to return to the design entity Count_3_Bit and develop yet another architecture body for it. The illustrated magnification of this new architecture, called Model_Gates_1, indicates that we are about to embark on a journey into a gate level decomposition of our 3-bit counter. Your company's top-down design methodologies should include a synthesis strategy so that you will not have to rely on time costly bottom-up techniques to manually build and document gate level implementations. The only reason why I am pursuing such a manual course of action is that the ensuing models serve

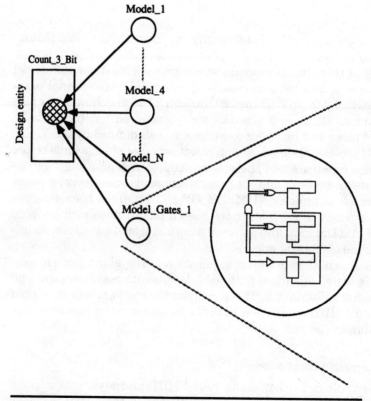

Figure 2.52 Gate level decomposition of the design entity Count_3_Bit.

as an excellent medium in which to introduce and discuss additional VHDL constructs, techniques, and caveats.

Actually, we have already encountered this intended gate level decomposition. It appeared in Fig. 2.33 (Sec. 2.13) during one of our previous journeys. Back then we used VHDL operators to capture the functionality of the combinational gates. The flip-flops, on the other hand, did not rely on any predefined, built-in VHDL operator. Rather, their registerlike behavior was implied by the synchronous manner in which their corresponding signals were updated. Our current game plan is to simulate a 3-bit counter at the gate level using VHDL models for the respective gates and flip-flops. This simulation effort will require compiled design entities for the following primitives: AND gate, NOT gate, Exclusive-OR gate, and D flip-flop. Recall that these compiled models have to exist only by the time we go to simulation. So we can optionally develop these models anytime between now and then. But let's do it now.

In our past journeys all the VHDL models were compiled into a single depository that was designated as the current working library. To make things more realistic I decided to compile the upcoming gate level models into a library that is different from this previously designated current working library. Recall that the creation and management of VHDL libraries is outside the scope of the language. It is up to the VHDL vendor to decide the mechanics and user interfaces for these activities. To successfully implement our aforementioned stated library objectives for the gate models, you will have to carry out the following library-related tasks. First, you must associate a logical library name with the intended physical location of the compiled gates. For instance, Gates_Simple_Lib is the logical name that I have chosen to be associated with the physical location of the compiled gates. Recall that a VHDL library is generally implemented as a specially designated subdirectory of a computer's filing system. Next, you must inform your VHDL tool suite that Gates_Simple_Lib is also to be considered as your current working library. Consequently, Gates_Simple_Lib is now synonymous with the logical library WORK. The tricky part is to remember that after compiling all the gates, you must then reassign WORK to your next intended current working library. Otherwise, any subsequent compilation will continue to direct the compiled models into Gates_Simple_Lib, which definitely is not what you want. Furthermore, if your current model requires compiled design entities from Gates_Simple_Lib, then you must remember to notify your compilation environment of this need. This notification is a two-step operation. First, you must announce Gates_Simple_Lib's physical location using the vendor's proprietary library management scheme. And second, your model must contain a library clause to make the library's log-

```
-- File Name : and_gate_model_1.vhd
--
-- Author    : Joseph Pick
--
-- ******************************************************  LINE NUMBER
entity And_Gate_Simple is                                   --  6
   generic (DELAY : TIME);                                  --  7
   port (Ain : in BIT; Bin : in BIT; Cout : out BIT);       --  8
end And_Gate_Simple;                                        --  9
                                                            -- 10
architecture Model_1 of And_Gate_Simple is                  -- 11
                                                            -- 12
begin                                                       -- 13
   Cout <= '1' after DELAY when Ain = '1' and Bin = '1'     -- 14
          else                                              -- 15
          '0' after DELAY;                                  -- 16
end Model_1;                                                -- 17
```

Figure 2.53

ical name visible to your model. We will see examples of this library clause very shortly in our journeys. By the way, when Gates_Simple_Lib is referenced in such an auxiliary manner, it is said to serve as a resource library.

The VHDL source code for an AND gate is shown in Fig. 2.53. Though the entity declaration and architecture body have been included in a single file, it is operationally sounder to keep them in separate files. The reason for this suggestion has to do with compilation dependencies and will reduce the number of time-consuming recompilations.

Since we are now down at the gate level, it makes sense to concern ourselves with propagation delays. This AND gate (and all our other gate models) will use a formal generic parameter DELAY to pass in a propagation delay. In Parts 2 and 3 of this book I will show you how to achieve a finer propagation delay granularity. Practical real-world techniques will be illustrated to incorporate high-to-low (Phl), low-to-high (Plh), minimum, maximum, and typical (min-max-typ) propagation delays.

The only new topic introduced by this AND gate model occurs on lines 14 through 16. Observe that the signal assignment on these lines is outside of any sequential region of code, such as a process or subprogram. So we can safely assume that this statement is some form of a concurrent signal assignment. In actuality it is a conditional (concurrent) signal assignment. The full details of this construct will be given in the Techniques and Recommendations part of this book but, as a first pass, it is sufficient for our immediate purposes to note the following features of this statement. During initialization if the BOOLEAN expression (Ain = '1' and Bin = '1') is TRUE, then '1' will be scheduled to be assigned to Cout at time 0 ns plus DELAY. Otherwise, Cout will be scheduled to get '0' at 0 ns plus DELAY. Henceforth, whenever Ain or Bin has an event then the BOOLEAN expression (Ain = '1' and Bin = '1') will be reevaluated with the updated value(s). Here again if this BOOLEAN expression is TRUE, then Cout will be scheduled to get '1'. Otherwise, it will get the value '0'. Note how a reading of the given conditional signal assignment statement seems to intuitively agree with this description. By the way, I could have substituted the given conditional signal assignment with the following concurrent signal assignment that applies the VHDL predefined operator and to the two input ports:

```
Cout <= Ain and Bin after DELAY;
```

Next, consider the VHDL model for a trailing edge-triggered flip-flop that is shown in Fig. 2.54. Its design entity name also contains the word *Simple* to reflect that, just like the aforementioned AND gate, this model is relatively quite simple and straightforward. This flip-flop model does not contain any tests to check for either setup times, hold times, or minimum pulse-width violations. Furthermore, unlike real D flip-flops, this model does not even contain a Preset or a Clear input port. Nonetheless, the model given in Fig. 2.54 quite readily satisfies our current needs for a simple D flip-flop design entity. To have incorporated the previously listed features would have drastically slowed down its simulation. However, if you wish to model at that fine level of granularity, then VHDL definitely gives you the capability to do so. Our current objective though is to just have a rudimentary D flip-flop model.

Suppose that the comment delimiter (--) would be removed from line 17 and placed onto line 18. In essence, my objective is to remove the

```
-- File Name : d_ff_trl_model_1.vhd
--
-- Author    : Joseph Pick
--
-- *************************************************   LINE NUMBER
entity D_FF_Trl_Simple is                              --   6
   generic (DELAY : TIME);                             --   7
   port (Clock : in BIT; D : in BIT; Q : out BIT);     --   8
end D_FF_Trl_Simple;                                   --   9
                                                       --  10
architecture Model_1 of D_FF_Trl_Simple is             --  11
                                                       --  12
begin                                                  --  13
   P_D_FF_Trl:                                         --  14
      process (Clock)                                  --  15
      begin                                            --  16
         --    if Clock = '0' then                     --  17
         if Clock = '0' and Clock'EVENT then           --  18
            Q <= D after DELAY;                        --  19
         end if;                                       --  20
      end process;                                     --  21
                                                       --  22
end Model_1;                                           --  23
```

Figure 2.54

check for Clock'EVENT and then to investigate what might potentially go wrong. Keep in mind that this design entity will always be just a subcomponent of a larger model. I have absolutely no control over what the port Clock will be connected to. So it is conceivable that Clock's initial input value might possibly be '0'. Let's therefore assume that it is. Recall that during the initialization phase this process will be executed until its first wait statement is sequentially reached. Since this process has a static sensitivity list, there is an implicit wait statement at its bottom just before line 21. In this case the uncommented line 17 will be executed during the initialization phase. Since Clock = '0' is assumed to be TRUE, the if statement test will pass and the simulator will continue on and execute line 19. Consequently, the output Q of our D flip-flop model will be scheduled to be updated with D's initial value. But wait a minute! Isn't there something wrong with this updating? Has the first trailing edge of Clock arrived yet? No it has not. And yet this model has just updated the output Q. So what went wrong? The root of the problem is a combination of the initialization phase and the implicit wait statement being at the process' bottom. The key lesson provided by this counterexample is that the VHDL engineer must always be aware of the ramifications of each process' execution during the simulation initialization phase.

To correct this initialization problem, you must create a situation whereby the if test initially fails. An appropriate workaround is shown on line 18 of Fig. 2.54. The additional check for Clock'EVENT will do the job. Remember that Clock'EVENT will be TRUE whenever the signal Clock just had an event. But during the initialization phase Clock'EVENT will always be FALSE. Assigning an initial value to Clock is not considered to be an event. In fact, according to the LRM, Clock is assumed to already have had its initially assigned value since the dawn of time, and so an event definitely did not occur at the start of the simulation. Hence even if Clock is initially '0' the additional test for Clock'EVENT means that the compound test (Clock = '0' and Clock'EVENT) will yield a FALSE value during the initialization phase. Consequently, the assignment to Q will now be bypassed during the initialization phase, and this process' execution will go directly to its bottom where it will then wait for events on Clock. So it seems like the initialization problem is solved. Unfortunately, this proposed workaround has introduced an inefficiency into our model. Just think about these new developments. The process is waiting at its bottom for an event on Clock. Whenever that happens its simulation will loop around and continue at its top. Suppose that Clock has a 50 percent duty cycle. Therefore the test (Clock = '0') will be TRUE 50 percent of the time. But Clock'EVENT will always be TRUE since the process will be reactivated only whenever Clock just had an event during the current simu-

```
-- File Name : d_ff_trl_model_2.vhd
--
-- Author    : Joseph Pick
--
-- ****************************************   LINE NUMBER
architecture Model_2 of D_FF_Trl_Simple is   --  6
                                             --  7
begin                                        --  8
   P_D_FF_Trl:                               --  9
      process                                -- 10
      begin                                  -- 11
         wait until Clock = '0';             -- 12
         Q <= D after DELAY;                 -- 13
      end process;                           -- 14
                                             -- 15
end Model_2;                                 -- 16
```

Figure 2.55

lation cycle. The test for Clock'EVENT is now redundant. So here is the dilemma. Clock'EVENT is only required to block the potential ill effects of the initialization phase, but henceforth it is no longer needed at all. Figure 2.55 illustrates a technique that circumvents all of the above difficulties. It uses our old friend wait until Clock = '0'. So the upshot of this D flip-flop example is that you must always be concerned with the ramifications of the initialization phase.

Our next design entity is a model for the Exclusive-OR gate. The if...then...else logic of lines 16 through 20 of Fig. 2.56 agrees with our understanding of an Exclusive-OR gate's behavior. If either of the inputs Ain or Bin change (have an event), then the output will be updated as follows: If Ain and Bin are now the same then the output Cout is scheduled to receive '0' after the specified propagation delay

```
-- File Name : xor_gate_model_1.vhd
--
-- Author    : Joseph Pick
--
-- *****************************************   LINE NUMBER
entity Xor_Gate_Simple is                     --  6
   generic (DELAY : TIME);                    --  7
   port (Ain : in BIT; Bin : in BIT; Cout : out BIT);  --  8
end Xor_Gate_Simple;                          --  9
                                              -- 10
architecture Model_1 of Xor_Gate_Simple is    -- 11
begin                                         -- 12
   P_Xor_Simple:                              -- 13
   process (Ain, Bin)                         -- 14
   begin                                      -- 15
      if Ain = Bin then                       -- 16
         Cout <= '0' after DELAY;             -- 17
      else                                    -- 18
         Cout <= '1' after DELAY;             -- 19
      end if;                                 -- 20
   end process P_Xor_Simple;                  -- 21
end Model_1;                                  -- 22
```

Figure 2.56

```
-- File Name : not_gate_model_1.vhd
--
-- Author    : Joseph Pick
--
-- *******************************************  LINE NUMBER
entity Not_Gate_Simple is                        -- 6
  generic (DELAY : TIME);                        -- 7
  port (Ain : in BIT; Cout : out BIT);           -- 8
end Not_Gate_Simple;                             -- 9
                                                 -- 10
architecture Model_1 of Not_Gate_Simple is       -- 11
begin                                            -- 12
  Cout <= not Ain after DELAY;                   -- 13
end Model_1;                                     -- 14
```

Figure 2.57

time. Otherwise, if these inputs differ, then Cout will be scheduled to receive '1' after the specified propagation. At first glance there does not seem to be anything new to discuss about this design entity. However, in light of our in-depth analysis of the D flip-flop model, there is something worthwhile to note regarding this Exclusive-OR gate model. Line 14 shows that this process has a static sensitivity list consisting of Ain and Bin. Recall that the D flip-flop model also had a process with a static sensitivity list. But in the D flip-flop model we had to concern ourselves with the signal attribute 'EVENT. Here in our current Exclusive-OR model there is no need to reference either Ain'EVENT or Bin'EVENT. So what is the difference between these models? Keep in mind that the D flip-flop is a synchronous device, whereas the Exclusive-OR gate is not. Hence the former should not be updated until the first appropriate edge of its clocking input signal. The latter, on the other hand, represents a combinational device and so at simulation startup it makes sense to just let the initial input values ripple through the model. And so, unlike the D flip-flop model, the Exclusive-OR process should not apply a 'EVENT attribute in order to block its execution during the initialization phase.

The final gate that we require a design entity for is the NOT gate. Its VHDL source code is given in Fig. 2.57. Because of our previous VHDL journeys there is nothing new to discuss about this model.

2.22.1 Information checkpoint

Before you venture further along your VHDL journeys, please make certain that you are familiar with the following topics that we have established via the models in this section:

- Conditional signal assignment.
- The importance of being aware of and, if necessary, counteracting the ramifications of the initialization phase.

2.23 File: count_3_bit_model_gates_1.vhd

Figure 2.58 illustrates the component instantiations for a gate level decomposition of our 3-bit counter. A quick glance at Fig. 2.52 will remind you how this design entity Count_3_Bit fits into our top-down journey.

Several interesting points immediately jump out at you from Fig. 2.58. First, the rectangles representing the gates and flip-flops are dotted even though we have just written and presumably compiled their corresponding VHDL models. This dotted representation is still used to emphasize the fact that even though the gates exist, they really do not have to at the time that this architecture is being compiled. The dotted rectangles represent only component templates and do not even have to correspond to any of the gate models of our previous section. Another worthwhile observation is that this is the first time in our journeys that an instantiated component's port will be connected to a port of the architecture's parent entity. Previously, all the actual parameters connected

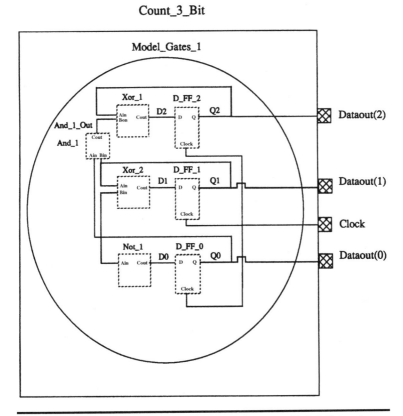

Figure 2.58 Component instantiations in the architecture Model_Gates_1.

to a component's port parameters had been signals declared within the architecture. Just for comparison you should review the contents of Fig. 2.24. In that diagram all the components' ports were connected to each other via signals declared in the architecture Structural_Decomposition. Our current Fig. 2.58 depicts yet another available option. Note how the Clock input ports of the three instantiated D flip-flops (D_FF_0, D_FF_1, and D_FF_2) are directly connected to the input port Clock of Model_Gates_1's associated entity Count_3_Bit. Though such direct port-to-port connections are a viable option you must always keep in mind that the ports must have the same data type and that they must also have the appropriate dataflow directions. For instance, it is incorrect to connect a component's in port to an entity's out port because the respective dataflows would then be clashing. Recall though (Fig. 2.24) that it is permissible to connect an instantiated component's out port to another instantiated component's in port via an explicitly declared signal. The term *explicitly declared signal* is used to differentiate it from entity declared formal port parameters. Though ports are signals they are implicitly declared signals since the reserved word *signal* does not exist in their declaration. By the way, before you get any ill-conceived notions, it is just a coincidence that the connecting components and entity ports have the same name Clock. This does not have to be so and, in general, the names are different.

Another neat feature of this architecture body is that, as the diagram shows, this will be the first time in our journeys that multiple instances are created for the same component. There are three instances of the D flip-flop component and two instances of the Exclusive-OR component. Our model is now finally beginning to have some degree of complexity. And this complexity is what makes this model so educational. Once all the necessary models are compiled, you can then use this gate level model to teach yourself all about the ins and outs of your VHDL simulation tool. Learn the menu pull-down commands that will permit you to traverse down into the hierarchy of your model. Discover how to place simulation breakpoints on any line of any of the multiple instances contained in your overall model. Experiment with selecting any signal within the total hierarchy. Then determine how to examine the current value of these selected signals. Also, learn how to monitor these signals during a simulation run so that their values may be displayed on a time versus value waveform graph. And of course practice how to create and manage multiple libraries. Remember that at one point the library Gates_Simple_Lib should be designated as a current working library and then later it should serve as a resource library.

Having laid all the aforementioned groundwork, let's now dive into the architecture Model_Gates_1 that is shown in Fig. 2.59. In form this model is similar to the architecture Structural_Decomposition of Sec.

```
-- File Name : count_3_bit_model_gates_1.vhd
--
-- Author    : Joseph Pick
--
-- ************************************************************ LINE NUMBER
architecture Model_Gates_1 of Count_3_Bit is              --  6
                                                          --  7
  component And_Gate_Simple                               --  8
          generic (DELAY : TIME);                         --  9
          port    (Ain : in BIT; Bin : in BIT; Cout : out BIT); -- 10
  end component;                                          -- 11
                                                          -- 12
  component Xor_Gate_Simple                               -- 13
          generic (DELAY : TIME);                         -- 14
          port    (Ain : in BIT; Bin : in BIT; Cout : out BIT); -- 15
  end component;                                          -- 16
                                                          -- 17
  component Not_Gate_Simple                               -- 18
          generic (DELAY : TIME);                         -- 19
          port    (Ain : in BIT; Cout : out BIT);         -- 20
  end component;                                          -- 21
                                                          -- 22
  component D_FF_Trl_Simple                               -- 23
          generic (DELAY : TIME);                         -- 24
          port    (Clock : in BIT; D : in BIT; Q : out BIT); -- 25
  end component;                                          -- 26
                                                          -- 27
  signal D0, D1, D2, Q0, Q1, Q2, And_1_Out : BIT := '1';  -- 28
                                                          -- 29
begin                                                     -- 30
                                                          -- 31
  And_1: And_Gate_Simple                                  -- 32
          generic map (5 ns)                              -- 33
          port map (Ain => Q0, Bin => Q1, Cout => And_1_Out); -- 34
                                                          -- 35
  Xor_1: Xor_Gate_Simple                                  -- 36
          generic map (5 ns)                              -- 37
          port map (Ain => And_1_Out, Bin => Q2, Cout => D2); -- 38
                                                          -- 39
  Xor_2: Xor_Gate_Simple                                  -- 40
          generic map (4 ns)                              -- 41
          port map (Ain => Q1, Bin => Q0, Cout => D1);    -- 42
                                                          -- 43
  Not_1: Not_Gate_Simple                                  -- 44
          generic map (5 ns)                              -- 45
          port map (Ain => Q0, Cout => D0);               -- 46
                                                          -- 47
  D_FF_0: D_FF_Trl_Simple                                 -- 48
          generic map (DELAY => 10 ns)                    -- 49
          port map (Clock, D0, Q0);                       -- 50
                                                          -- 51
  D_FF_1: D_FF_Trl_Simple                                 -- 52
          generic map (10 ns)                             -- 53
          port map (Clock => Clock, D => D1, Q => Q1);    -- 54
                                                          -- 55
  D_FF_2: D_FF_Trl_Simple                                 -- 56
          generic map (10 ns)                             -- 57
          port map (Clock => Clock, D => D2, Q => Q2);    -- 58
                                                          -- 59
  Dataout <= Q2 & Q1 & Q0 ;                               -- 60
                                                          -- 61
end Model_Gates_1;                                        -- 62
```

Figure 2.59

2.6. The architecture declarative part contains component declarations that are later instantiated in the architecture statement region. Note that the component Xor_Gate_Simple is instantiated twice (Xor_1 and Xor_2) and that the component D_FF_Trl_Simple is instantiated three times (D_FF_0, D_FF_1, and D_FF_2). Observe that each instance of the same component must have a different label. Recall that component instantiations are VHDL's equivalent of a netlist. So, essentially, what we have here is the VHDL version of an optimized gate level schematic for a 3-bit counter. The port connections were all done manually and are based on the schematic previously shown in Fig. 2.33. Clearly, it is preferable to rely on a synthesis tool to automatically generate such a netlist instead of laboriously doing it by hand.

The generic map constructs (lines 33, 37, 41, 45, 49, 53, and 57) are used to pass in propagation delay values to be used during the simulation of these instances. Lines 37 and 41 point out that different instances of the same component may be assigned different actual generic values.

For illustrative purposes the format of the port and generic mappings highlight your option to use either named or positional associations. Lines 50, 54, and 58 show that it is possible to connect an instantiated component's port parameter to a port of the encapsulating architecture's associated entity. But as stated earlier their dataflow directions have to be compatible. For example, in the instance labeled Xor_1 (lines 36 through 38) it would have been incorrect to write

```
port map (Ain => And_1_Out, Bin => Dataout(2), Cout => D2); -- WRONG !
```

The reasoning behind this error is that Dataout(2) is a formal port of mode out, whereas Bin is a local port of mode in. Recall that ports of mode out can only be on the left-hand side of a signal assignment, whereas ports of mode in can only appear on the right-hand side. Alternatively, one can say that out ports can only be written to, whereas in ports can only be read. An analogy can be made with VHDL's port-mapping rules. Hence it would be contradictory to map (directly connect) a local port that is only readable onto a formal port that is only writable. Instead, VHDL forces us to declare intermediate signals to do the connecting in an indirect way. Recall that explicitly declared signals can be both read and written to. Lines 38 and 60 illustrate how the signal Q2 is used to indirectly connect a local in port (Bin) to a formal out port (Dataout(2)). While we are discussing line 60 let me remind you that, in general, concurrent signal assignments are examples of a dataflow modeling style. Since component instantiations are indicative of a structural model we see that, as previously advertised, VHDL permits the mixture of these two modeling paradigms. Additionally, I could also have included a process containing a complex behavioral algorithm.

And in that case we definitely would have had all three modeling styles (structural, dataflow, and behavioral) in one VHDL model.

One final operational remark about this model. Line 28 shows that you can collectively declare several signals at the same time, each signal separated by a comma. Of course they must all be of the same type, and they are all going to be initialized to the same value. This same collective notation can be used with variable declarations as well. Though this option exists, I recommend that it not be used because you will then be locked into assigning the same initial value to all the collectively declared objects. Also, that technique makes your code more difficult to read and to manage during its full life cycle.

2.23.1 Information checkpoint

Before you venture further along your VHDL journeys, please make certain that you are familiar with the following topics that we have established via this model:

- The same declared component may be multiply instantiated.
- An actual parameter referenced in the *port map* of a *component* instantiation may also be a *port* named in the entity declaration associated with the architecture body in which this instantiation occurs.
  - But it should be noted that the respective *port* modes must satisfy VHDL's *port*-mapping rules.
- VHDL objects may be declared collectively in a list, but each element of the list must be separated by a comma.

2.24 File: cfg_cnt3_model_gates_1.vhd

So here we are again with our standard presimulation choice. Should we rely on VHDL's default binding when simulating our gate level 3-bit counter. Or should we use a configuration declaration to explicitly pronounce the intended constituents of our simulation run. Though the default-binding option exists, it is, nonetheless, highly recommended that you write a configuration declaration. In general, this recommendation is very sound but it is even more so here in light of the fact that the test bench (term fully explained in Sec. 2.6) for our 3-bit counter now has two layers of hierarchy: one for the device under test (Count_3_Bit) and a second tier for the gate level decomposition of this device under test.

Line 9 (Fig. 2.60) states that we are now going to create a configuration called Universe_Model_Gates_1 for our old friend Encapsulate_3_Bit_Count. If you do not recall this old friend, then please renew

```
-- File Name : cfg_cnt3_model_gates_1.vhd
--
-- Author    : Joseph Pick
--
-- ************************************************************ LINE NUMBER
--                                                              -- 6
library Gates_Simple_Lib;                                       -- 7
                                                                -- 8
configuration Universe_Model_Gates_1 of Encapsulate_3_Bit_Count is -- 9
                                                                -- 10
   for Structural_Decomposition                                 -- 11
                                                                -- 12
      for Synch: Clock                                          -- 13
         use entity WORK.Clock(Model_1);                        -- 14
      end for;                                                  -- 15
                                                                -- 16
      for Logic_Analyzer: Count_Monitor                         -- 17
         use entity WORK.Count_Monitor(Model_1);                -- 18
      end for;                                                  -- 19
                                                                -- 20
      for Gen_Cnt: Count_3_Bit                                  -- 21
         use entity WORK.Count_3_Bit(Model_Gates_1);            -- 22
                                                                -- 23
         for Model_Gates_1                                      -- 24
                                                                -- 25
            for And_1: And_Gate_Simple                          -- 26
               use entity Gates_Simple_Lib.And_Gate_Simple(Model_1); -- 27
            end for;                                            -- 28
                                                                -- 29
            for all: Xor_Gate_Simple                            -- 30
               use entity Gates_Simple_Lib.Xor_Gate_Simple(Model_1); -- 31
            end for;                                            -- 32
                                                                -- 33
            for Not_1: Not_Gate_Simple                          -- 34
               use entity Gates_Simple_Lib.Not_Gate_Simple(Model_1); -- 35
            end for;                                            -- 36
                                                                -- 37
            for all: D_FF_Trl_Simple                            -- 38
               use entity Gates_Simple_Lib.D_FF_Trl_Simple(Model_1); -- 39
            end for;                                            -- 40
                                                                -- 41
         end for;                                               -- 42
                                                                -- 43
      end for;                                                  -- 44
                                                                -- 45
   end for;                                                     -- 46
end Universe_Model_Gates_1;                                     -- 47
```

Figure 2.60

your friendship by reviewing Figs. 2.15, 2.20, 2.21, 2.24, 2.25, 2.27, 2.37, 2.42, and 2.46.

The for construct on line 11 says that of all the architecture bodies (there are five) associated with the design entity Encapsulate_3_Bit_Count this specific configuration is going to concentrate on the architecture called Structural_Decomposition. Recall (see Sec. 2.6) that this architecture contained three instantiated components labeled Synch, Logic_Analyzer, and Gen_Cnt, respectively. Lines 13 through 15 define an explicit binding for the component instance labeled Synch. There is nothing new here and, in fact, we have already seen this exact binding in Sec. 2.18. The component instance labeled Logic_Analyzer is explicitly bound on lines 17 through 19, and we have also seen this binding in Sec. 2.18.

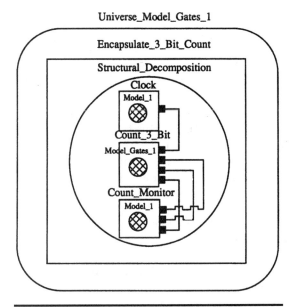

Figure 2.61 Upper level entity/architecture bindings specified in the configuration Universe_Model_Gates_1.

The for construct on line 21 initiates our intention to bind the instantiation of the component Count_3_Bit that was labeled Gen_Cnt. The specific binding is given on the following line, and it will use the entity/architecture pair Count_3_Bit and Model_Gates_1 from our current working library. Figure 2.61 depicts the bindings that we have established so far. Comparing this diagram to the one in Fig. 2.46 shows that up to this point we have really not done anything new. The primary novelty of this configuration declaration occurs between lines 24 and 42 where a second level hierarchical binding is formulated.

The for construct of line 24 announces our intentions to bind those component instantiations that were made in the architecture Model_Gates_1. The for construct on lines 26 through 28 will give a binding for the instantiation of the component And_Gate_Simple that was labeled And_1. A slight twist to an already familiar concept occurs on line 27. Remember that the design entity for And_Gate_Simple was deliberately compiled into a library that is now no longer the current working library. Hence we cannot reference the logical library WORK during the binding of the design entity And_Gate_Simple. Instead, we have to use the logical name of the library into which this design entity was previously compiled. That is why we have to use Gates_Simple_Lib on line 27. But to be able to refer to this logical library VHDL requires that we make its name visible. This visibility is achieved on line 7 by using a library clause.

The remaining new feature highlighted by this configuration declaration occurs on lines 30 and 38. Since every instance of the component Xor_Gate_Simple is to be bound to the same design entity, it is possible to compactly express this collective binding in a single for statement using the reserved word *all*. The multiple instantiations of the component D_FF_Trl_Simple are handled similarly using this for all technique.

By the way, in this configuration declaration the component names and their respective generic and port parameters match those of their binding design entities. Consequently, there is no need for either a generic or a port map association to account for any differences. Recall that the binding of our N-bit counter required an explicit mapping between the local component and formal design entity parameters (see Sec. 2.21).

Let's conclude the analysis of this configuration declaration with an important VHDL style guide remark. The source code indentations are a very powerful way to visually capture the intended hierarchy of our configuration. Moreover, these indentations make the VHDL source code more readable and user friendly. Figure 2.62 illustrates the complete two-tier hierarchy established and traversed via the configuration Universe_Model_Gates_1.

2.24.1 Information checkpoint

Before you venture further along your VHDL journeys, please make certain that you are familiar with the following topics that we have established via this model:

- Usage of a library clause to make the logical name of a library visible.
- A complete hierarchy may be both specified and traversed via a sequence of nested for...end for binding statements.
- All instantiations of a component may be collectively bound to the same design entity via the for all construct.

2.25 File: cfg_cnt3_cfg.vhd

The purpose of this section is to show you another configuration option that you may find beneficial in your top-down modular coding methodologies.

All of our previous configuration declarations dealt with self-contained top-level design entities that did not have any ports. Figure 2.63 highlights another capability available to the VHDL modeler. Line 7 states that Config_To_Bind_Gates is a configuration for the lower level design entity Count_3_Bit. In actuality, what I did was take the gate level bind-

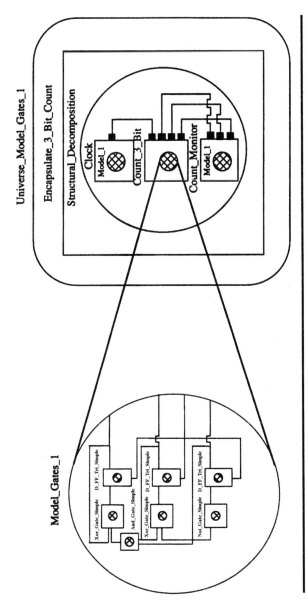

Figure 2.62 Complete top-down entity/architecture bindings specified in the configuration Universe_Model_Gates_1.

```
-- File Name : cfg_to_bind_gates.vhd
--
-- Author    : Joseph Pick
--
-- *********************************************************** LINE NUMBER
library Gates_Simple_Lib;                                       -- 6
configuration Config_To_Bind_Gates of Count_3_Bit is            -- 7
                                                                -- 8
   for Model_Gates_1                                            -- 9
                                                                -- 10
      for And_1: And_Gate_Simple                                -- 11
         use entity Gates_Simple_Lib.And_Gate_Simple(Model_1);  -- 12
      end for;                                                  -- 13
                                                                -- 14
      for all: Xor_Gate_Simple                                  -- 15
         use entity Gates_Simple_Lib.Xor_Gate_Simple(Model_1);  -- 16
      end for;                                                  -- 17
                                                                -- 18
      for Not_1: Not_Gate_Simple                                -- 19
         use entity Gates_Simple_Lib.Not_Gate_Simple(Model_1);  -- 20
      end for;                                                  -- 21
                                                                -- 22
      for all: D_FF_Trl_Simple                                  -- 23
         use entity Gates_Simple_Lib.D_FF_Trl_Simple(Model_1);  -- 24
      end for;                                                  -- 25
                                                                -- 26
   end for;                                                     -- 27
                                                                -- 28
end Config_To_Bind_Gates;                                       -- 29
```

Figure 2.63

ings for the components instantiated in the architecture Model_Gates_1 and place them into their own configuration. This configuration Config_To_Bind_Gates may then later be referenced in a hierarchically higher level configuration. Line 20 of Fig. 2.64 is an example of this usage. Note that the reserved word *configuration* is applied where we had previously used the word entity.

2.25.1 Information checkpoint

Before you venture further along your VHDL journeys, please make certain that you are familiar with the following topics that we have established via these models:

- Any design entity may be configured even if it is not a top level entity.
- *use configuration* may also be applied in a binding statement.

2.26 File: mvl4_pkg_.vhd

By now you are thoroughly convinced about the educational merits of our VHDL journeys. Well, here comes the icing on the cake.

Consider the block diagram in Fig. 2.65. It provides a visual background for the next and final leg of our VHDL journeys. Suppose that

```
-- File Name : cfg_cnt3_cfg.vhd
--
-- Author    : Joseph Pick
--
-- ************************************************** LINE NUMBER
configuration Universe_Model_Gates_Config
                        of Encapsulate_3_Bit_Count is  --  7
                                                       --  8
   for Structural_Decomposition                        --  9
                                                       -- 10
      for Synch: Clock                                 -- 11
         use entity WORK.Clock(Model_1);               -- 12
      end for;                                         -- 13
                                                       -- 14
      for Logic_Analyzer: Count_Monitor                -- 15
         use entity WORK.Count_Monitor(Model_1);       -- 16
      end for;                                         -- 17
                                                       -- 18
      for Gen_Cnt: Count_3_Bit                         -- 19
         use configuration WORK.Config_To_Bind_Gates;  -- 20
      end for;                                         -- 21
                                                       -- 22
   end for;                                            -- 23
                                                       -- 24
end Universe_Model_Gates_Config;                       -- 25
```

Figure 2.64

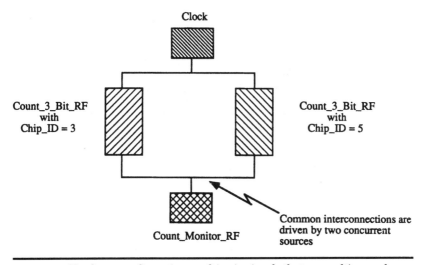

Figure 2.65 Hardware configuration resulting in signals that are multisourced.

there are two 3-bit counters each connected to a common clock. On each trailing edge they will both simultaneously increment their own internal counters. Whenever this incremented value is equal to its predesignated chip identification, then the respective counter will output this current value onto a common bus. Based on all our past VHDL adventures I am sure that you have already determined some of the mechanics for the VHDL implementation of Fig. 2.65. There really is only a need for one enhanced 3-bit counter. Its chip identification will be assigned to it via the generic parameter CHIP_ID. An architecture will be written that will declare a component template having the same name and description as our enhanced 3-bit counter. This component will then be instantiated twice and, as per Fig. 2.65, the values 3 and 5 will be, respectively, mapped onto these instances' CHIP_ID generic parameter. If you have understood all that has just been read then congratulations! You have definitely mastered the essence of our past VHDL journeys. Let's now continue along on our new adventure.

Observe from the diagram that the output of both instances of this enhanced counter are connected to the same set of wires. Hence the VHDL signals corresponding to these physical wires will be multiply driven from two concurrent constructs. By itself, VHDL does not directly handle a multisourcing environment. The concept of a tristated (tristate is a trademark of National Semiconductor Corporation), wired-and, or wired-or bus is not hard-wired into the VHDL language. And why should it be? VHDL plans to be around for a long time, and so it should not commit itself to any present-day technology. Instead, VHDL serves as a robust tool that provides the capability to model present and future technological developments. This adaptability is achieved by a two-pronged attack. First, the VHDL user has the freedom to develop new data types that can abstract a technology's physical states. And second, VHDL allows its users to write a specially dedicated resolution function that captures the essence of a technology's multisourcing characteristics.

Very shortly, more will be said about resolution functions, but right now let's explore why the standard data type BIT is not good enough for our intended tristate model. In addition to driving a '0' or a '1', our 3-bit counter must also be able to create the look and feel of high impedance. Furthermore, consideration must be given to the tristate bussing conflict that occurs when one source drives a '0' while the other source drives a '1'. These tristate characteristics cannot be handled by BIT's two-valued logic system. So our first task at hand is to define a new enhanced data type that will incorporate additional symbols for the high impedance and unknown states. But where do we declare it? Should its declaration appear in the architecture declarative part as was done earlier for our state machine enumeration type. Actually, there is a problem with this location. It is too local and restrictive. Other design entities will not have access to it. To make this data type globally accessible, you must define it in a VHDL package. A package may intuitively be viewed as a repository for certain definitions that may be exported to other models and/or projects. Figure 2.66 contains the package that will be referenced later by our tristated counters.

```
--   File Name  :  mvl4_pkg_.vhd                              --
--                                                            --
--   Author     :  Joseph Pick                                --
--                                                            --
--   ************************************************* LINE NUMBER
package MVL4_Pkg is                                           --  6
                                                              --  7
   type Logic4 is ('X', '0', '1', 'Z');                       --  8
                                                              --  9
   type Logic4_Vector is array (NATURAL range <>) of Logic4;  -- 10
                                                              -- 11
   -- Resolution functions (RF)                               -- 12
   function Tristate_RF (V : Logic4_Vector) return Logic4;    -- 13
   function Wired_And_RF (V : Logic4_Vector) return Logic4;   -- 14
   function Wired_Or_RF  (V : Logic4_Vector) return Logic4;   -- 15
                                                              -- 16
   -- Declare those resolved subtypes (RS) that are to be     -- 17
   -- associated with a resolution function.                  -- 18
   subtype Tristate_RS is Tristate_RF Logic4;                 -- 19
   subtype Wired_And_RS is Wired_And_RF Logic4;               -- 20
   subtype Wired_Or_RS is Wired_Or_RF Logic4;                 -- 21
                                                              -- 22
   type Tristate_RS_Vector is array (NATURAL range <>)        -- 23
                                       of Tristate_RS;        -- 24
   type Wired_And_RS_Vector is array (NATURAL range <>)       -- 25
                                       of Wired_And_RS;       -- 26
   type Wired_Or_RS_Vector is array (NATURAL range <>)        -- 27
                                       of Wired_Or_RS;        -- 28
                                                              -- 29
   -- Overloading of the predefined logical operators         -- 30
   --    (Scalar version).                                    -- 31
   function "and"  (L, R : Logic4) return Logic4;             -- 32
   function "nand" (L, R : Logic4) return Logic4;             -- 33
   function "or"   (L, R : Logic4) return Logic4;             -- 34
   function "nor"  (L, R : Logic4) return Logic4;             -- 35
   function "xor"  (L, R : Logic4) return Logic4;             -- 36
   function "not"  (S    : Logic4) return Logic4;             -- 37
                                                              -- 38
   -- Overloading of the predefined logical operators         -- 39
   --    (Vector version).                                    -- 40
   function "and"  (L, R : Logic4_Vector) return Logic4_Vector; -- 41
   function "nand" (L, R : Logic4_Vector) return Logic4_Vector; -- 42
   function "or"   (L, R : Logic4_Vector) return Logic4_Vector; -- 43
   function "nor"  (L, R : Logic4_Vector) return Logic4_Vector; -- 44
   function "xor"  (L, R : Logic4_Vector) return Logic4_Vector; -- 45
   function "not"  (V    : Logic4_Vector) return Logic4_Vector; -- 46
                                                              -- 47
   -- Conversion functions (Overloaded scalar version):       -- 48
   function To_Bit    (S : Logic4) return BIT;                -- 49
   function To_Logic4 (S : BIT) return Logic4;                -- 50
                                                              -- 51
      -- Conversion functions (Overloaded vector version):    -- 52
   function To_Bit    (V : Logic4_Vector) return BIT_VECTOR;  -- 53
   function To_Logic4 (V : BIT_VECTOR) return Logic4_Vector;  -- 54
                                                              -- 55
   -- Subprograms to sense/identify a specific value          -- 56
   -- in a bit stream:                                        -- 57
                                                              -- 58
   -- TRUE is returned if X is sensed/identified              -- 59
   -- in input bit stream.                                    -- 60
   function Sense_X (Bit_Stream : Logic4_Vector) return BOOLEAN; -- 61
                                                              -- 62
   -- TRUE is returned if Z is sensed/identified in input     -- 63
   -- bit stream.                                             -- 64
   function Sense_Z (Bit_Stream : Logic4_Vector) return BOOLEAN; -- 65
                                                              -- 66
   -- Input bit stream in simultaneously checked for X or Z.  -- 67
   -- Returned value will be:                                 -- 68
   --         '0'  when there are no Xs or Zs                 -- 69
   --         'X'  when there is an X                         -- 70
   --         'Z'  when there is a Z but no X                 -- 71
   --              (hence X has higher priority than Z)       -- 72
   function Sense_XZ (Bit_Stream : Logic4_Vector) return Logic4; -- 73
                                                              -- 74
end MVL4_Pkg;                                                 -- 75
```

Figure 2.66

The interesting point about this package is that in form it is typical of any MVLN package, where MVL stands for *multivalued logic,* and N indicates the *number* of elements in its base type. Consequently, if you understand the form and substance of this package, then you will understand every MVLN package since they all have the same look and feel. The package contained in Fig. 2.66 may be considered to be an MVL4 package since its base data type consists of four elements: 'X', '0', '1', and 'Z'. The new IEEE standard STD_LOGIC_1164 package contains a base type consisting of nine elements, and so it may be viewed as an MVL9 package. The following detailed analysis of our simpler MVL4 package will superbly prepare you to read through and utilize the standard IEEE STD_LOGIC_1164 package.

Figure 2.66 contains the package declaration for the package MVL4_Pkg. Analogous to a design entity a package has two constituents: a package declaration and a package body. A package declaration is similar to an entity declaration in that both are primary design units that present an external view to the outside world. A package body is similar to an architecture body in that both are secondary design units that contain internal behavioral descriptions that will be executed during a simulation. In the architecture's case concurrent constructs such as processes are being executed, whereas in a package body's case it is subprograms that are being executed. Any declaration made within an architecture body is not available (visible) to other design entities. Likewise, any declaration made within a package body is not available (visible) to the user of the package. Another similarity is that the package declaration and body may be in separate files, just like the entity declaration and its associated architecture body. Package is the shorthand collective term for a package specification and its body. Similarly, entity is the shorthand collective term for an entity declaration and its architecture body. These design entity and package analogies break down on one key point. Whereas an entity can have multiple architecture bodies, a package can have at most only one body associated with it.

The reserved word *package* on line 6 initiates our intentions to develop a package declaration. Package specification is another term often used to denote a package declaration. In this particular case our package declaration and body are in separate files, and it is recommended that they always be so. The name of this package is MVL4_Pkg, where the significance of the acronym MVL4 has already been explained. It is optional to repeat this package's name at the termination of the package specification, but it is recommended that you do so for readability as I have done on line 75.

All MVLN packages begin with a definition of a base type. In our particular package this is done on line 8 with an enumeration data type definition. The enumeration literal 'X' will correspond to our concept of an unknown value. A little refresher now about the attribute 'LEFT. Since 'X' is the leftmost element listed in the enumeration list, it will correspond to Logic4'LEFT. Hence any signal or variable of type Logic4 that is not initialized by the user will automatically be initialized to 'X' by the simulation engine. Be careful in regards to 'X'. The single quotes imply that it is a character literal. Hence lower case 'x' is not the same as upper case 'X'. This distinction must always be remembered. The enumeration literal 'Z' will correspond to our concept of high impedance. Please keep in mind we are not going to be disconnecting a signal by assigning 'Z' to it. Rather, the interactions of 'Z' with other Logic4 values will yield the look and feel of a tristated bus. One driven 'Z' combined with another driven 'Z' will yield a 'Z'. And a 'Z' combined with a '1' will yield a '1'. Incidentally, VHDL does not allow the ordinary, regularly defined signals that we have been working with to be disconnected. Only a special kind of signals called guarded signals can be disconnected. But do not get your hopes up too high. Guarded signals are very tedious, restrictive, and inefficient. Most VHDL modelers do not rely on them. Consequently, the topic of guarded signals will not be developed in this book.

After the definition of the package's base type, most MVLN packages continue onto a type definition for an unconstrained array where each element is of the package's base type. In our particular MVL4_Pkg this definition is made on line 10. The definition for Logic4_Vector is very similar to that of BIT_VECTOR, which we have already encountered in one of our previous journeys. As with BIT_VECTOR there is a great utility in defining Logic4_Vector to be an unconstrained array. This unconstrained status is achieved via the usage of the box (<>) notation.

Lines 13 through 15 contain the specification for three resolution functions. As their names suggest they are, respectively, supposed to create the environment for a tristate, wired-and, or a wired-or bussing scheme. But at this point in time the compiler just views these three lines as specifications for three ordinary functions. The compiler is not yet aware that these functions are supposed to be used as resolution functions. This realization only comes about with the subtype definitions on lines 19 through 21.

Let's closely examine the subtype definition on line 19. Tristate_RS is the name of a subtype that is going to be associated with the function Tristate_RF, and Logic4 is the parent type of this newly defined subtype. It would be an error to use the reserved word *type* in this resolution function context. Only subtype can be used. Having compiled this statement the compiler now realizes that the previous function specification for Tristate_RF is intended to be used as a resolution function. The compiler will then reexamine this function's format to confirm that it has the correct resolution function profile. A resolution function can have only one input parameter that must be an unconstrained array. Furthermore, the data type of the object being returned by the resolution function must match the data type of its input array's base element. Since Tristate_RF satisfies both these constraints, the compiler is appeased and it will allow Tristate_RF to be used as a resolution function. Note well that a typical function cannot take on the responsibilities of a resolution function since it can have numerous formal input parameters that do not necessarily have to be of an unconstrained data type. Moreover, a typical function can return an object having any arbitrary previously defined data type.

All right, it is about time that we looked at the big picture of resolution functions. Consider our multisourced environment depicted in Fig. 2.65. The tristated bus is three lines wide and each line has two sources, one coming from each of the counters. When it is time to update one of these lines, the resolution function will be called by the simulation engine to resolve the various sources on this line and to derive an effective value to be assigned to this line. In our case since there are two sources, the resolution function will be called with an input array of two elements. These two array elements will correspond to the two sources on this line. Now suppose that this line had 100 sources. Then the resolution function would be called with an input array of 100 elements. So now you can see the importance and significance of VHDL's demand that a resolution function's input be an unconstrained array. A resolution function must be able to handle any number of multiple sources and hence the length of its input array must be unbounded. Otherwise, the resolution function would only work for a fixed number of sources. Such a modeling approach would be impractical and very limited in its scope. Returning to our two-counter environment, suppose that all three lines of our tristated bus had to be simultaneously updated (i.e., during the first half of the same simulation cycle). Then the resolution function would be called three times by the simulation engine, and on each call its input array would consist of exactly two elements. The most important observation that you should have made regarding the resolution function is that you do not invoke it. Rather, it is the simulation engine that calls this function. Your only responsibilities are to associate the multisourced signal with a resolution function. It is then up to the simulation engine to call this resolution function whenever its associated resolved signal is to be updated. The calling of a resolution function is not your responsibility. In this respect resolution functions are totally different from the ordinary functions that you must explicitly invoke in the body of your model.

Lines 20 and 21 define subtypes to be associated with the resolution functions Wired_And_RF and Wired_Or_RF, respectively. As their names suggest, these subtypes should be used when modeling a wired-and and a wired-or bus, respectively.

Lines 23 through 28 define unconstrained array types where the base elements are of the subtype Tristate_RS, Wired_And_RS, and Wired_Or_RS, respectively. Using these new types we can collectively declare a group of signals where each individual signal is associated with the same resolution function. In other words, we are now in a position to model tristated, wired-and, or wired-or buses.

To best understand what is being done on lines 32 through 37 recall that VHDL has several built-in predefined logical operators that can be applied to objects of type BIT, BOOLEAN, and BIT_VECTOR. For consistency and readability, it would be very worthwhile if the same logical operators could be applied to objects that are of type Logic4 and Logic4_Vector. The purpose of the function specifications on lines 32 through 37 is to give us the capability to write assignments such as

```
C_Logic4 := A_Logic4 and B_Logic4;
```

where the names of the objects reflect their data type. Note that all the functions' names on lines 32 through 36 are in double quotes. VHDL requires this because the functions are overloading VHDL's predefined operators. The term overloading is used in VHDL whenever the same name is used but in different contexts. By the way, the function's formal input parameters L and R are abbreviations for Left input and Right input, respectively. Lines 41 through 46 are the Logic4_Vector versions of these same overloaded predefined operators. By the way, the VHDL'93 version of this package should include overloaded operators for the new shift and rotate operators that have been added to the language.

The typical MVLN package will also include conversion functions between the newly defined data types and the VHDL standard data types BIT and BIT_VECTOR. In our package the specifications for these conversion functions are given on lines 49 through 54. Both the scalar (single element) and the vector (array) versions are given. Note that the same function names are used, but they are not in double quotes. The difference between these overloaded conversion function names and the previously encountered overloaded names is that these current names are not overloading any of the predefined VHDL operators. Whenever predefined VHDL operators are being overloaded, the overloading function name must be in double quotes.

The parameters S and V in these conversion functions are abbreviations for Scalar and Vector, respectively. Note well that there is no conversion function between Logic4 and Tristate_RS. The fact that one is a subtype (subset) of the other means that the compiler views them as compatible data types. Hence it is possible to assign a Logic4 declared signal to a Tristate_RS declared signal and vice versa. The same can be said of Wired_And and Wired_Or. But what about assignments between Logic4_Vector signals and Tristate_RS_Vector signals? This one is a bit tricky since Logic4_Vector and Tristate_RS_Vector are each defined as a type. One is not a subtype of the other. However, each element of Tristate_RS_Vector is a subtype of Logic4, which is the base type of each element of Logic4_Vector. Hence these two array types are said to be closely related. VHDL has a special technique for converting between closely related arrays and so explicitly declared conversion functions between Logic4_Vector and Tristate_RS_Vector are not required. A later model will show you the mechanics to convert between closely related arrays.

This package concludes with several sensing utility functions (lines 57 through 73) that search for the occurrence of a specific Logic4 value in an input data stream.

Special utility packages such as this one should be compiled into their own uniquely identifiable library. Recall that we already carried out a similar maneuver when we compiled the logic gates into the logical library Gates_Simple_Lib. In our current journey the MVL4_Pkg package (declaration and body) will be compiled into the logical library MVL_Packages_Lib.

And now we are done with this package declaration. Let's move onto MVL4_Pkg's body.

2.26.1 Information checkpoint

Before you venture further along your VHDL journeys, please make certain that you are familiar with the following topics that we have established via this model:

- Format of a package declaration.
- A package declaration and its corresponding body may be in separate files.
- Only declarations made in a package declaration are available (visible) to the users of this package.
- The format to define a subtype to be associated with a resolution function.
- The role played by the resolution function in determining the effective value of a signal when that signal is multiply driven.
- The restrictions placed on a resolution function's profile.
- The possibility of overloading a subprogram's name.
- The double quote format required when overloading a predefined VHDL operator.

- The key point of this section is that the package MVL4_Pkg is, in format, similar to a typical MVLN package, where N is the number of elements in the package's base enumeration type.

2.27 Excerpt of File: mvl4_pkg.vhd

The package body for MVL4_Pkg is rather lengthy. Consequently, we will not examine it line by line. Instead, only its salient features will be discussed. Figure 2.67 contains an excerpt of the file that contains the complete package body.

Line 8 announces our intentions to write a package body for the package MVL4_Pkg. The reserved word *body* informs the compiler that this

```
-- File Name  : mvl4_pkg.vhd  (*** EXCERPT ONLY ***)
--
-- ********** COMPLETE FILE IS NOT SHOWN **********
--
-- Author     : Joseph Pick
--
-- *************************************************************** LINE NUMBER
package body MVL4_Pkg is                                              -- 8

    -- Declare a two dimensional table type that will be used
    -- by the resolution functions.
    type Logic4_Table is array (Logic4, Logic4) of Logic4;            -- 12

    -- ***** LINES HAVE BEEN OMITTED *****

    -- Look-up table for the resolution function, Tristate_RF.
    constant TRISTATE_RF_TABLE : Logic4_Table := (('X','X','X','X'),  -- 26
                                                  ('X','0','X','0'), -- 27
                                                  ('X','X','1','1'), -- 28
                                                  ('X','0','1','Z')  -- 29
                                                 );                   -- 30

    -- This function resolves multiply driven tristated signals.
    function Tristate_RF (V : Logic4_Vector) return Logic4 is         -- 31
              variable Result : Logic4       := 'Z';                  -- 32
    begin                                                              -- 33
        for I in V'RANGE loop                                         -- 34
            Result := TRISTATE_RF_TABLE(Result, V(I));                -- 35
            exit when Result = 'X';                                   -- 36
        end loop;                                                     -- 37
        return Result;                                                -- 38
    end Tristate_RF;                                                  -- 39

    -- Look-up table for the resolution function, Wired_And_RF.
    constant WIRED_AND_RF_TABLE : Logic4_Table := (('X','0','X','X'), -- 42
                                                   ('0','0','0','0'),-- 43
                                                   ('X','0','1','1'),-- 44
                                                   ('X','0','1','Z')-- 45
                                                  );                  -- 46

    -- This function resolves multiply driven wired_and signals.
    function Wired_And_RF (V : Logic4_Vector) return Logic4 is        -- 49
              variable Result : Logic4       := 'Z';                  -- 50
    begin                                                              -- 51
        for I in V'RANGE loop                                         -- 52
            Result := WIRED_AND_RF_TABLE (Result, V(I));              -- 53
            exit when Result = '0';                                   -- 54
        end loop;                                                     -- 55
```

Figure 2.67

```
           return Result;                                          -- 56
        end Wired_And_RF;                                          -- 57

        -- ***** LINES HAVE BEEN OMITTED *****
        -- Overloading of the predefined logical operators.

        -- Look-up table for the overloaded "and" function.
        constant AND_TABLE : Logic4_Table :=                       -- 94
                            (('X', '0', 'X', 'X'),                 -- 95
                             ('0', '0', '0', '0'),                 -- 96
                             ('X', '0', '1', 'X'),                 -- 97
                             ('X', '0', 'X', 'X'));                -- 98

        -- Overloaded "and" function (scalar version).
        function "and"  (L, R : Logic4) return Logic4 is           --101
        begin                                                      --102
           return AND_TABLE(L,R);                                  --103
        end "and";                                                 --104

        -- Overloaded "and" function (vector version).
        function "and"  (L, R : Logic4_Vector) return Logic4_Vector is  --107
           alias L_Alias    : Logic4_Vector(L'LENGTH - 1 downto 0) is L; --108
           alias R_Alias    : Logic4_Vector(R'LENGTH - 1 downto 0) is R; --109
           variable Result  : Logic4_Vector(L'LENGTH - 1 downto 0);     --110
        begin                                                           --111
           assert L'LENGTH = R'LENGTH                                   --112
              report "Length mismatch during overloaded vector and"     --113
              severity ERROR;                                           --114
                                                                        --115
           for I in L_Alias'RANGE loop                                  --116
              Result(I) := AND_TABLE(L_Alias(I),R_Alias(I));            --117
           end loop;                                                    --118
           return Result;                                               --119
        end "and";                                                      --120

        -- ***** LINES HAVE BEEN OMITTED *****

        end MVL4_Pkg;                                                   --371
```

Figure 2.67 (*Continued*)

design unit currently being compiled must have associated with it a package declaration that has already been compiled into the current working library. If this is not so, then the compiler will generate an error message and terminate its activities. Incidentally, recall from our previous section that MVL4_Pkg's declaration was compiled into a library having the logical name MVL_Packages_Lib. So when compiling this body the current working library will have both the logical names WORK and MVL_Packages_Lib.

Recall that the package declaration contained numerous subprogram specifications. VHDL requires that their respective bodies be contained in the package's body. It is forbidden for these subprogram bodies to be in the package declaration. We are now going to look at the bodies for some of the functions that were specified earlier. Let's begin with the resolution function Tristate_RF. Consider how it could be implemented. One approach is to use a sequence of nested if...then...else logic to pairwise compare the input array's elements. For example, if the first two elements are 'Z', then their combination would yield a 'Z'. An 'X' and a '1' combination will yield an 'X'. So you can see that we would be obligated to test for all of the possible combination pairings between Logic4 elements. But this approach is very inefficient. Most modern computers consist of a pipelined architecture. At the assembly language level our nested if...then...else logic will result in a series of conditional branching statements. Pipeline architectures view conditional branches as a major catastrophe for their interstage dataflow. But every path of this if...then...else logic results in a constant value that is even known at compile time. Hence it would be more efficient to use a lookup table to derive these fixed combinational results. And, in fact, this lookup table approach is a standard trick that is often used in the VHDL community.

Like every object in VHDL, this lookup table must also be associated with a data type. Line 12 contains the type definition for the resolution function's lookup table. This type called Logic4_Table is a two-dimensional array where each element of this matrix is of type Logic4. At first glance the usage of the enumeration type Logic4 as the index of an array looks somewhat peculiar. The appearance of this notation is somewhat contrary to our usage of numerical indices such as (5, 7) to access points in the two-dimensional Euclidean plane. But I contend that you have already seen something analogous to this type definition, although you did not embellish it with the degree of formalism that we have on line 12.

Pick up any road map and you will see a table that provides the distance between the major cities. A sample of such a table is given at the top of Fig. 2.68. The abstraction of this table is a three-step strategy. First, you declare an enumeration data type Cities that lists the cities in the left-to-right order exactly as they appear on the top of the map's distance table. The next step is to define a two-dimensional data type called Inter_City_Dist_Type, where each element of the matrix is of type NATURAL. An enumeration type always defines an ascending order for its listed members. This ordering implies that the notation (Cities, Cities) is really shorthand for the following:

```
(Cities range Baltimore to Montreal,
                    Cities range Baltimore to Montreal)
```

This two-dimensional table type can now be used to produce the constant lookup table called Inter_City_Dist_Table that is shown in Fig. 2.68. Do not be concerned with the parentheses within parentheses. They are merely VHDL syntax requirements. Instead, you should focus on the order of the listed numbers and realize that I simply lifted the numbers straight from the map's table and placed them, as is, into this constant lookup matrix. To really crystallize matters, what you should do is write Baltimore, Montreal, and Toronto across the top and down the side of this matrix. You now have a two-dimensional grid where the axis values are derived from an enumeration list instead of from some

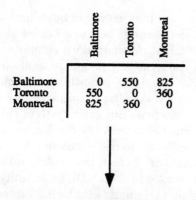

```
type Cities is (Baltimore, Toronto, Montreal);
type Inter_City_Dist_Type is array (Cities, Cities) of NATURAL;
constant Inter_City_Dist_Table : Inter_City_Dist_Type
    := ((0,550,825),
        (550,0,360),
        (825,360,0));
```

Distance_Travelled := Inter_City_Dist_Table(Montreal, Toronto);

Figure 2.68 Real-world example of a two-dimensional array indexed via enumeration types.

numbering system. Figure 2.68 concludes with an application of this lookup table using Montreal and Toronto to index into the first and second dimensions of this table. Using the city names that you wrote along the top and side of the constant matrix, you can deduce that the expression Inter_City_Dist_Table(Montreal, Toronto) will yield the value 360. So, with the aid of your past road map experiences you now totally understand the meaning of the type definition that was given on line 12. This excellent road map analogy was created by Ray Popp, one of my former colleagues at Westinghouse. One more remark about the data-type Logic4_Table. Recall that the users of a package can only reference those definitions that are made in the package declaration. Since Logic4_Table was defined in the package body and not in the package declaration, it is not available to the users of the package. But the type Logic4_Table is available anywhere in the package body following its definition. Let's now apply this two-dimensional array type to create a lookup table for the Tristate_RF resolution function.

Lines 26 through 30 show the declaration for a constant two-dimensional array of type Logic4_Table. To better understand this constant lookup table you should write 'X', '0', '1', and 'Z' along its top and sides just as you did for the road map analogy. The contents of this matrix captures our concept of how a tristated bus behaves. A 'Z' combined with a 'Z' will yield a 'Z'. A 'Z' combined with a '1' will yield a '1'. A '1' combined with a '0' will yield an 'X'. Use the vertical and horizontal labels that you appended to the constant matrix to confirm these results. Incidentally, this lookup table may also be viewed as a truth table for our tristate resolution function.

Lines 31 through 39 contain the body for the resolution function Tristate_RF. It is optional to repeat a subprogram's name at its physical conclusion and I always do so for readability (see line 39). Observe that line 31 is almost identical to this function's specification as it was given in the corresponding package declaration. The only difference is that the trailing semicolon was replaced by the reserved word *is*. The region between the reserved words *is* and *begin* is known as the function's declarative part. I could have placed the constant TRISTATE_RF_TABLE in the functions declarative but to do so would have been very inefficient. Being in the function's declarative part means that this constant lookup table would be reconstructed on the computer's stack every time the function would be called. By placing this constant table outside of the function guarantees that it will be stored on the computer's heap and will never have to be recreated on the stack every time the function is called. On line 32 the variable Result is declared and initialized to 'Z'. This variable will be used when cycling through the elements of the function's input array. 'Z' was deliberately chosen as the initial value for Result since, according to TRISTATE_RF_TABLE, 'Z' has a neutral effect on the other Logic4 elements.

Lines 34 through 37 contain a for loop statement. We have already seen this construct before even with the conditional exit option that occurs on line 36. The only new topic is the array attribute 'RANGE. The formal input parameter of this function is defined to be of type Logic4_Vector. Because Logic4_Vector is an unconstrained array, the run time bounds for the input V will be determined by the range of V's corresponding to the actual array when this function is called. The purpose of 'RANGE is to adjust the for loop bounds so that they will accommodate any input array size. As its name suggests, V'RANGE will expand into the range bounds of V. For instance, if V's indices have the ascending range of 0 to 31, then V'RANGE is equivalent to the expression 0 to 31. Similarly, if V's indices have the descending range of 31 down to 0, then V'RANGE will be equivalent to the expression 31 down to 0. In general, the usage of the attribute 'RANGE allows subprograms to be elastic and to adjust to any input array size. This capability is very powerful and allows subprograms to be reused from one project to the next. On line 35 the variable Result is updated via a table lookup using the previous value of Result and the Ith element of the input array. Since 'X' combined with any other value yields an 'X', it makes sense to jump out of the for loop whenever an 'X' is derived from the table. This jumping out is the purpose of the conditional exit statement on line 36. Whether this exit path is taken or the for loop is allowed to complete all of its iterations, the next statement to be executed occurs on line 38. Here the value of Result is returned to the caller of this resolution function. As a friendly reminder, keep in mind that the simulation engine will call the resolution function. It is not the VHDL engineer's responsibility to do so. Before leaving this resolution function, let me add that when debugging VHDL models it is sometimes necessary to travel into a resolution function to determine the various values driven by a signal's multiple sources. Some VHDL tool suites will just let you step right into the execution of a resolution function. Others require a special, proprietary debugging command to do so. In your preliminary debug iterations, it is very beneficial to step into a called resolution function and examine its input array. This technique will allow you to observe the set of values that are driving a multisourced signal.

Lines 42 through 46 contain the constant lookup table to be used by the Wired_And_RF resolution function. Inspection will show that its contents agree with our concept of how signals interact on a wired-and bus. The Wired_And_RF resolution function is very similar to Tristate_RF. The only difference is that it references the constant WIRED_AND_TABLE instead of TRISTATE_RF_TABLE. The Wired-Or_RF resolution function is similarly implemented.

The constant lookup table given on lines 94 through 98 can be thought of as a truth table that captures the behavior of the overloaded "and" operator. For your own interest, you should contrast this table with the one that was presented earlier for a wired-and bussing environment. Lines 101 through 104 contain the scalar version for the overloading of the VHDL predefined "and" operator. The only interesting observation is that, as line 103 shows, it is possible to return an expression instead of just a variable, which is what we have been doing so far.

The vector version of the overloaded "and" operator introduces two very important constructs. Let's first talk about alias. The alias construct has many useful applications. One of them is to rename specific elements of an array. Suppose that bit number five of a computer's status register is the overflow bit. Instead of having to remember the index value five whenever referencing this overflow bit, it is possible to use the alias construct to rename this specific bit of the status register to OVERFLOW. The beauty of this alias maneuver is that it causes OVERFLOW to inherent all of the VHDL intrinsic characteristics of whatever object it is aliasing. For instance, if the overflow bit is a signal, then OVERFLOW is also a signal. Moreover, aliasing guarantees that scheduling a value to be assigned to OVERFLOW is equivalent to making the same scheduled assignment to the fifth bit of the status register. You can see the great utility in such a usage of alias since remembering specific bit numbers is highly error prone.

Let's now return to the alias construct given on lines 108 and 109. The exact background necessitating its usage will be given in the Techniques and Recommendations segment of this book. Suffice it to say that without this alias construct, a range mismatch between the function's two inputs will result in a simulation run time error. In effect, the alias construct is used to normalize the inputs' array bounds so that all the inputs will have the exact same range indices. The Techniques and Recommendations segment of this book will also point out some practical run time efficiency techniques that may be applied when making declarations such as those that are made in this function's declarative part.

The assert statement on lines 112 through 114 works like this. L'LENGTH is another attribute associated with arrays and it is the number of elements that the array contains. The assert condition L'LENGTH = R'LENGTH is checking to confirm that the two inputs have the same size so that the overloaded "and" operator may be correctly applied to them bitwise. When this assert expression is FALSE, then the ASCII string following the reserved word *report* is printed onto your screen. Most vendors will allow you to optionally pipe this message into a file for later examination. Furthermore, if the assert condition is FALSE, then, as per line 114, a severity level of ERROR will be reported to the simulation engine. The severity level options are NOTE, WARNING, ERROR, and FAILURE. This listed order defines an ascending order such that NOTE < WARNING < ERROR < FAILURE. By default, simulators will halt their model execution whenever a severity of ERROR or greater is reported. Most VHDL tool suites allow their users to set the simulator's default halting severity level. The great utility of this halting environment is that once an unwanted condition is reported, the simulation can be stopped on a dime. Variables and signals may then be examined to determine the perpetrator of the error. Observe that the semicolon appears only after the severity level is announced and not before. But what happens if the assert condition is TRUE? Then the report and severity clauses will be completely bypassed.

There is nothing new in the remainder of this overloaded "and" operator. But pay close attention to see how the attributes 'LENGTH and 'RANGE are used so that this function can adjust to any input array size.

2.27.1 Information checkpoint

Before you venture further along your VHDL journeys, please make certain that you are familiar with the following topics that we have established via this model:

- Format of a package body.
- Declarations made in a package body are not available (visible) to the package's user.

- Subprogram optimization techniques such as
  - Constant tables defined outside of a subprogram declarative part.
  - Lookup tables used instead of nested if...then...else logic.
- Subprogram reusability achieved via
  - Unconstrained formal parameters.
  - Flexible and dynamic looping bounds.
- Behavior of the assert statement.
- Typical applications of the alias construct.

2.28   File: count_3_bit_rf_.vhd

Figure 2.69 contains the entity declaration for our enhanced 3-bit counter. The CHIP_ID declared on line 10 will be used for autoidentification purposes by the corresponding architecture body. The input Clock is of the standard type BIT since it is not tristated. On the other hand, since Dataout is to model a tristated output it is declared on line 12 as Tristate_RS_Vector (2 down to 0). Recall from our analysis of the package MVL4_Pkg that each element of Tristate_RS_Vector (2 down to 0) is associated with the resolution function Tristate_RF. Hence this resolution function will be called by the simulation engine whenever Dataout is to be updated, and so Dataout will have the characteristics of a tristated line. To make the identifier (name) Tristate_RS_Vector visible requires both a library clause and a use clause. The library clause on line 6 makes the logical library name MVL_Packages_Lib visible. Recall from our previous two sections that the package MVL4_Pkg was compiled into this logical library. Line 7 makes both the name of the package MVL4_Pkg visible along with all of its contents. The visibility of all of the package's contents is achieved via the reserved word *all*. Keep in mind that all of the package's contents refers only to

```
--   File Name  :  count_3_bit_rf_.vhd
--
--   Author     :  Joseph Pick
--
--   ************************************************** LINE NUMBER
library MVL_Packages_Lib;                                --  6
use MVL_Packages_Lib.MVL4_Pkg.all;                       --  7
                                                         --  8
entity Count_3_Bit_RF is                                 --  9
       generic (CHIP_ID : NATURAL);                      -- 10
       port    (Clock   : in BIT;                        -- 11
                Dataout : out Tristate_RS_Vector(2 downto 0)  -- 12
                         := (others => 'Z')              -- 13
               );                                        -- 14
end Count_3_Bit_RF;                                      -- 15
```

Figure 2.69

those names appearing in the package's declaration part. Any name (data type, constant, subprogram, etc.) introduced in the package body is not visible to the users of the package. As a prelude to what will happen in the next section, be aware that any name made visible to the entity will automatically be visible to all of its associated architecture bodies. Consequently, the following names will also be visible to any of Count_3_Bit_RF's associated architecture bodies: MVL_Packages_lib, MVL4_Pkg, and all of the names appearing in MVL4_Pkg's package declaration part.

Line 13 shows that for the first time in our journeys we are going to explicitly assign an initial value to an out port. Though not necessary, I wanted this model to start up in a high impedance state. Consequently, I initialized each element of Dataout with the value 'Z'. You have already encountered the usage of the reserved word *others* as a shorthand way to assign a single value to multiple objects. If I would not have assigned any initial value to Dataout, then the simulation engine would have automatically assigned Tristate_RS'LEFT to each of Dataout's elements. Because of the manner in which the subtype Tristate_RS was defined, we have that Tristate_RS'LEFT is equal to Logic4'LEFT, which, in turn, is equal to 'X'. Hence if I would not have explicitly initialized Dataout, then the simulation engine would have initialized each of its elements to 'X'.

2.28.1 Information checkpoint

Before you venture further along your VHDL journeys, please make certain that you are familiar with the following topics that we have established via this model:

- Method to make a library's logical name visible to your VHDL model.
- Method to make all of the contents of a package visible to your VHDL model.
- Technique to explicitly assign an initial value to a port.
- Usage of generic parameters to pass in a chip identification number.

2.29 File: count_3_bit_rf_model_1.vhd

Figure 2.70 contains an architecture body for the design entity Count_3_Bit_RF. Before we delve into its implementation fine points, let's first review its functionality. On each trailing edge of the input Clock, the current value of an internal counter will be incremented. This updated value will then be compared to the design entity's CHIP_ID. If there is a match, then this updated value will be exported out of the design entity via the port Dataout. If there is no match, then

An Excursion into VHDL

```
--  File Name  :  count_3_bit_rf_model_1.vhd
--
--  Author     :  Joseph Pick
--
--  ************************************************************* LINE NUMBER
--  use MVL_Packages_Lib.Math_Bit_Vector_Pkg.all;                  --  6
architecture Model_1 of Count_3_Bit_RF is                          --  7
                                                                   --  8
use MVL_Packages_Lib.Math_Bit_Vector_Pkg.all;                      --  9
                                                                   --10
begin                                                              -- 11
                                                                   --12
   Bit_Incr:                                                       -- 13
   process                                                         -- 14
      variable Current_Count : BIT_VECTOR(2 downto 0);             -- 15
                                                                   --16
   begin                                                           -- 17
                                                                   --18
      wait until Clock = '0';                                      -- 19
                                                                   --20
      Current_Count := Inc (Current_Count);                        -- 21
                                                                   --22
      if CHIP_ID = To_Natural (Current_Count) then                 -- 23
         Dataout <= Tristate_RS_Vector (To_Logic4 (Current_Count));-- 24
      else                                                         -- 25
         Dataout <= "ZZZ";                                         -- 26
      end if;                                                      -- 27
                                                                   --28
   end process Bit_Incr;                                           -- 29
                                                                   --30
end Model_1;                                                       -- 31
```

Figure 2.70

---

the design's output should have the look and feel of high impedance. Implementing this specification item by item we know that the first thing we must have is the visibility of MVL4_Pkg's contents. But this visibility was already established via the library and use clauses given in the entity declaration Count_3_Bit_RF that corresponds to this architecture body. Hence MVL4_Pkg's visibility will automatically migrate over to this architecture. And so there is nothing explicit that has to be done regarding this matter.

To make this model more real-worldlike and modular I placed a bitwise incrementation algorithm into a function and then made a call to it when I needed to increment the 3-bit counter. Furthermore, I placed this algorithm into the package called Math_Bit_Vector_Pkg and compiled this package into the same logical library that contains MVL4_Pkg. Since the architecture's corresponding entity declaration did not make the contents of Math_Bit_Vector_Pkg visible, it is up to us to explicitly do so in this architecture body. Suppose that we would attempt to use the contents of line 6 to achieve this visibility. The fact that this line is commented out is a clue that it does not work. Here is what goes wrong. Suppose that you are a compiler and you come across the uncommented version of this line 6. Since the library name MVL_Packages_Lib is not yet visible to you, an error message will be generated and you will terminate your compilation activities. One solution is to precede the use clause on line 6 with the following library clause: library MVL_Packages_Lib. Another option is to take advantage of the fact that the architecture's corresponding entity declaration already contained this library clause (line 6 of Fig. 2.69). Hence it is sufficient to place the appropriate use clause in the architecture declarative part as shown on line 9 of Fig. 2.70. A library clause is now no longer required since on line 7 this architecture is associated with the design entity Count_3_Bit_RF. Hence by the time line 9 is reached, all of Count_3_Bit_RF's visible identifiers are also visible to this architecture. In particular, the library name MVL_Packages_Lib is now visible as well.

Line 15 is a variable declaration for the architecture's internal incrementor. Note well that this variable was declared as a constrained BIT_VECTOR instead of a constrained Tristate_RS_Vector. There are two main reasons for this choice. First, it forces us to get involved with some practical real-world conversion techniques. And second, it shows you that even though the port interfaces may have to be of a specific type, internally your model may use any data type that you feel is appropriate in terms of efficiency speed or even readability.

Line 19 is our old friend that will put this process to sleep until the occurrence of Clock's trailing edge. When this trailing edge arrives (is identified), then this process will be reactivated and its execution will continue at line 21. The variable Current_Count will immediately be assigned the value of the expression on the right-hand side of the assignment. The right-hand side of this assignment contains a call to a function named Inc that will increment the current value of the variable Current_Count. By the way, this function Inc was created by cutting and pasting the bit stream incrementation algorithm given in Fig. 2.28. The formal input parameter to this function was declared to be of type BIT_VECTOR, thus allowing an input of any array size. The original for loop fixed bounds (0 to 2) was replaced by Input'RANGE where Input is the name of Inc's formal parameter. Recall from our journey into MVL4_Pkg that the attribute 'RANGE allows the for loop to adjust to any array length. With these changes the Inc function is now able to increment any sized BIT_VECTOR array.

The if test on line 23 is comparing this recently incremented value of Current_Count with the generic parameter CHIP_ID. Since CHIP_ID and Current_Count are of different type (NATURAL versus BIT_VECTOR), it would have been erroneous to write the following comparison test:

```
if CHIP_ID = Current_Count then  -- **** ERROR ****
```

Consequently, a conversion function is required to convert from one type to another. The function To_Natural is a function also contained in the package Math_Bit_Vector_Pkg that will convert an array of type

156   An Excursion into VHDL

BIT_VECTOR into a NATURAL number. Let us suppose that this comparison test on line 23 yields the Boolean value FALSE. Then execution will continue into the else clause beginning at line 26. Observe that the string "ZZZ" will, under these circumstances, be scheduled to be assigned to Dataout 1 delta time unit from now. So because of the way 'Z' interacts with the other elements 'X', '0', and '1', it follows that we have just conceptually set these three output lines to high impedance, and we will be floating any line that is attached to them. Observe that the resolution function Tristate_RF is not called anywhere in this architecture. All that I had to do was to declare the array Dataout to be of type Tristate_RS_Vector. Because of this declaration each element of Dataout was, in effect, declared to be of type Tristate_RS. And hence, each element of the array Dataout automatically became associated with the resolution function Tristate_RF. Based on what was stated earlier, it follows that anytime an element of Dataout is to be updated, the simulation engine will call this resolution function instead of you having to do it.

Now suppose that CHIP_ID is equal to the NATURAL equivalent of Current_Count. Then line 24 will be executed. This line is a signal assignment, the source (right-hand side) of which seems slightly complicated. Let's carefully ease into this expression. The basic problem is this. On the left-hand side we have Dataout, which is of type Tristate_RS_Vector, and we want to assign the value of Current_Count to it. But Current_Count is of type BIT_VECTOR. These two data types are not compatible and so our immediate objective is to convert from one data type into the other. Since BIT_VECTOR is on the right-hand side of the assignment, we will convert it into Tristate_RS_Vector. To_Logic4 is a function that is contained in the package MVL4_Pkg. Recall from going over this package's declaration part that To_Logic4 is an overloaded function that comes in two flavors. The scalar version converts BIT to Logic4 while the vector (array) version converts BIT_VECTOR to Logic4_Vector. Since the input Current_Count is of type BIT_VECTOR, the latter overloaded function will be called. But we are not yet out of the woods. Since Logic4_Vector and Tristate_RS_Vector are two distinct data types, it would be incorrect to write:

```
Dataout <= To_Logic4(Current_Count); -- **** ERROR ****
```

In Sec. 2.26 I mentioned that the package MVL4_Pkg deliberately did not include any conversion functions between Logic4_Vector and Tristate_RS_Vector since these two types are closely related arrays (term defined in Sec. 2.26). I also indicated that VHDL has a built-in technique to implement conversions between closely related arrays. This method is illustrated by our next conversion maneuver. To convert from Logic4_Vector to Tristate_RS_Vector it is sufficient to place parentheses around the object of type Logic4_Vector and then to simply write Tristate_RS_Vector in front of this parenthesized expression. And that is it. Presto! We are done. The resulting expression looks like a function call to convert from Logic4_Vector to Tristate_RS_Vector, and some VHDL tool suites might internally implement it this way. But as far as I am concerned, because Logic4 is the base type of each of these closely related arrays, I conjecture that an efficient compiler will not generate any extra assembly code or make a call to a function. Instead, it will simply do nothing at all. I intuitively believe that the purpose of this type conversion technique between Logic4_Vector and Tristate_RS_Vector is to merely acquiesce the compiler because now both sides of the assignment statement are of the same type. Enough said. Let's move on.

2.29.1 Information checkpoint

Before you venture further along your VHDL journeys, please make certain that you are familiar with the following topics that we have established via this model:

- Libraries and packages made visible to the entity declaration are also visible to any of its corresponding architecture bodies.
- Usage of functions from user-defined packages.
- Conversion technique between closely related arrays.

2.30 File: count_monitor_rf.vhd

This section will present a monitor for our enhanced 3-bit counter that is a slight variation of our previously discussed monitor (Sec. 2.17). Back then the incremented values were converted into a NATURAL number and then printed out. Our present monitor will wait for an event on the tristated bus and then print out the updated tristate value instead of first converting it to a NATURAL number. Figure 2.71 shows the file that contains both the entity declaration and the architecture body for this tristated version of the monitor.

Lines 6 and 7 together make the contents of the package MVL4_Pkg visible. This visibility is immediately required on line 10 where Datain's declaration references the data type Tristate_RS_Vector.

Line 14 makes the contents of the package MVL4_Textio visible. This package contains procedures that facilitate the writing and reading of those objects that are of type Logic4 or Logic4_Vector. This package is required because the VHDL TEXTIO package is only applicable to objects that belong to the standard VHDL data types such as INTE-

158   An Excursion into VHDL

```
-- File Name : count_monitor_rf.vhd
--
-- Author    : Joseph Pick
--
-- *********************************************************** LINE NUMBER
library MVL_Packages_Lib;                                       -- 6
use MVL_Packages_Lib.MVL4_Pkg.all;                              -- 7
                                                                -- 8
entity Count_Monitor_RF is                                      -- 9
    port (Datain : in Tristate_RS_Vector (2 downto 0));         -- 10
end Count_Monitor_RF;                                           -- 11
                                                                -- 12
architecture Model_1 of Count_Monitor_RF is                     -- 13
    use MVL_Packages_Lib.MVL4_Textio.all;                       -- 14
    use STD.TEXTIO.all;                                         -- 15
                                                                -- 16
begin                                                           -- 17
                                                                -- 18
  Display:                                                      -- 19
  process                                                       -- 20
    variable Current_Count_RF : Logic4_Vector(2 downto 0);      -- 21
                                                                -- 22
    file Outfile : TEXT is out "counter_rf_values";             -- 23
--  In VHDL'93 the above line MUST be replaced by               -- 24
--  file Outfile : TEXT open WRITE_MODE is "counter_rf_values"; -- 25
                                                                -- 26
    variable Outline : LINE;                                    -- 27
                                                                -- 28
  begin                                                         -- 29
                                                                -- 30
    write (Outline, STRING'(" TIME   VALUE OF TRI-STATED BUS"));-- 31
    writeline (Outfile, Outline);                               -- 32
                                                                -- 33
    Main_Loop:                                                  -- 34
    loop                                                        -- 35
                                                                -- 36
      Current_Count_RF := Logic4_Vector(Datain);                -- 37
                                                                -- 38
      write (L      => Outline, Value => Now,                   -- 39
             Field => 6);                                       -- 40
      write (Outline,                                           -- 41
             Current_Count_RF,                                  -- 42
             Field => 8);                                       -- 43
      writeline (Outfile, Outline);                             -- 44
                                                                -- 45
      wait on Datain;                                           -- 46
                                                                -- 47
    end loop Main_Loop;                                         -- 48
                                                                -- 49
  end process;                                                  -- 50
                                                                -- 51
end Model_1;                                                    -- 52
```

Figure 2.71

GER, BIT, BIT_VECTOR, BOOLEAN, TIME, CHARACTER, STRING. By the way, it was very straightforward to design the procedures in MVL4_Textio. To write objects of type Logic4 and Logic4_Vector into an ASCII file I first converted them into objects of type CHARACTER and STRING, respectively. I then wrote them out using the VHDL's TEXTIO procedures for these standard data types. To read objects of type Logic4 or Logic4_Vector from an ASCII file I did the reverse. I read in a CHARACTER or a STRING and then converted them into Logic4 and Logic4_Vector, respectively. Later in our journeys I will show you how to apply the new VHDL'93 attribute 'IMAGE to avoid having to write such a pseudo-Textio package for your user-defined data types.

The use clause on line 15 is required because this architecture body is also going to reference some of the procedures from the standard TEXTIO package. Though the logical library STD is named in this use clause, there is no need for a library clause since STD is a special library that, as per the LRM, is always visible. Consequently, you never have to write:

```
library STD; -- **** NOT REQUIRED ****
```

Though VHDL'93 is, on the whole, upward compatible with VHDL'87, there are some discrepancies that cannot be resolved. One problem area is file declarations. We already saw this in the previous monitor model. There commented lines were added that would have to replace file declarations in the VHDL'87 compliant model. Similarly, this section's monitor model also contains such file declarations that are only suitable for VHDL'87. Hence to compile this source code on a VHDL'93 complaint compiler you must remove the comment dashes from line 25 and instead add them to line 23.

This current tristated monitor is similar in both form and spirit to the previous BIT_VECTOR monitor. A header is printed out via lines 31 and 32. Then an infinite loop is entered that writes out the current simulation time and Datain's values. This loop is executed once during the initialization phase and henceforth every time that Datain has an event.

On line 37 Datain's current value is converted from Tristate_RS_Vector type into a Logic4_Vector type. As in the previous section this is a conversion between closely related arrays and is required because the MVL4_Textio write procedure only works with Logic4 and Logic4_Vector type objects.

It is interesting to compare the procedure calls on lines 39 and 41. Although they both have the same name, their parameter profiles are different. This difference is used by the compiler to determine which of these similarly named procedures is going to get called. The word *write* on line 39 will come from the standard VHDL TEXTIO package, whereas the *write* on line 41 will come from the MVL4_Textio package.

2.30.1 Information checkpoint

Before you venture further along your VHDL journeys, please make certain that you are familiar with the following topics that we have established via this model:

- The logical library name STD is always visible and does not require a library clause.

160   An Excursion into VHDL

- In a VHDL'87 compliant environment the VHDL engineer must design a utility package that will contain subprograms to both read and write objects of any newly defined enumeration type.
- Usage of a subprogram's unique parameter profile to distinguish it from other overloaded subprograms having its same name.

2.31   File: encapsulate_rf_components_pkg_.vhd

Let's now create a test bench for our tristated 3-bit counter environment. To make things more interesting I am going to include the original BIT_VECTOR version of the 3-bit counter along with its monitor. The purpose of this inclusion is to use the original 3-bit BIT_VECTOR counter as a reference to verify the correctness of the data on the tristated bus. I am anticipating to see "ZZZ" on the bus except when the BIT_VECTOR incrementer has the decimal equivalent of 3 or 5. In this case the tristated bus should contain "011" and "101", respectively. Figure 2.72 describes the overall hardware configuration and their interconnections. Based on our previous journeys you know that we need a top-level design entity

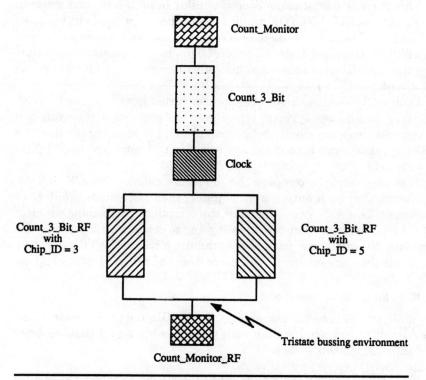

Figure 2.72   Hardware configuration for the architecture Encapsulate_RF. Design entities Count_3_Bit and Count_Monitor included as a reference.

that will define the appropriate component templates and then instantiate them. Herein lies the purpose of this section.

Suppose that these component declarations would be required by several models spanning over several projects. Would you want to type in these component declarations over and over again into the declarative part of your various architectures? This is what we have been doing so far in our journeys. Even having the luxury to cut and paste these components with a text editor does not seem satisfactory. VHDL acknowledges that this tedious, repetitive task should be streamlined, and so it allows component declarations be placed into a package. This package may then be reused repeatedly as needed. Figure 2.73 is an example of such a package. It contains all the component declarations that we will need in our top-level test bench. Storing commonly declared components in a package helps to improve both the modularity of your design and your overall modelling productivity.

Two observations about this package. The library and use clauses at the top of this package emphasize that, once visibility is established, one package may reference the contents of another package. Another feature to observe is that this package does not have a body associated with it.

```vhdl
--   File Name :   encapsulate_rf_components_pkg_.vhd
--
--   Author    :   Joseph Pick
--
library MVL_Packages_Lib;
use MVL_Packages_Lib.MVL4_Pkg.all;

package Encapsulate_RF_Components_Pkg is

    component Clock
            generic (PULSE_WIDTH : TIME);
            port    (Clock : out BIT);
    end component;

    component Count_Monitor
            port    (Datain : in BIT_VECTOR(2 downto 0));
    end component;

    component Count_3_Bit
            port    (Clock   : in BIT;
                     Dataout : out BIT_VECTOR(2 downto 0));
    end component;

    component Count_Monitor_RF
            port (Datain : in Tristate_RS_Vector (2 downto 0));
    end component;

    component Count_3_Bit_RF
            generic (CHIP_ID : NATURAL);
            port    (Clock   : in BIT;
                     Dataout : out Tristate_RS_Vector(2 downto 0)
                    );
    end component;

end Encapsulate_RF_Components_Pkg;
```

Figure 2.73

2.31.1 Information checkpoint

Before you venture further along your VHDL journeys, please make certain that you are familiar with the following topics that we have established via this model:

- VHDL allows you to place component declarations into a package declaration for future reuse by other models or projects.
- It is not necessary for all packages to possess a corresponding package body.

2.32 File: encapsulate_cnt_rf.vhd

Figure 2.74 contains a top-level design entity that fulfills the requirements of the block diagram given in Fig. 2.72. By now you should be very familiar with the look and feel of this design entity. The only difference is that the component declarations are contained in a package instead of being included in the architecture's declarative part. The use clause on line 12 will make available all the component declarations kept in the package Encapsulate_RF_Components. As is the case with the logical library STD, the logical library WORK also does not require a library clause to make it visible. As promised, the component Count_3_Bit_RF is instantiated twice. In one instance, 3 (line 38) is assigned to the generic parameter CHIP_ID, while in the other instance 5 (line 44) is assigned.

And now comes an interesting topic. On lines 19 and 20 the signal Data_Cnt_RF is deliberately initialized to "111" instead of to the nice neutral value "ZZZ". But because this signal is "connected" to the local out port Dataout (line 40), its initial value will be influenced by the output value of Dataout's corresponding formal port. On line 13 of Fig. 2.69 the formal out port Dataout was initialized to "ZZZ". Therefore the two instances Gen_Cnt_3 and Gen_Cnt_5 will both initially drive "ZZZ" on their respective outputs Dataout. Since Data_Cnt_RF is being driven by both these values (lines 40 and 46), its initial value will be the effective value of these multiple sources as determined by the resolution function Tristate_RF. Consequently, Data_Cnt_RF will begin with the initial value "ZZZ." This overriding of its user-specified initial value will be implemented by the VHDL tool suite during the elaboration phase. Recall that the elaboration phase is the first activity that is executed by the VHDL simulator. We have already seen several significant operations that are conducted during this important presimulation (even preinitialization) stage. Additionally, it is during the elaboration phase when the effective initial values of all signals and ports are determined. In our particular example it is during the elaboration phase that the initial "111" value of the signal Data_Cnt_RF will be

```
-- File Name : encapsulate_cnt_rf.vhdl
--
-- Author    : Joseph Pick
--
-- ********************************************************* LINE NUMBER
entity Encapsulate_Cnt_RF is
end Encapsulate_Cnt_RF;

library MVL_Packages_Lib;
use MVL_Packages_Lib.MVL4_Pkg.all;

use WORK.Encapsulate_RF_Components_Pkg.all;              -- 12

architecture Encapsulate_RF of Encapsulate_Cnt_RF is

    signal Tic_Toc     : BIT := '0';
    signal Data_Cnt    : BIT_VECTOR(2 downto 0);
    signal Data_Cnt_RF : Tristate_RS_Vector(2 downto 0)  -- 19
                        := (others => '1');              -- 20

begin

    Synch: Clock
             generic map (50 ns)
             port map    (Tic_Toc);

    Logic_Analyzer: Count_Monitor
                port map (Data_Cnt);

    Logic_Analyzer_RF: Count_Monitor_RF
                port map (Data_Cnt_RF);

    Gen_Cnt: Count_3_Bit
             port map (Tic_Toc, Data_Cnt);

    Gen_Cnt_3: Count_3_Bit_RF
             generic map (CHIP_ID => 3)                  -- 38
             port    map (Clock   => Tic_Toc,
                          Dataout => Data_Cnt_RF         -- 40
                         );

    Gen_Cnt_5: Count_3_Bit_RF
             generic map (CHIP_ID => 5)                  -- 44
             port    map (Clock   => Tic_Toc,
                          Dataout => Data_Cnt_RF         -- 46
                         );

end Encapsulate_RF;
```

Figure 2.74

overridden by the value of "ZZZ". Though resolution function techniques entered this discussion the real underlying principle is that there is going to be an overriding of the initial value assigned to a signal whenever that signal is connected to an out port. I am giving you this "under the VHDL hood" information so that you will not be surprised when sometimes the initial values you assign to signals will seem to mysteriously disappear.

2.32.1 Information checkpoint

Before you venture further along your VHDL journeys, please make certain that you are familiar with the following topics that we have established via this model:

- The logical library WORK is always visible and does not require a library clause.
- Initial value assigned to a signal will be overridden when the signal is used as a port map actual to connect to a local component's out port. If this signal is single sourced, then its initial value will become the initial value of the formal out port to which it is connected. If this signal is multisourced, then the effective value of all the driving sources will be used to override the signal's initial value.

2.33 File: cfg_cnt_rf.vhd

Figure 2.75 contains the configuration declaration for our test bench Encapsulate_Cnt_RF. There is really nothing new to say about the contents of this file. You have already seen it all.

2.34 Baggage Reduction with 1164

The previously described tristate environment required you, the modeler, to create three packages: MVL4_Pkg, Math_Bit_Vector_Pkg, and MVL4_Textio. Aside from the workload needed to design, code, and test these packages there is yet another burden that you have to contend with. The problem of data type portability has to be addressed and

```
-- File Name : cfg_cnt_rf.vhd
--
-- Author    : Joseph Pick
--
configuration Universe_RF of Encapsulate_Cnt_RF is

  for Encapsulate_RF

    for Synch: Clock
      use entity WORK.Clock(Model_1);
    end for;

    for Logic_Analyzer: Count_Monitor
      use entity WORK.Count_Monitor(Model_1);
    end for;

    for Gen_Cnt: Count_3_Bit
      use entity WORK.Count_3_Bit(Model_1);
    end for;

    for Logic_Analyzer_RF: Count_Monitor_RF
      use entity WORK.Count_Monitor_RF(Model_1);
    end for;

    for all: Count_3_Bit_RF
      use entity WORK.Count_3_Bit_RF(Model_1);
    end for;
  end for;

end Universe_RF;
```

Figure 2.75

resolved so that models from different projects and companies can communicate with each other. Portability issues can come about in a number of ways. One company might be modeling with a four-element data type while another might be doing so with seven. Even if both companies are using a four-element data type, the names might be different. One company might declare the enumeration type with the name Logic4 while another might call it MVL4. Even if they both use the same number of elements and the same name, there is no guarantee that all is going to be well. Suppose that the two companies have models that rely on the data type Logic4 but their respective declarations are different:

```
type Logic4 is ('X', '0', '1', 'Z'); -- Company A

type Logic4 is ('Z', '0', '1', 'X'); -- Company B
```

Clearly, there will be portability problems between Company A and Company B. For instance, in Company A's case Logic4'LEFT is 'X' while in Company B's case it is 'Z'. In all these variations different data typing roads are being traveled. To transport models freely between these multitudes of potential data typing roads requires data type conversion functions to serve as a bridge between the many roads being traversed. And so there are even more utility functions that you would have to design, code, and test. But what is even worse is that these conversion functions have introduced additional run time execution activities. Precious CPU time will be gobbled up by these conversion functions.

To overcome these portability issues the IEEE mandated in 1992 a standard data type package called std_logic_1164. Aside from solving the portability problem this IEEE package has the extra benefit of relieving you of the burden of having to write your own MVLN package to suit the needs of your current project. Moreover, now that an industrywide standard data type exists, many of the VHDL vendors have added value to their products by bundling in a wide assortment of auxiliary packages all based on this new standard. The availability of all these packages have, in essence, lightened your modeling workload. You no longer have to deal with the excess baggage of having to write your own data type and associated utilities package.

The structural format of this std_logic_1164 package is very similar to our MVL4_Pkg. But in content there are two significant differences, both of which make the std_logic_1164 package pedagogically more difficult to use as the first example of an MVLN package. Enough new ground was being covered during our initial introduction to an MVLN package that I wanted to avoid the further complications that are implied by these differences. But having made the journey into MVL4_Pkg you now have a better understanding of the big picture. So

166 An Excursion into VHDL

now is a good time for an introductory exploration of both the content and also some applications of the std_logic_1164 package.

One key difference between std_logic_1164 and MVL4_Pkg is the notion of strengths that is embedded into std_logic_1164's base data type. The base data type defined in MVL4_Pkg had no implied strengths. Another significant difference is that std_logic_1164 has only one resolution function, whereas MVL4_Pkg has three, one for each bussing scheme. In the std_logic_1164 environment different busses are modeled by the multisourced assignments of strength values. The sole resolution function will then use the strengths of the various signal sources to derive a value that is indicative of the bussing scheme being modeled. More will be said about the implications of both these strength values and this single resolution function in Chap. 9 of this book.

The salient definitions made in the std_logic_1164 package are listed below. Because of our previous journeys the format and potential applications of these definitions should seem very familiar to you.

```
type std_ulogic is ('U',    -- Uninitialized
                    'X',    -- Forcing Unknown
                    '0',    -- Forcing 0
                    '1',    -- Forcing 1
                    'Z',    -- High Impedance
                    'W',    -- Weak Unknown
                    'L',    -- Weak 0
                    'H',    -- Weak 1
                    '-'     -- Don't care
                   );

type std_ulogic_vector is array (NATURAL range <>) of std_ulogic;
function resolved (S : std_ulogic_vector) return std_ulogic;
subtype std_logic is resolved std_ulogic;
type std_logic_vector is array (NATURAL range <>) of std_logic;
```

The convention in the VHDL community is to compile the std_logic_1164 package into a library having the logical name IEEE. Typically, the 1164-related auxiliary packages supplied by a VHDL vendor are also resident in this IEEE library.

Now let's go back to our multisourced counting environment and redo the models with this IEEE standard data type. The availability of both the IEEE std_logic_1164 package and a vendor's auxiliary packages together create a user-friendly environment in which to model multisourced signals. Since the models are very similar to our MVL4_Pkg referencing models, we can breeze through them very quickly and concentrate only on the differences. It should be noted that our modeling activities have now been drastically streamlined. It is no longer necessary to design, code, and test any new MVLN package. But it is required that you familiarize yourself with the vendor-supplied package specifications so that you will know what data types and subprograms are

```
-- File Name :  count_rf_1164_.vhd
--
-- Author    :  Joseph Pick
--
library IEEE;
use IEEE.std_logic_1164.all;

entity Count_3_Bit_RF_1164 is
      generic (CHIP_ID : NATURAL);
      port    (Clock   : in BIT;
               Dataout : out std_logic_vector(2 downto 0)
                            := (others => 'Z')
              );
end Count_3_Bit_RF_1164;
```

Figure 2.76

available to you. Later on, when you have more leisure time, you can then enhance your VHDL repertoire by studying the techniques used in these vendor-supplied package bodies. There always is something new to be learned.

Figure 2.76 shows the entity declaration for the enhanced 3-bit counter. The only significant difference is the usage of the std_logic_vector instead of Tristate_RS_Vector. The library and use clauses were required to make this data type visible (available).

Figure 2.77 contains the architecture body associated with the entity Count_3_Bit_RF_1164. The package std_logic_unsigned is a Synopsys

```
-- File Name :  count_rf_1164_model_1.vhd
--
-- Author    :  Joseph Pick
--
architecture Model_1 of Count_3_Bit_RF_1164 is

use IEEE.std_logic_unsigned.all;

begin

   Incr:
   process
      variable Current_Count : std_logic_vector(2 downto 0) := "000";

   begin

      -- Wait until the trailing edge of the clock.
      wait until Clock = '0';

      -- Increment the counter.
      Current_Count := Current_Count + "001";

      -- Determine what value is to be placed on the tri-stated bus.
      if CHIP_ID = Current_Count then
         Dataout <= Current_Count;
      else
         Dataout <= "ZZZ";
      end if;

   end process Incr;

end Model_1;
```

Figure 2.77

std_logic_1164 auxiliary package that treats std_logic_vector arrays as representations of unsigned numbers. The standard VHDL operators, including equality, have also been overloaded in this package. Consequently, it is now valid to increment Current_Count by adding "001" to it. But you have to remember to initialize Current_Count to "000" or else it will default to "UUU" (why?). Adding "001" to "UUU" will then not be appropriate. An interesting technique to observe is the direct comparison that can be made between CHIP_ID and Current_Count. In our previous journeys this could not be done since these two objects were of different data types. But there exists a Synopsys function that overloads the equality operator to allow this comparison between dissimilar data types. An advantage of working exclusively with std_logic_vector is that

```
-- File Name : count_monitor_rf_1164.vhd
--
-- Author    : Joseph Pick
--
library IEEE;
use IEEE.std_logic_1164.all;

entity Count_Monitor_RF_1164 is
    port (Datain : in std_logic_vector(2 downto 0));
end Count_Monitor_RF_1164;

architecture Model_1 of Count_Monitor_RF_1164 is
    use IEEE.std_logic_textio.all;
    use STD.TEXTIO.all;
begin

   Display:
   process

      file Outfile : TEXT is out "counter_rf_values";
      -- In VHDL'93 the above line MUST be replaced by
      --   file Outfile : TEXT open WRITE_MODE is "counter_rf_values";
      variable Outline : LINE;
   begin

      write (Outline, STRING'(" TIME    VALUE OF TRI-STATED BUS"));
      writeline (Outfile, Outline);

      Main_Loop:
      loop
         write (L      => Outline, Value => Now,
                Field => 6);
         write (Outline,
                Datain,
                Field => 8);
         writeline (Outfile, Outline);

         wait on Datain;

      end loop Main_Loop;

   end process;

end Model_1;
```

Figure 2.78

conversions do not have to be made when assigning the incremented value to the output Dataout. Recall how in our previous model (Fig. 2.70) we had to convert from BIT_VECTOR to Tristate_RS_Vector.

Figure 2.78 shows the counter's monitor in an std_logic_1164 environment. The only worthwhile note to be made is that you no longer have to design, code, and test a package that reads and writes std_logic_vector data-type objects. Instead, all you have to do is reference your vendor's auxiliary std_logic_1164-related input/output package, as I have done with the Synopsys std_logic_textio package. The file declarations used in this model concur with the VHDL'87 format. When VHDL'93 tool suites become available, then the commented line must be used instead.

There is nothing significant in the remaining models, and they are included in Figs. 2.79, 2.80, and 2.81 just for completion.

Our excursion into VHDL is now complete. Welcome home.

2.35 Journey's Epilogue

At the end of each journey there is always a feeling of sadness. All that is left are fond memories and photos. And yet we feel spiritually richer

```
--   File Name :   encap_components_pkg_.vhd
--
--   Author    :   Joseph Pick

library IEEE;
use IEEE.std_logic_1164.all;

package Encapsulate_Components_Pkg is

   component Clock
             generic (PULSE_WIDTH : TIME);
             port    (Clock : out BIT);
   end component;

   component Count_Monitor
             port    (Datain : in BIT_VECTOR(2 downto 0));
   end component;

   component Count_3_Bit
             port    (Clock : in BIT;
                      Dataout : out BIT_VECTOR(2 downto 0));
   end component;

   component Count_Monitor_RF_1164
             port (Datain : in std_logic_vector (2 downto 0));
   end component;

   component Count_3_Bit_RF_1164
             generic (CHIP_ID : NATURAL);
             port    (Clock   : in BIT;
                      Dataout : out std_logic_vector(2 downto 0)
                     );
   end component;

end Encapsulate_Components_Pkg;
```

Figure 2.79

An Excursion into VHDL

```
-- File Name : encap_cnt_rf_1164.vhd
--
-- Author    : Joseph Pick
--
entity Encapsulate_Cnt_RF_1164 is
end Encapsulate_Cnt_RF_1164;

library IEEE;
use IEEE.std_logic_1164.all;

use WORK.Encapsulate_Components_Pkg.all;

architecture Model_1 of Encapsulate_Cnt_RF_1164 is

    signal Tic_Toc      : BIT := '0';
    signal Data_Cnt     : BIT_VECTOR(2 downto 0);
    signal Data_Cnt_RF  : std_logic_vector(2 downto
                          := (others => '1');
begin

   Synch: Clock
          generic map (50 ns)
          port map    (Tic_Toc);

   Logic_Analyzer: Count_Monitor
                   port map (Data_Cnt);

   Logic_Analyzer_RF: Count_Monitor_RF_1164
                      port map (Data_Cnt_RF);

   Gen_Cnt: Count_3_Bit
            port map (Tic_Toc, Data_Cnt);

   Gen_Cnt_3: Count_3_Bit_RF_1164
              generic map (CHIP_ID => 3)
              port    map (Clock   => Tic_Toc,
                           Dataout => Data_Cnt_RF
                          );

   Gen_Cnt_5: Count_3_Bit_RF_1164
              generic map (CHIP_ID => 5)
              port    map (Clock   => Tic_Toc,
                           Dataout => Data_Cnt_RF
                          );

end Model_1;
```

Figure 2.80

for having made the journey. Henceforth we will view all future activities from a more mature perspective and use our traveling experiences to better understand ourselves and our surroundings. We are all the wiser for what we have accomplished and eager for our next adventures.

The remainder of this book is essentially a collection of VHDL tricks, techniques, and caveats based on real-world VHDL coding scenarios. They are intended to enhance your VHDL problem solving skills and to make you more productive. But always keep in mind this far-reaching VHDL voyage that we have just taken together. The global view that you acquired during this trip will serve as a solid foundation for all your future VHDL activities.

Thank you for sharing this VHDL journey with me.

```
-- File Name : cfg_cnt_rf_1164.vhd
--
-- Author    : Joseph Pick
--
configuration Universe_RF_1164 of Encapsulate_Cnt_RF_1164 is

  for Model_1

    for Synch: Clock
      use entity WORK.Clock(Model_1);
    end for;

    for Logic_Analyzer: Count_Monitor
      use entity WORK.Count_Monitor(Model_1);
    end for;

    for Gen_Cnt: Count_3_Bit
      use entity WORK.Count_3_Bit(Model_1);
    end for;

    for Logic_Analyzer_RF: Count_Monitor_RF_1164
      use entity WORK.Count_Monitor_RF_1164(Model_1);
    end for;

    for all: Count_3_Bit_RF_1164
      use entity WORK.Count_3_Bit_RF_1164(Model_1);
    end for;
   end for;
  end Universe_RF_1164;
```

Figure 2.81

Part 2

Driving the Simulation Flow

Chapter

3

Signal-Updating Algorithms

The purpose of Part 2 is to describe the main forces that are driving and directing a VHDL simulation. One such force is the algorithm that VHDL applies when updating signals. Knowledge of this internal algorithm will not only improve your VHDL debugging skills, but it will also give you a deeper appreciation of how VHDL is able to model the signal-updating characteristics of real hardware.

In the Excursion you saw that signals are always scheduled to be updated at a later simulation time. This user-specified delay can be either an absolute amount of time or it may be an infinitesimal amount known as a delta time unit. These two options have already been introduced and developed in the Excursion. But there is even more to the story. In actuality, VHDL has two signal-updating schemes. One is called the inertial delay model, and unless specified otherwise, it serves as the default. Hence all the signal updates in the Excursion were implemented using this inertial delay model. The other signal-updating scheme is called the transport delay model, and its application requires the usage of the reserved word *transport*.

Inertial delay is used to model devices that respond only to signals that persist for a given amount of time. This signal-updating method is useful in modeling devices that ignore spikes in their inputs. Any pulse that has a width less than the user-specified delay time will not be transmitted to the targeted signal. On the other hand, transport delay is analogous to the delay incurred by passing a current through a wire. Hence there is no minimum amount of time that the signal must persist in order that it propagate through to the targeted signal. Both these definitions may be best understood with the aid of a timing diagram.

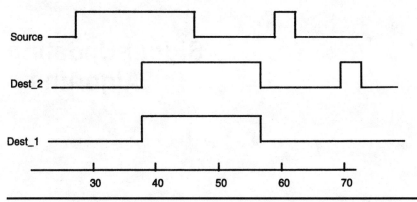

Figure 3.1 Timing diagram illustrating the difference between the inertial and transport delay models used to update signals.

Consider the waveforms drawn in Fig. 3.1. Assume that the assignments to Dest_1 and Dest_2 are both concurrent signal assignments. The reserved word *transport* on the right-hand side of the signal assignment will cause the updating of Dest_2 to comply with a transport delay updating model. Consequently, Dest_2's waveform will be identical to Source's waveform except that it will be delayed (shifted to the right) by the specified delay time of 10 ns. On the other hand, the inertially updated Dest_1 will not honor any of Source's pulses that are less than the specified delay time of 10 ns. This explains why the second pulse on Source is not propagated to Dest_1, whereas the first one is. If this second pulse width had been greater than or equal to the specified delay time of 10 ns, then it too would have been part of Dest_1's waveform.

Though these explanations are somewhat satisfactory they, nonetheless, represent only an external user's perspective. To better understand the VHDL mechanics of signal updating, one must travel down to the VHDL algorithmic level and observe, first hand, how the aforementioned described inertial and transport models are actually implemented. Let's begin with a definition.

The driver for a signal consists of two parts: the signal's current value and its projected scheduled waveform. A signal's projected scheduled waveform can be thought of as a set of ordered pairs (V_N, T_N), where the value V_N is to be assigned to the signal at simulation time T_N. Recall from the Excursion segment that the term *simulation time* refers to the combination of both an absolute and a delta time. For example, a signal might be updated at 30 ns plus 2 delta. And an assignment at 30 ns plus 0 ns is simply referred to as an update at 30 ns. With this arsenal of nomenclature conventions at our disposal we are now ready to explore the transport and inertial update algorithms and, in particular, apply them toward the updating of Dest_1 and Dest_2. First, let's do the simpler one, which is the transport delay model.

As per the timing diagram the initial values of both Source and Dest_2 are '0'. Hence at simulation startup both Source and Dest_2 will each be driving '0', and their projected drivers will both be empty since nothing has yet been scheduled for them. In the Excursion you learned that every concurrent signal assignment has a process that is theoretically equivalent to it. This equivalent process will have a static sensitivity list consisting of all the signals referenced on the right-hand side of the corresponding signal assignment. Furthermore, at the bottom of this equivalent process is an implicit wait statement that is waiting for events on those signals listed in the static sensitivity list. Consequently, during the initialization phase, Dest_2 will be scheduled to receive '0' 10 ns from the current simulation. Since the initialization phase occurs at 0 ns, this updating of Dest_2 will occur at the simulation time 10 ns (really 10 ns plus 0 delta). So at the completion of the initialization phase Dest_2's projected driver will be {('0', 10 ns)}. When the simulation engine jumps to 10 ns, this value '0' will be assigned to Dest_2, after which its projected driver will once again be empty. As per the timing diagram Source will transition from '0' to '1' at simulation time 28 ns. Based on our Excursion experiences we can now say more precisely that Source will be updated during the first half of the simulation cycle that occurs at time 28 ns plus 0 delta. This change in Source's value will cause Source to be tagged as just having received an event. This event on Source will cause the equivalent process to be awakened during the second half of the simulation cycle at 28 ns plus 0 delta. The signal assignment to Dest_2 will then be executed and, as a result, Dest_2 will be scheduled to receive the value '1' at 10 ns from now, which will be at 38 ns (28 ns plus 10 ns). Therefore the projected driver for Dest_2 will become {('1', 38 ns)}. When simulation time advances to 38 ns, this '1' will be assigned to Dest_2, and the net result will be that Dest_2 will have a leading edge exactly 10 ns after Source's leading edge. Using the same aforementioned algorithm it can similarly be shown that Source's trailing edge at 46 ns will be propagated to Dest_2 at 56 ns (really 46 ns plus 10 ns). Let's now continue to Source's second pulse, which has a width that is less than the specified propagation delay of 10 ns. It is interesting to observe the effects of this small pulse width on Dest_2's projected driver. At time 58 ns Source has an event when it goes from '0' to '1'. As before, the signal assignment to Dest_2 will be scheduled to occur 10 ns later and, as a result, Dest_2's projected driver will become {('1', 68 ns)}. Now comes an interesting twist. At time 62 ns Source will go back down to '0'. This event will once again schedule the occurrence of a signal assignment to Dest_2 just as before. Therefore the ordered pair ('0', 72 ns) will have to be added to Dest_2's projected driver. But unlike any of the previous inclusions Dest_2's projected driver will already contain an ordered pair when this new ordered pair is to be placed into it. A decision will have to be made regarding the fate of the ordered pair that already exists in the projected driver. Since Dest_2 is being updated using the transport delay model, the only check that has to be conducted is the confirmation that the time element of this new ordered pair is greater than the time element of every ordered pair that already exists in the projected driver. If this new time had been less than or equal to the time element of any of the preexisting ordered pairs, then these preexisting ordered pairs would all have to be deleted from the projected driver. In our current situation the times are in ascending order, and so the new ordered pair will merely be appended to the projected driver. Consequently, Dest_2's projected driver will be the set {('1', 68 ns), ('0', 72 ns)}. Since there are no more events on Source, Dest_2's projected driver will no longer be appended to. All that remains is to carry out those assignments implied by the Dest_2's projected driver. This means that Dest_2 will be assigned '1' and '0' at 68 ns and 72 ns, respectively. The net effect of these last two assignments is that Dest_2 will follow Source's second pulse width, even though this pulse has a pulse width that is shorter than the user-specified delay time of 10 ns. Figure 3.2 summarizes this chronological flow of Dest_2's projected driver.

Let's now explore the inertial updating algorithm. Up until 62 ns, Dest_1's projected driver will be exactly the same as Dest_2's projected driver. And as in the transport case it will be necessary at 62 ns to add the ordered pair ('0', 72 ns) to the preexisting projected driver {('1', 68 ns)}. Recall that in the earlier transport case only the ascending order of the times had to be confirmed before the ordered pair ('0', 72 ns) could be appended to the projected driver. This time comparison test must also be conducted by the inertial signal-updating algorithm. But once this test passes, then an additional check must also be conducted. If the new projected value is different from any of the preexisting projected values, then their corresponding preexisting ordered pairs must be deleted from the projected driver. This is precisely the situation at 62 ns when the ordered pair ('0', 72 ns) is to be added to the projected driver {('1', 68 ns)}. Because the new scheduled value '0' is different from the previously scheduled value '1', it follows from the aforementioned inertial signal-updating algorithm that the previously existing

Time	Projected Driver for Dest_2
0 ns	{('0', 10 ns)}
10 ns	Empty
28 ns	{('1', 38 ns)}
38 ns	Empty
46 ns	{('0', 56 ns)}
56 ns	Empty
58 ns	{('1', 68 ns)}
62 ns	{('1', 68 ns), ('0', 72 ns)}
68 ns	{('0', 72 ns)}
72 ns	Empty

Figure 3.2 Chronological flow of Dest_2's projected driver.

ordered pair ('1', 68 ns) must be deleted from Dest_1's projected driver. The upshot of this deletion is that Dest_1's projected driver will now only consist of ('0', 72 ns). And so '0' will be assigned to Dest_1 at 72 ns. Observe that unlike Dest_2, the signal Dest_1 will not be assigned '1' at 68 ns. Hence the second pulse on Source will not be transmitted to Dest_1. As per Dest_1's perspective this second pulse will be interpreted as a glitch and will be filtered out. Figure 3.3 summarizes this chronological flow of Dest_1's projected driver.

From the previous inertial updating algorithm it follows that the minimum duration of a nonrejected pulse width must be greater than or equal to the user-specified propagation delay. In general, this restriction is acceptable, but every once in a while one would like to model a rejection time that is less than the specified propagation delay. In VHDL'87 this modeling requirement can be achieved via two concurrent signal assignments. For instance, a rejection time of 3 ns and a propagation delay of 10 ns may be modeled via:

```
Temp_Signal <= A after 3 ns;

B <= transport Temp_Signal after 7 ns;
```

VHDL'93 has a more compact and elegant solution that references the two newly introduced reserved words *reject* and *inertial*. Conse-

Time	Projected Driver for Dest_1
0 ns	{('0', 10 ns)}
10 ns	Empty
28 ns	{('1', 38 ns)}
38 ns	Empty
46 ns	{('0', 56 ns)}
56 ns	Empty
58 ns	{('1', 68 ns)}
62 ns	{~~('1', 68 ns)~~, ('0', 72 ns)}
72 ns	Empty

Figure 3.3 Chronological flow of Dest_1's projected driver.

quently, in VHDL'93 the above two concurrent signal assignments may be singly written as:

```
B <= reject 3 ns inertial A after 10 ns;
```

Two points must be made regarding this new VHDL'93 feature. First, the rejection time must be less than or equal to the user-specified propagation delay time. And second, it readily follows that the earlier detailed inertial updating algorithm must be slightly enhanced to incorporate this new pulse-rejection capability.

When followed to the letter, the aforementioned transport and inertial updating algorithms may be systematically applied to better understand a model's data flow, irrespective of the model's complexity. These algorithms are further explored in the Experiments segment of this book (Test_18B, Test_18C, Test_18D, Test_113, and Test_18E).

Chapter 4

Predefined Data Type Attributes

The previous chapter showed how VHDL's inertial and transport algorithms control a simulation's behavior. A simulation flow can also be directed via branching decisions based on VHDL's predefined attributes.

The predefined VHDL attributes may be associated with either data types or signals. You can think of these predefined attributes as windows into the simulation engine that provide key information about what is currently going on. Such information is very important. Remember that your overall game plan is to model reality within the constraints of the VHDL language. The predefined attributes are specifically designed to help you to achieve this modeling objective. This chapter discusses those attributes that are associated with data types. The next chapter will both introduce and explore those attributes that are associated with signals. Let's now start off with the attributes associated with the data type INTEGER, which is defined as follows in the VHDL predefined package called STANDARD:

```
type INTEGER is range -2147483647 to 2147483647;
```

Typically, the INTEGER-related attributes that you will probably use in your models are 'LEFT, 'RIGHT, 'HIGH, 'LOW, and 'IMAGE. The single quote in each of these attributes is read as tick, and this naming convention is true for all other VHDL attributes as well. The value of INTEGER'LEFT (read as INTEGER tick LEFT) is −2147483647 since this number is the leftmost range bound used in INTEGER's definition. The significance of this INTEGER'LEFT value is that the VHDL simulation engine will use it to initialize any INTEGER signal or variable object that is not given an initial value when it is formally declared. For example, consider the following declaration:

```
variable Count : INTEGER;
```

Since an initial value was not assigned to Count, the VHDL simulation engine will initialize it to −2147483647. Keep this fact in mind whenever you are simulating with INTEGER data types. If this negative number unexpectedly (and unwantingly) pops up in your simulation, then you must have forgotten to assign an appropriate initial value to it. INTEGER'LOW is the smallest of the two bounds specified by the defined range. Since INTEGER is defined with an ascending range, it follows that INTEGER'LEFT is equal to INTEGER'LOW.

INTEGER'RIGHT is equal to 2147483647 since this number is on the rightmost range bound used in INTEGER's definition. Since INTEGER's range is ascending, we will have that INTEGER'RIGHT is the same as INTEGER'HIGH. The significance of INTEGER'HIGH is that it is used in the following subset definition:

```
subtype NATURAL is INTEGER range 0 to INTEGER'HIGH;
```

As per the aforementioned pattern for 'LEFT, we have that NATURAL'LEFT is 0. Consequently, any signal or variable declared to be of type NATURAL that is not initialized by you, the modeler, will automatically be initialized by the VHDL simulation engine to 0 (= NATURAL'LEFT).

Now, suppose that the object Integer_Var is a variable of type INTEGER. Then the attribute INTEGER'IMAGE(Integer_Var) will generate the string representation of the number Integer_Var. This attribute comes in very handy when working with assert statements. Here is how you might want to use it in conjunction with the concatenation operator &:

```
assert Integer_Var < 5555
    report "Integer_Val's unexpected value is " & INTEGER'IMAGE(Integer_Var)
    severity NOTE;
```

Let's now look at the most common predefined attributes that you will probably reference when working with enumeration data types. This data typing capability was introduced in the Excursion and is a powerful technique that can be applied to model devices at any desirable abstraction level.

Consider the following user-defined enumeration type:

```
type States is (IDLE, TRANSFER, HALT, BUSY);
```

Each member of this enumeration data type is called an enumeration literal. Since IDLE is the leftmost element in the list, we have that States'LEFT is equal to IDLE. Analogously, States'RIGHT is equal to BUSY. In general, the order in which the elements are listed in an enumeration data type definition determines an ascending order relationship so that States'LEFT is the smallest and States'RIGHT is the highest. In our particular example we will have that:

```
IDLE < TRANSFER < HALT < BUSY
```

Consequently, States'LOW is equal to IDLE and States'HIGH is equal to BUSY. By the way, while we are on the subject of enumeration literal ordering, let me point out that it is valid to write the following for loop construct:

```
for K in States loop
    -- sequential code
end loop;
```

The usage of the type mark (term has already been defined) States in this context implies that the for loop bounds should be interpreted as:

```
for K in IDLE to BUSY loop
```

'POS and 'VAL are two other VHDL attributes that can be applied to enumerated data types. Let's first look at 'POS. The left-to-right positioning of each enumeration literal defines a unique number. The leftmost element is said to be in position 0. The element immediately to the right of this leftmost element is in position 1. This incrementing numbering scheme is continued until the rightmost element is reached. The attribute 'POS can be used to derive the position number of an enumeration literal. For example, States'POS(IDLE) is equal to 0 and States'POS(HALT) is equal to 2. On the other hand, the attribute 'VAL may be thought of as the inverse of 'POS. Given a number K then 'VAL will derive the enumeration literal that is in position K. For instance, States'VAL(1) is equal to TRANSFER. The attributes 'POS and 'VAL are very powerful tools in your arsenal of VHDL tricks and techniques. They will be used in very innovative ways during the remaining two parts of this book.

Here are some of the other attributes that you may want to reference when working with enumeration data types. States'SUCC(TRANSFER) will yield that enumeration element that is the immediate successor of TRANSFER. Consequently, States'SUCC(TRANSFER) is equal to HALT. Note well that requesting the successor of BUSY via States'SUCC(BUSY) will result in an error since BUSY is the last enumeration literal in the list. The attribute 'SUCC does not wrap around to the beginning of the list. The immediate predecessor of an enumeration literal can be derived using the attribute 'PRED. For example, States'PRED(HALT) is the enumeration literal TRANSFER. Here again, it is worthwhile to note that States'PRED(IDLE) will result in an error since no enumeration literal comes before IDLE.

This user-defined enumeration data type States now poses an interesting problem. Suppose that I wish to read the various enumeration literals from an ASCII file and use them as test vectors for my simulation. Or what if I wish to write into an ASCII file the enumeration literal that represents the machine state that my simulation currently is in. Unfortunately, I cannot readily accomplish either of these tasks. The VHDL TEXTIO package will only allow us to read and write the standard VHDL data types such as INTEGER, REAL, BOOLEAN, BIT, BIT_VECTOR, CHARACTER, STRING, and TIME. This TEXTIO package cannot be used to either read or write any new data type that is defined by a VHDL modeler. Consequently, in VHDL'87 the only alternative was to write your own subprograms that will read and write the elements of your newly defined data type. Recall that this is exactly what I had to do in the Excursion with elements of type Logic4 and Logic4_Vector. Fortunately, VHDL'93 offers two new data-type attributes so that we will no longer have to write our own tedious subprograms. The attribute 'IMAGE will provide a string character for any data-type element. This attribute was already used earlier to incorporate an INTEGER's current value into an assert's report string. Clearly, there was no need to use 'IMAGE to write an INTEGER typed value out to an ASCII file since there already exists an overloaded write subprogram in the TEXTIO package that can readily accomplish this. But there is no such subprogram for our newly defined enumeration data type States. Instead, we can use 'IMAGE to convert any enumeration literal into a STRING data type and then rely on the predefined VHDL TEXTIO package to write this STRING into an ASCII file. Here is how it works. Suppose that you wish to write the variable States_Var, which is of type States, into the file that has the logical name Dataout. This can be accomplished via the following two steps:

```
write (Buffer_Line, States'IMAGE(States_Var));
writeline (Dataout, Buffer_Line);
```

The first procedure writes the STRING States'IMAGE(States_Var) into Buffer_Line. Actually, what really happens is that the pointer Buffer_Line will point to this STRING. The second procedure writes Buffer_Line into the ASCII file that has the logical name Dataout. Here again some qualification is in order. The contents of Buffer_Line are not really being written out. Instead, the STRING that Buffer_Line is pointing to gets written out to the file. But from an operational point of view, what is most important to us is the manner in which the attribute 'IMAGE may be applied. We no longer have to create our own subprogram when we wish to write an ASCII representation of the enumeration literals that constitute the data type States.

The attribute 'VALUE (not to be confused with 'VAL) is the inverse of 'IMAGE. Given an ASCII string, this attribute may be applied to convert this STRING-typed object into a user-defined enumeration literal. A typical application of 'VALUE is to read in an enumeration literal that is resident in an ASCII file. Suppose that such a file has the logical name Datain and the STRING pointer has been named Buffer_Line. Then the following statements will achieve our reading objectives:

```
readline (Datain, Buffer_Line);
read (Buffer_Line, Input_String);
Enumerated_Value := States'VALUE(Input_String);
```

The procedures readline and read are from the predefined VHDL TEXTIO package. Readline will access one line from the ASCII file Datain and point Buffer_Line to it. The second procedure will read the first N characters pointed to by Buffer_Line and assign them to the variable Input_String. As its name suggests, this variable Input_String had previously been defined to be of type STRING. Note well that this assumes that the ASCII representation of the enumeration literal was the first item to occur on the line that has just been read in. If this is not the case, then the preceding items must be read before you can get to this STRING representation of the enumeration literal. The third line merely applies the attribute 'VALUE to derive the enumeration literal that has the string representation of Input_String. Enumerated_Value is a variable of type States. By the way, the attribute 'VALUE is very friendly since it will ignore leading and trailing blank spaces. Consequently, Input_String may be defined as a STRING from 1 to the length of the longest enumerated literal. Consequently, you can read in "IDLE " and the expression States'VALUE("IDLE ") will correctly convert it into IDLE. The trailing blanks in the string will be ignored. Without this capability you would not know how wide to make the variable Input_String. The real culprit is the overloaded TEXTIO STRING read procedure that considers blanks to be valid CHARACTERs and hence does not skip over them.

Let's now move up the data type complexity chain and discuss the predefined VHDL attributes that are associated with arrays. Consider the following type and object declarations:

```
type Bits_2D is array (NATURAL range <>, NATURAL range <>) of BIT;
type Logic4 is ('X', '0', '1', 'Z');
type Logic4_To_Bit is array (Logic4) of BIT;

variable Input_To         : Bits_2D(12 to 31, 5 to 14);
variable Input_Mixed      : Bits_2D(12 to 31, 14 downto 5);
variable Input_Enum_Index : Logic4_To_Bit;
```

Before we explore the VHDL predefined attributes that can be associated with these variable objects, let's first make some remarks about the declarations themselves. The type declaration for Logic4_To_Bit is shorthand for the following:

```
type Logic4_To_Bit is array (Logic4 range 'X' to 'Z') of BIT;
```

Observe how the range direction and bounds were automatically implied by the order in which the enumeration literals were listed in Logic4's definition. The variable declaration for Input_Mixed shows that the index ranges of the various array dimensions do not have to be in the same direction. The first dimension has an ascending range, whereas the second dimension has a descending range. Since the data type Logic4_To_Bit is defined to be a constrained array, it would have been invalid to reference a slice of it in an object's declaration. Hence it would have been incorrect to have written:

```
variable Input_Enum_Index_Slice :
                    Logic4_To_Bit('0' to 'Z'); -- **** ERROR
```

The only time that a slice of an array data type can be referenced in an object's declaration is when the data type was declared as an unconstrained array via the box notation <>. Enough said. Let's now explore the VHDL's predefined array attributes via some concrete examples.

The attribute Input_To'LEFT(2) refers to the left bound of the range that was defined for Input's second dimension. If a number is not given, as in Input_To'LEFT, then the attribute refers to the left bound of the range that was defined for Input's first dimension. This omission rule is true for all multidimensional arrays. Consequently, we have that Input_To'LEFT(2) is equal to 5 and Input_To'LEFT is equal to 12. Similarly, Input_Mixed'LEFT(2) has the value 14 since this number is the leftmost bound of the range that was defined for Input_Mixed's second dimension. As for Input_Enum_Index'LEFT, its value is 'X' since the range defined for its sole dimension spans the values 'X' to 'Z'. Analogously, we will have that Input_To'RIGHT(2), Input_Mixed'RIGHT(2), and Input_Enum_Index'RIGHT will, respectively, be equal to 14, 5, and 'Z'. The attribute 'HIGH when applied to arrays yields the larger of the two bounds used to define the index range for a specific dimension of an array. Consequently, Input_To'HIGH(2) and Input_Mixed-'HIGH(2) both have the same value 14 and Input_Enum_Index'HIGH is equal to 'Z'. Analogously, Input_To'LOW(2), Input_Mixed'LOW(2), and Input_Enum_Index are equal to 5, 5, and 'X', respectively. The attribute 'LENGTH refers to the number of values in the specified dimension of an array. Consequently, Input_To'LENGTH(2), Input_Mixed-'LENGTH(2), and Input_Enum_Index'LENGTH are equal to 10, 10, and 4, respectively. The attribute 'RANGE refers to the range specification of an array's dimension. Both the bounds and the direction are honored by this attribute. So, for instance, Input_To'RANGE(2) is equal to the range specification 5 to 14. Likewise Input_Mixed-'RANGE(2) is equal to 14 downto 5 and Input_Enum_Index'RANGE is equal to 'X' to 'Z'. The attribute 'REVERSE_RANGE is true to its name in that the user-specified dimension ranges are reversed. Therefore Input_To'REVERSE_RANGE(2) is equal to 14 downto 5, Input_Mixed'REVERSE_RANGE(2) is equal to 5 to 14, and Input_Enum_Index'REVERSE_RANGE is equal to 'Z' downto 'X'.

These array attributes are very powerful modeling tools that will allow you to create elastic subprograms that can internally adjust to any array input length. This flexibility means that you only have to design and test such elastic subprograms once and then you can use them indefinitely on successive projects. Also, if an algorithm can be found that may be applied to model a device irrespective of its input and output bandwidth, then these elasticizing array attributes may be applied to create a reusable generic hardware part. In the Excursion we already saw how some of these array attributes may be applied to design both elastic subprograms and generic hardware components. The remaining segments of this book will point out some potential traps that you might fall into when working with subprograms that have unconstrained parameters.

VHDL'93 introduced the array attribute 'ASCENDING, which has the value TRUE if the range of a specified dimension is declared to be in an ascending order. For instance, Input_To'ASCENDING(2) is TRUE, whereas Input_Mixed'ASCENDING(2) is FALSE. The problem with this attribute is that it serves as an invitation for the VHDL engineer to create models that are memory inefficient. The availability of this attribute might be used to determine which of two for loops should be executed, depending on whether the array's range is ascending or descending. Consequently, the same for loop might be written twice except for the different range bounds. This repetition of code is unnecessary and, even worse, it is an inefficient usage of precious computer memory. Actually, there is even more to the story. Experience has shown that whenever I had a need to know an input array's range direction ('RANGE was not suitable), I also had a need to control its index range bounds. Putting these two facts together makes me believe that, in the long run, it will be better to rely on alias techniques to gain complete control of an array's indices. You have already come across the alias construct during the Excursion, and its uses will be further discussed in the remaining parts of this book.

Chapter 5

Predefined Signal Attributes

This chapter describes the various attributes that may be associated with a signal. A later subsection will show how these attributes may be applied to control the simulation flow of a VHDL model.

Signal attributes come in two flavors. Some are signals and some are not. Let's first begin with those signal attributes that are also signals. In all the descriptions that follow, the attribute prefix S must be a signal and the attribute argument T must be an object of type TIME.

S'DELAYED(T) defines a signal having the same type as S, and its value is the value of S delayed by the time T. If T is not given, as in S'DELAYED, then T is assumed to be 0 ns. S'DELAYED has the same value as S but delayed by 1 delta time unit. Another way to say this is that S'Delayed has the same value as S but is delayed by one simulation cycle. The terms delta time and simulation cycle have already been developed in the Excursion.

S'STABLE(T) defines a BOOLEAN signal having the value TRUE if S has not had an event for the length of time T. Recall that if a signal is updated with a value that is different from its current value, then an event is said to have occurred on this signal. If T is not given, as in S'STABLE, then T is assumed to be 0 ns. S'STABLE will be TRUE if S has not had an event during the current simulation cycle.

S'QUIET(T) defines a BOOLEAN signal having the value TRUE if S has not had a transaction for the length of time T. If a signal is updated with a value that is the same or different from its current value, then a transaction is said to have occurred on this signal. Note well that an event is always a transaction, but a transaction does not always imply an event. If T is not given, as in S'QUIET, then T is assumed to be 0 ns. S'QUIET will be TRUE if S has not had a transaction during the current simulation cycle.

S'TRANSACTION defines a BIT signal that toggles each time a transaction occurs on S.

The remaining signal attributes are not signals. In general, you must always know which attributes are signals and which are not. Otherwise, you might misuse them and your VHDL models will not behave as expected.

S'EVENT defines a BOOLEAN value that is TRUE if an event has occurred on S during the current simulation cycle.

S'ACTIVE defines a BOOLEAN value that is TRUE if a transaction has occurred on S during the current simulation cycle.

S'LAST_EVENT refers to the amount of absolute time that has elapsed since S last had an event. There are two warnings to be made about this attribute. First and foremost is the fact that this attribute does not give you the time at which the signal S last had an event. Rather, it provides the amount of time that has passed since S last had an event. And second, you must also be aware that this attribute does not honor delta time units. If S last had an event 15 ns and 3 deltas ago, then S'LAST_EVENT will yield only the value 15 ns.

S'LAST_ACTIVE refers to the amount of absolute time that has elapsed since S last had a transaction. The warnings given for S'LAST_EVENT can be analogously given for this attribute.

S'LAST_VALUE refers to the previous value of S when S is a single scalar object. However, when S is an array, VHDL'87 has a very silly definition for this attribute. As per the VHDL'87 rules S'LAST_VALUE is determined element-wise. The net result of this definition is a value that has no practical use. Let me explain with an example. Suppose that the array signal S is defined to have a range of 0 to 3 and that its initial value is "0000". If at time 10 ns S became "0001", then at this point in the simulation S'LAST_VALUE would be equal to "0000" because the previous value of S(3) was '0'. Now suppose that at time 20 ns S would take on the value "0011". Because this attribute attacks the signal S element-wise, we will have that S'LAST_VALUE is equal to "0000". Note that the previous values of S(2) and S(3) have been merged together to derive this totally useless value "0000". But as per your modeling needs what you really wanted was the previous value of the array as a whole. The way around this problem in VHDL'87 is to save the previous value of S into a variable whenever S is updated. Fortunately, the VHDL'93 definition for S'LAST_VALUE will always yield the previous value regardless of whether or not S is an array.

S'DRIVING_VALUE is equal to the current driving value of the signal S. There are two caveats regarding this attribute. First, it is only available in processes, concurrent statements that have equivalent processes, and subprograms. And second, if the signal S is multi-sourced, then this attribute will not equal the signal's effective value as derived by the resolution function. Rather, this attribute will equal that single value that its encapsulating process is driving.

5.1 Signal Attribute Updates

To better understand how to use these signal attributes you must first learn about their updating characteristics. A good place to start is with their initial values. Suppose that the object Signal_A is declared as follows:

```
signal Signal_A : NATURAL := 5555;
```

As per the VHDL rules it is assumed that Signal_A has had this initial value of 5555 for an infinite amount of time prior to the start of the simulation. Figure 5.1 summarizes the initial values for the various attributes associated with Signal_A.

Since VHDL presumes that Signal_A had its initial value of 5555 for a very long time, then surely it definitely must have had this value 1 delta before the start of simulation. Consequently, the initial value of Signal_A'DELAYED is also 5555. Likewise, 25 ns before the initialization phase Signal_A was still 5555, and so Signal_A'DELAYED(25 ns) must also have the initial value 5555. The assignment of an initial value to Signal_A is not viewed as an event or as a transaction. Therefore at simulation startup Signal_A'STABLE and Signal_A'QUIET must both be TRUE, and Signal_A'EVENT and Signal_A'ACTIVE must both be FALSE. Moreover, Signal_A did not have an event or a

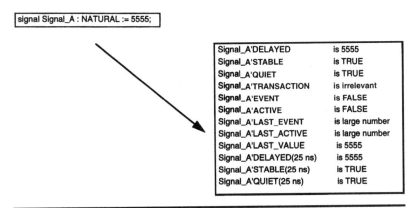

Figure 5.1 Example illustrating the initial signal attribute values.

transaction 25 ns before the initialization phase. Hence Signal_A'STABLE(25 ns) and Signal_A'QUIET(25 ns) must both be TRUE. Furthermore, Signal_A'LAST_VALUE will be 5555 since Signal_A is assumed to have never had an event in its past prior to the simulation startup. The initial value of Signal_A'TRANSACTION is not defined by the VHDL language. In fact, the LRM even states that a VHDL model must not be dependent on whether this initial value is a '0' or a '1'. All that really matters is the toggling of this signal once the initialization phase has transpired. The attribute Signal_A'LAST_EVENT is asking for the amount of time that has transpired since Signal_A last had an event. But Signal_A is assumed to have always been 5555. Hence Signal_A last had an event an infinite time ago. To capture a sense of infinity, this attribute is set to a very large number, the actual value of which is not specified by the VHDL language. Similarly, Signal_A'LAST_ACTIVE also has a very large number as its initial value.

The updating of these signal attributes depend on the update values assigned to Signal_A. Suppose that this Signal_A, which we had previously, initialized to 5555, is scheduled at time 0 ns to be updated via the following concurrent waveform signal assignment:

```
Signal_A <= 66 after 10 ns, 77 after 15 ns,
            88 after 60 ns, 88 after 100 ns;
```

Inspection of this waveform assignment to Signal_A shows that an event (and a transaction) will occur at 10, 15, and 60 ns. The assignment of 88 at 100 ns will be a transaction but not an event since Signal_A already will be 88 at that time. Figures 5.2 through 5.7 summarize the updating of Signal_A's signal attributes. Each figure contains the original waveform assignment and Signal_A's progressive update values for quick and easy reference. Let's begin with Figure 5.2, which focuses on the attributes 'DELAYED and 'DELAYED(25 ns). By the way, there is nothing special about 25 ns. Any other absolute time value could have been used instead.

The initial values on the 0 ns row have already been established. At 10 ns Signal_A has an event going from 5555 to 66. Both 25 ns and 1 delta time earlier Signal_A was still 5555. Consequently, Signal_A'DELAYED and Signal_A'DELAYED(25 ns) are both equal to 5555 on the 10 ns row. At 10 ns plus 1 delta Signal_A'DELAYED will take on the value that Signal_A had 1 delta time earlier at 10 ns. Consequently, Signal_A'DELAYED will become 66 at 10 ns plus 1 delta. But Signal_A'DELAYED(25 ns) will still be 5555 because 25 ns ago Signal_A had this value. Similar reasoning will yield that Signal_A'DELAYED will still be 66 at 15 ns, even though at this time Signal_A will be updated to 77. Signal_A'DELAYED(25 ns) will still be 5555 at this

Signal_A <= 66 after 10 ns, 77 after 15 ns, 88 after 60 ns, 88 after 100 ns;

Time	Signal_A	Signal_A'DELAYED	Signal_A'DELAYED(25 ns)
0 ns	5555	5555	5555
10 ns	66	5555	5555
10 ns + 1Δ	66	66	5555
15 ns	77	66	5555
15 ns + 1Δ	77	77	5555
35 ns	77	77	66
40 ns	77	77	77
60 ns	88	77	77
60 ns + 1Δ	88	88	77
85 ns	88	88	88
100 ns	88	88	88
100 ns + 1Δ	88	88	88
125 ns	88	88	88

Figure 5.2 Progressive updates of Signal_A'DELAYED and Signal_A'DELAYED(25 ns).

time. Then 1 delta time later, at 15 ns plus 1 delta, Signal_A'DELAYED will become 77 because 1 delta earlier (at 15 ns) Signal_A was 77. But Signal_A'DELAYED(25 ns) will still be 5555. Finally, at 35 ns Signal_A'DELAYED(25 ns) will become 66 because 35 ns is 25 ns after the time (10 ns) at which Signal_A went to 66. Similarly, Signal_A'DELAYED(25 ns) will become 77 at 40 ns, which is 25 ns after Signal_A went to 77. The remaining rows of this figure may now be analogously deduced. Inspection of the time durations will confirm that the 'DELAYED attribute is updated using the transport delay algorithm.

Figure 5.3 shows the progressive updating for Signal_A'STABLE and Signal_A'STABLE(25 ns). The 0 ns row displays the initial values that these two attributes take on. Keep in mind that Signal_A'STABLE is, in essence, asking if Signal_A has been stable (i.e., no event) during the current simulation cycle. Likewise, Signal_A'STABLE(25 ns) is asking if Signal_A has been stable for the last 25 ns. With this information in hand it follows that every time Signal_A has an event, Signal_A'STABLE and Signal_A'STABLE(25 ns) both become FALSE. If you do not feel comfortable with Signal_A'STABLE(25 ns) being FALSE whenever Signal_A has an event, then you should look at the situation this way. If Signal_A just had an event, then it is currently not stable, and if it currently is not stable, then surely it has not been stable for the inclusive, previous 25 ns. Another point to observe is that in this example Signal_A is kept constant for at least 1 delta after its events at 10, 15, and 60 ns. Consequently, Signal_A'STABLE will tran-

Signal_A <= 66 after 10 ns, 77 after 15 ns, 88 after 60 ns, 88 after 100 ns;

Time	Signal_A	Signal_A'STABLE	Signal_A'STABLE(25 ns)
0 ns	5555	TRUE	TRUE
10 ns	66	FALSE	FALSE
10 ns + 1Δ	66	TRUE	FALSE
15 ns	77	FALSE	FALSE
15 ns + 1Δ	77	TRUE	FALSE
35 ns	77	TRUE	FALSE
40 ns	77	TRUE	TRUE
60 ns	88	FALSE	FALSE
60 ns + 1Δ	88	TRUE	FALSE
85 ns	88	TRUE	TRUE
100 ns	88	TRUE	TRUE
100 ns + 1Δ	88	TRUE	TRUE
125 ns	88	TRUE	TRUE

Figure 5.3 Progressive updates of Signal_A'STABLE and Signal_A' STABLE(25 ns).

sition from FALSE to TRUE at 10 ns plus 1 delta, 15 ns plus 1 delta, and 60 ns plus 1 delta. The attribute Signal_A'STABLE(25 ns) will finally become TRUE at 40 ns and 85 ns since Signal_A did not have an event for the 25 ns duration following 15 ns and 60 ns, respectively.

Figure 5.4 displays the progressive updates of Signal_A' QUIET and Signal_A'QUIET(25 ns). Recall that 'QUIET deals with transactions, whereas 'STABLE zeros in on events. Therefore Signal_A'QUIET and Signal_A'QUIET(25 ns) will have the same value as Signal_A'STABLE and Signal_A'STABLE(25 ns) whenever Signal_A has an event. Differences will arise when the nonevent transaction occurs at 100 ns. At this point in simulation time both Signal_A'QUIET and Signal_A'QUIET(25 ns) will both become FALSE. One delta time later Signal_A'QUIET will return to TRUE, and 25 ns later Signal_A' QUIET(25 ns) will return to TRUE.

Figure 5.5 shows the update values for Signal_A'EVENT and Signal_A'ACTIVE. These two attributes will have the same values except when a nonevent transaction occurs. The two columns are identical up until 100 ns when 88 will be assigned to Signal_A. This assignment is a nonevent transaction because at that time Signal_A will already have this value. And so Signal_A'EVENT will remain FALSE but Signal_A'ACTIVE will become TRUE. Then 1 delta later Signal_A' ACTIVE will return to FALSE since Signal_A did not receive a transaction at that simulation time.

Figure 5.6 shows the update values for Signal_A'LAST_EVENT and Signal_A'LAST_ACTIVE. The 0 ns row contains the infinity symbol for

Signal_A <= 66 after 10 ns, 77 after 15 ns, 88 after 60 ns, 88 after 100 ns;

Time	Signal_A	Signal_A'QUIET	Signal_A'QUIET(25 ns)
0 ns	5555	TRUE	TRUE
10 ns	66	FALSE	FALSE
10 ns + 1Δ	66	TRUE	FALSE
15 ns	77	FALSE	FALSE
15 ns + 1Δ	77	TRUE	FALSE
35 ns	77	TRUE	FALSE
40 ns	77	TRUE	TRUE
60 ns	88	FALSE	FALSE
60 ns + 1Δ	88	TRUE	FALSE
85 ns	88	TRUE	TRUE
100 ns	88	FALSE	FALSE
100 ns + 1Δ	88	TRUE	FALSE
125 ns	88	TRUE	TRUE

Figure 5.4 Progressive updates of Signal_A'QUIET and Signal_A'QUIET(25 ns)

Signal_A <= 66 after 10 ns, 77 after 15 ns, 88 after 60 ns, 88 after 100 ns;

Time	Signal_A	Signal_A'EVENT	Signal_A'ACTIVE
0 ns	5555	FALSE	FALSE
10 ns	66	TRUE	TRUE
10 ns + 1Δ	66	FALSE	FALSE
15 ns	77	TRUE	TRUE
15 ns + 1Δ	77	FALSE	FALSE
35 ns	77	FALSE	FALSE
40 ns	77	FALSE	FALSE
60 ns	88	TRUE	TRUE
60 ns + 1Δ	88	FALSE	FALSE
85 ns	88	FALSE	FALSE
100 ns	88	FALSE	TRUE
100 ns + 1Δ	88	FALSE	FALSE
125 ns	88	FALSE	FALSE

Figure 5.5 Progressive updates of Signal_A'EVENT and Signal_A'ACTIVE.

Signal_A <= 66 after 10 ns, 77 after 15 ns, 88 after 60 ns, 88 after 100 ns;

Time	Signal_A	Signal_A'LAST_EVENT	Signal_A'LAST_ACTIVE
0 ns	5555	∞	∞
10 ns	66	0 ns	0 ns
10 ns + 1Δ	66	0 ns	0 ns
15 ns	77	0 ns	0 ns
15 ns + 1Δ	77	0 ns	0 ns
35 ns	77	20 ns	20 ns
40 ns	77	25 ns	25 ns
60 ns	88	0 ns	0 ns
60 ns + 1Δ	88	0 ns	0 ns
85 ns	88	25 ns	25 ns
100 ns	88	40 ns	0 ns
100 ns + 1Δ	88	40 ns	0 ns
125 ns	88	65 ns	25 ns

Figure 5.6 Progressive updates of Signal_A'LAST_EVENT and Signal_A'LAST_ACTIVE.

both of these attributes. This notation symbolizes that, as per the VHDL language rules, these two attributes must have very large initial values. The remaining values are fairly straightforward, but you must always remember that both these attributes are asking for the amount of time that has elapsed. And so Signal_A'LAST_EVENT is concerned with the duration of time that has passed since Signal_A last had an event. The only tricky point here is that whenever Signal_A just had an event, then 0 ns have elapsed from the time that this event occurred. That is why 0 ns is listed at 10, 15, and 60 ns. Note that delta time units are not taken into consideration. This attribute only provides absolute simulation times. As for Signal_A'LAST_ACTIVE, it refers to the amount of time that has passed since Signal_A last had a transaction, and this attribute will be identical to Signal_A'LAST_EVENT until a nonevent transaction occurs. This is precisely what will happen at 100 ns and beyond.

Figure 5.7 shows the update values for Signal_A'TRANSACTION and Signal_A'LAST_VALUE. Let's deal with the former one first. As per this figure the BIT-valued Signal_A'TRANSACTION was initialized to '1' on the simulator that I was working with. As stated earlier this initial value could well have been '0' since VHDL does not specify an initial value for this attribute. In any case what is important is that Signal_A'TRANSACTION will toggle every time Signal_A has a transaction. Hence toggling can be observed at 10, 15, 60, and 100 ns. As for Signal_A'LAST_VALUE, this attribute will correspond to the previous value of Signal_A. Inspection of Fig. 5.7 will confirm this behavior.

Signal_A <= 66 after 10 ns, 77 after 15 ns, 88 after 60 ns, 88 after 100 ns;

Time	Signal_A	Signal_A'TRANSACTION	Signal_A'LAST_VALUE
0 ns	5555	'1'	5555
10 ns	66	'0'	5555
10 ns + 1Δ	66	'0'	5555
15 ns	77	'1'	66
15 ns + 1Δ	77	'1'	66
35 ns	77	'1'	66
40 ns	77	'1'	66
60 ns	88	'0'	77
60 ns + 1Δ	88	'0'	77
85 ns	88	'0'	77
100 ns	88	'1'	77
100 ns + 1Δ	88	'1'	77
125 ns	88	'1'	77

Figure 5.7 Progressive updates of Signal_A'TRANSACTION and Signal_A' LAST_VALUE.

5.2 Application of Signal Attributes

As a general rule you should use 'EVENT in a sequential region instead of 'STABLE since the former is not a signal, whereas the latter is. As for concurrent constructs, all I can recommend is that extensive experimentation should be carried out to determine which of these two attributes should be applied.

In a sequential region 'EVENT can be used as follows to identify a leading edge:

```
if Clock = '1' and Clock'EVENT then
```

This coding scenario works just fine when Clock is of type BIT. However if Clock would be any other data type, then the above if condition must be extended to include the following extra attribute qualification:

```
if Clock = '1' and Clock'LAST_VALUE = '0' and Clock'EVENT then
```

Another application of 'EVENT is to control the logic flow after a wait statement. Suppose that a wait construct is waiting for an event on any of several signals via the statement

```
wait on Signal_A, Signal_B;
```

Suppose that an event occurs on either Signal_A or Signal_B. The encapsulating process will be awakened and then will continue executing the first line following this wait statement. But what if your model must execute different code segments depending on which specific signal had the event? The attribute 'EVENT may then be applied, as follows, to distinguish between these two code segment options:

```
if Signal_A'EVENT then
    --- sequential statements
end if;


if Signal_B'EVENT then
    --- sequential statements
end if;
```

This solution works fine except when Signal_A and Signal_B both have an event during the same simulation cycle. Both code segments will then be executed. If this is how you want your model to behave, then all is well. But if you wish to execute only one of these two code segments, then you must use an if...then...else construct to prioritize the choices. Here is how it can be done with Signal_A having the higher priority:

```
if Signal_A'EVENT then
    --- sequential statements
else
    if Signal_B'EVENT then
        --- sequential statements
    end if;
end if;
```

This prioritization technique may also be applied to write a VHDL model for a device that has an asynchronous reset. Observe that in Fig. 5.8 Reset'EVENT is omitted since, irrespective of Clock's leading edge, the modeled device is considered to be in a reset state anytime that the signal Reset is active (= '1').

While on the topic of resets let's look at two other reset-related scenarios. Figure 5.9 illustrates a VHDL model for a synchronous reset environment.

Figure 5.10 returns to the topic of asynchronous resets but in a more complex environment. The illustrated model has numerous wait statements, each of which is waiting for events on several signals. Since the modeled device has an asynchronous Reset, anytime a wait condition occurs the Reset signal must also be included. Once Reset goes active

```
process (Clock, Reset)
    -- Variable declarations
begin
    -- NOTE that Reset'EVENT is deliberately omitted
    if Reset = '1' then
        -- Reset appropriate variables and signals
    else
        if Clock = '1' and Clock'EVENT then
            -- Sequential statements that capture  the functionality/purpose of
            -- this process
        end if;
    end if;
end process;
```

Figure 5.8 Modeling an asynchronous reset.

```
process
    -- Variable declarations
begin
    wait until Clock = '1';
    if Reset = '1' then
        -- Reset appropriate variables and signals
    else
        -- Sequential statements that capture  the functionality/purpose of
        -- this process
    end if;
end process;
```

Figure 5.9 Modeling a synchronous reset.

the model should stop executing the regular code and, instead, jump to a neutral region where the pertinent variables and signal can be reset. The key trick to achieving this jumping maneuver is to encapsulate the main portion of the code into an infinite loop. This loop is then selectively exited whenever the signal Reset has an event. There is no need to confirm a reset value of '1' since Main_Loop will only be entered when Reset takes on its inactive state of '0'.

The best place to use the new VHDL'93 attribute 'DRIVING_VALUE is in the coding scenario for which it was originally intended to support. Suppose that Clock is declared to be a port of mode out. Hence Clock can only appear on the left-hand side of a signal assignment. But suppose that you wish to write:

```
Clock <= not Clock;
```

```
process
    -- Variable declarations
begin

    if Reset = '1' then
        -- Reset appropriate variables and signals
        wait on Reset;   -- Continue when Reset = '0'
    end if;

    Main_Loop:
    loop
        -- Sequential statements excluding wait
        wait on A, B, C, Reset;
        exit Main_Loop when Reset'EVENT;
            -- Sequential statements excluding wait

        wait on D, E, Reset;
        exit Main_Loop when Reset'EVENT;
            -- Sequential statements excluding wait
    end loop Main_Loop;

end process;
```

Figure 5.10 Modeling a complex asynchronous reset environment.

Because the out port Clock is on the right-hand side of this assignment statement, you will get a compilation error. Quick and easy fixes to this problem may be achieved by declaring Clock to be either a buffer or an inout port. Both these solutions have weaknesses that are discussed in a later part of this book. The VHDL'93 solution is to keep Clock as an out port and to write:

```
Clock <= not Clock'DRIVING_VALUE;
```

'TRANSACTION is one of my favorite attributes. It is a valuable gem in my bag of VHDL tricks. Typically, it is used to identify a transaction on a real (as in hardware real) or on an imaginary (as in cute trick) signal as follows:

```
wait on Signal_A'TRANSACTION;
```

By the way, on the surface it appears as if either 'ACTIVE or 'QUIET may be substituted for 'TRANSACTION in this wait statement. However both are inappropriate. Whereas Signal_A'TRANSACTION is a signal, Signal_A'ACTIVE is not. Since the objects listed in a wait on construct must be signals, you will get a compile error if you were to write

```
wait on Signal_A'ACTIVE; -- **** COMPILE ERROR
```

As for Signal_A'QUIET, though it is a signal, it is still inappropriate because of its defined behavior. Suppose that Signal_A had a transaction at time 25 ns. Signal_A'QUIET would then transition from TRUE to FALSE and as a result the following wait statement would be triggered, which by the way, is what we want:

```
wait on Signal_A'QUIET; -- **** SIMULATION ERROR
```

However, at 25 ns plus 1 delta, assuming that Signal_A remained stable, the attribute Signal_A'QUIET would have an event going from FALSE to TRUE. The previous wait statement would then once again be triggered, despite the fact that Signal_A did not receive another transaction. This false triggering will play havoc with your VHDL model and might be difficult to debug. However, it is feasible to write:

```
wait until Signal_A'QUIET = FALSE; -- **** SIMULATION INEFFICIENCY
```

But as the comment suggests, this approach is inefficient. The inefficiency comes about because, as shown earlier, every transaction on Signal_A will cause the encapsulating process to be awakened twice, only one of which will satisfy the desired condition. The first awakening will occur at the time of the transaction (condition will be satisfied), and the second one will occur 1 delta time later (condition will not be satisfied).

VHDL's signal attributes may also be used to compactly capture and model a complex verbal description. Suppose that you wanted a model to wait for an event on the BIT-type signal called Signal_A. When this event occurs you then want to confirm that Signal_A just had a leading edge and that the previous '0' pulse width lasted for at least 40 ns prior to this current change to '1'. This detailed description may be modeled via the following wait statement:

```
wait on Signal_A until Signal_A = '1'
            and Signal_A'DELAYED'STABLE(40 ns);
```

Before dissecting this solution observe that because Signal_A'DELAYED is a signal any of VHDL's signal attributes may be applied to it. This technique is true in general and can go on indefinitely as long as the attribute prefix is a signal. Based on our earlier discussions it should come as no surprise that Signal_A's leading edge can be identified via

```
wait on Signal_A until Signal_A = '1' ......
```

But what is Signal_A'DELAYED'STABLE(40 ns) trying to achieve? Let's explore this expression in a piecemeal fashion. Recall that Signal_A'DELAYED has a waveform that is identical to Signal_A's waveform except that it is delayed by 1 delta time unit. If Signal_A has a leading edge during the current simulation cycle, then 1 delta time earlier Signal_A had the value '0'. So whenever Signal_A experiences a leading edge Signal_A'DELAYED will be equal to '0'. By applying 'STABLE(40 ns) to Signal_A'DELAYED we are in effect asking if the signal Signal_A'DELAYED was stable for at least 40 ns. In other words, when Signal_A has a leading edge the expression Signal_A'DELAYED'STABLE(40 ns) is checking that Signal_A had its previous value of '0' for at least 40 ns, which as per our previous verbal description, is exactly what we want. Incidentally, another equivalent condition that may have been used is:

```
wait on Signal_A until Signal_A = '1'
            and Signal_A'DELAYED'LAST_EVENT >= 40 ns;
```

The advantage of using 'LAST_EVENT instead of 'STABLE will be shown in a later example.

VHDL's predefined signal attributes may also be applied to check for setup and hold timing violations. Figure 5.11 shows a standard setup and hold timing diagram that, I am certain, you are already familiar with.

There are many VHDL modeling styles that you can use to verify these timing conditions. You can write a process, a subprogram, or a concurrent assert statement. Though the process method is more simulation efficient, I will instead apply concurrent assert statements since they require more usage of VHDL's predefined signal attributes. Let's begin with a quick lesson on the concurrent assert statement.

The sequential version of the assert statement was already discussed in the Excursion. The concurrent version exists by itself in a VHDL concurrent region. In other words, a concurrent assert statement cannot exist inside of either a process or a subprogram. A concurrent assert statement will be activated whenever any of the signals in its sensitivity list has an event. If the resulting assert condition is TRUE, then nothing will happen. But if it is FALSE then the assert's report string

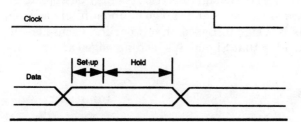

Figure 5.11 Setup and hold timing diagram.

will be printed onto the screen, and the user-specified severity level will be recorded by the simulation engine. This severity level could potentially halt the simulation.

A setup timing violation will be identified by the following concurrent assert statement where SET_UP_TIME is either a declared constant or a generic parameter:

```
assert (Clock = '0') or (not Clock'EVENT)
           or (Data'DELAYED'LAST_EVENT >= SET_UP_TIME)
    report "Set-up violation observed"
    severity ERROR;
```

The easiest way to understand how this concurrent assert statement accomplishes its objective is to apply Demorgan's law to the assert condition. Remember that Demorgan's law states that

```
not (A or B) = (not A) and (not B)
```

Therefore a setup violation will be observed when either Clock or Data'DELAYED have an event and the following conditions are all TRUE:

- Clock is equal to '1'.
- Clock has just had an event.
- Data'DELAYED has changed within SET_UP_TIME.

Extensive simulation has shown that using Data'DELAYED' LAST_EVENT instead of just Data'LAST_EVENT will allow for the correct handling of the situation when Clock and Data are simultaneously updated, even at time 0 ns plus 1 delta. The possibility of these extreme predicaments occurring in the real world are pretty slim. And so you may be able to get away with using just Data'LAST_EVENT instead of the more complex expression Data'DELAYED'LAST_ EVENT.

A hold timing violation will be identified by the following concurrent assert statement where HOLD_TIME is either a declared constant or a generic parameter:

```
assert (Clock = '0') or (Clock'DELAYED'LAST_EVENT >= HOLD_TIME)
           or (Data'STABLE)
    report "Hold violation observed"
    severity ERROR;
```

Here again Demorgan's law may be applied to better understand the validity of this concurrent assert statement. Whenever Clock, Clock-

'DELAYED, or Data'STABLE has an event, a hold violation will be reported if the following conditions are all TRUE:

- Clock is '1'.
- Clock'DELAYED has changed with HOLD_TIME.
- Data has just had an event.

Analogous to the previous setup violation, using Clock'DELAYED' LAST_EVENT instead of Clock'LAST_EVENT allows for the correct handling of the situation when Clock and Data are simultaneously updated, even at 0 ns plus delta. Here again the reality of either of these situations is questionable.

Extensive simulation has shown that this given hold time assertion check will erroneously fire when the following conditions are all true:

- The '0' level pulse width of Clock is less than HOLD_TIME.
- Data has an event during the same simulation cycle that Clock changes to '1'.

Hence, when conducting error checks at the hold time granularity level, it is also required to include minimum pulse width assertion monitors. A pulse-width timing violation will be identified by the following concurrent assert statement where MINIMUM_PULSE_WIDTH is either a declared constant or a generic parameter:

```
assert (Clock'STABLE) or
       (Clock'DELAYED'LAST_EVENT >= MINIMUM_PULSE_WIDTH)
       report "Minimum pulse width violation observed"
       severity ERROR;
```

Here again Demorgan's law may be applied to better understand the validity of this concurrent assert statement. Whenever Clock'STABLE or Clock'DELAYED has an event, a pulse width violation will be reported if the following conditions are all TRUE:

- Clock just had an event.
- Clock'DELAYED has changed within MINIMUM_PULSE_WIDTH.

In this situation there is no choice in regards to referencing Clock'DELAYED'LAST_EVENT instead of just simply Clock'LAST_EVENT. Suppose that Clock just had an event. The attribute Clock'LAST_EVENT would then be 0 ns, and the comparison to MINIMUM_PULSE_WIDTH would be false. A false minimum pulse width violation would then be reported.

While on the topic of setup and hold checks, let me show you another trick. In essence, all the check times have been fixed. But what if we would like a VHDL model to support a min-max-typical timing environment? This modeling granularity can be achieved via the following methodology. First, you will need the following declarations:

```
type Delay_Type is (MIN, MAX, Typical);
type Delay_Type_Array is array (Delay_Type) of TIME;
constant SET_UP_TIME : Delay_Type_Array := (5 ns, 15 ns, 10 ns);
```

Note that the enumeration literal MAX may be applied as an index to the array SET_UP_TIME to yield the user-specified maximum time of 15 ns (SET_UP_TIME(MAX) = 15 ns). If you wish even more flexibility, you could have defined the constant SET_UP_TIME values via a function call that will get the timing data from an input file. This way you will not have to recompile your model whenever the min-max-typical times are redefined. For instance, a user-designed function called Get_Times could be referenced as follows, where Data_File is the name of the input file containing the min-max-typical timings:

```
constant SET_UP_TIME : Delay_Type_Array := Get_Times(Data_File);
```

The following concurrent assert construct incorporates a min-max-typical option when checking for a set-up timing violation. The identifier CURRENT_DELAY_TIME is a generic parameter.

```
assert (Clock = '0') or (not Clock'EVENT)
        or (Data'DELAYED'LAST_EVENT >=
              SET_UP_TIME(CURRENT_DELAY_TIME))
    report "Set-up violation observed"
    severity ERROR;
```

Incidentally, it might be tempting to replace the test Data'DELAYED'LAST_EVENT >= SET_UP_TIME(CURRENT_DELAY_TIME) with Data'DELAYED'STABLE(SET_UP_TIME(CURRENT_DELAY_TIME)). But there is a major VHDL'87 portability issue here. VHDL'87 is very vague about allowing such a construct. Consequently, some VHDL tool suites allow it while others do not. This ambiguity has been amended in VHDL'93 so that such an application of 'STABLE must be supported by all VHDL'93 compliant tools.

Part

3

VHDL Techniques and Recommendations

Chapter 6

Compilation Caveats

The purpose of this chapter is to enhance your VHDL problem-solving skills by exposing and solving a broad spectrum of nontrivial compilation errors that you will eventually encounter.

Error. Case selector expression contained the & operator as shown in the following code segment:

```
case A & B & C & D is -- ERROR

    when "0000" =>
```

This error comes about because the & operator really is a function that returns an unconstrained array, and the case selector expression cannot be an unconstrained array. Test_162 confirms the compilation failure of this coding scenario.

For the same reason the following will also not compile where both W and V are declared as BIT_VECTOR(0 to 1):

```
case W or V is -- ERROR

    when "00" =>
```

Here again, the or logical operator is a function that returns an unconstrained array. However, the following will compile since the or operator applied to the single BITs J and K returns a single BIT that is not an array, so the constrained issue does not even come into play:

```
case J or K is -- OK

    when '0' =>
```

Recommended solution. Declare an appropriate subtype and then use it to qualify the case selector expression as in the following:

```
subtype Nibble is BIT_VECTOR(0 to 3);
```

....

```
case Nibble'(A & B & C & D) is
```

In general, type qualification should be attempted whenever the compiler seems confused about an object's data type.

Incidentally, since BIT_VECTOR(0 to 3) is not a data type, the following attempted qualification will not compile:

```
case BIT_VECTOR(0 to 3)'(A & B & C & D)  -- ERROR
```

Error. In a VHDL'87 environment the case selector expression contained an array slice.

```
case Instruction_Register(31 downto 26) is  -- ERROR in VHDL'87
    when "0000" =>
```

Note well that this is an error only in VHDL'87.

Recommended solution. Here again, type qualification will solve the problem.

```
subtype Opcode is BIT_VECTOR(31 downto 26);
```

....

```
case Opcode'(Instruction_Register(31 downto 26)) is  -- OK
```

Incidentally, VHDL'93 accepts array slices in the case selector expression.

```
case Instruction_Register(31 downto 26) is  -- OK in VHDL'93
```

Error. Double quoted string was used as the actual to the overloaded TEXTIO write procedure.

```
write (Current_Line, "HEADER FOR OUTPUT DATA");  -- ERROR
```

The peculiarities of this error have already been discussed in the Excursion segment (Sec. 2.17) of this book.

Recommended solution. Here again, type qualification is required to direct the compiler in overcoming its ambiguity with regards to the data type of the literal between the double quotes.

```
write (Current_Line, STRING'("HEADER FOR OUTPUT DATA"));
```

Error. The actual in a port map association was not a signal in a VHDL'87 environment.

```
Instance_1: Component_1
        port map (Data_In => '0',  -- ERROR in VHDL'87
```

As indicated by the comment this is only an error in VHDL'87.

Recommended solution. VHDL'87 requires the following silly workaround:

```
signal Ground : BIT := '0';
    . . . . . . . . . . . . . .
port map (Data_In => Ground,  -- OK
```

This restriction is very annoying, and, fortunately VHDL'93 does allow ports of mode in to be associated with a constant literal.

```
port map (Data_In => '0',  -- OK in VHDL'93
```

Error. Interpretation of identifiers defined in a package was unintentionally overridden.

This scope and visibility error is addressed in Test_158 and Test_190 of the Experiments segment of this book.

Error. An out port was read (process version).

During the Excursion I pointed out several times that a port of mode out cannot be read. There are several coding scenarios in which this error may occur. The signal assignment version is illustrated in Fig. 6.1.

This compilation error has two quick and easy fixes, but they are both unsatisfactory in the long run. Clock's out mode may be replaced by either the mode inout or buffer.

The problem with such an artificial usage of an inout port is explored in the Experiments segment of this book via Test_195.

The mode buffer solution has many hierarchical pitfalls. Buffers may only be directly connected to explicitly declared signals or to other buffer ports. Such a restriction is very unsatisfactory because some-

```
entity Clock is
    port (Clock : out BIT);
end Clock;

architecture Behave_1 of Clock is
begin
    process
    begin
        wait for 50 ns;
        Clock <= not Clock;    -- **** ERROR
    end process;
end Behave_1;
```

Figure 6.1 Signal assignment version of the VHDL fact that an out port cannot be read.

where along the hierarchical chain it will eventually be desirable to directly connect a buffer port to an out port as shown in Fig. 6.2.

Incidentally, to say that an out port cannot be read also implies that an out port's value cannot be examined as per the following code segment, where Dataout is a port of mode out:

```
if Dataout = '1' then -- ERROR
```

Recommended solution. The preferred VHDL'87 solution to the signal assignment version of this compilation error is to use an intermediate variable as was done in the Excursion for the design entity named Clock. An even better solution exists in VHDL'93. This VHDL update contains a new attribute specifically created to solve this problem of reading an out port. The VHDL'93 solution allows Clock to be declared with an out mode. The assignment to Clock can now be achieved via the new VHDL'93 attribute 'DRIVING_VALUE as follows:

```
Clock <= not Clock'DRIVING_VALUE;  -- OK in VHDL'93
```

This new attribute may also be used to solve the out port examination problem as follows:

```
if Dataout'DRIVING_VALUE = '1' then -- OK in VHDL'93.
```

Error. An out port was read (component instantiation version).

Figure 6.3 illustrates the component instantiation version of the inability to read an out port. Since the component declaration specified Bin as an in port, it cannot be directly connected to the out port Q_Bar. Technically speaking, the formal out port Q_Bar cannot be associated as an actual parameter to the local in port parameter Bin.

Compilation Caveats   213

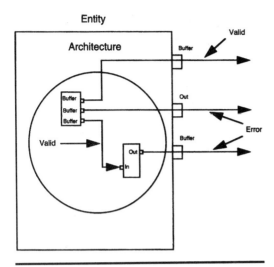

Figure 6.2 Hierarchical weaknesses of the mode buffer.

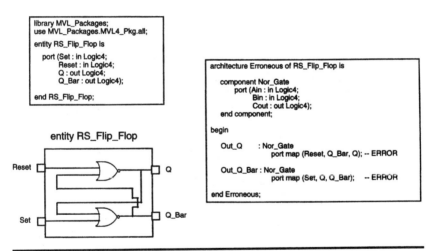

Figure 6.3 Component instantiation version of the VHDL restriction that an out port cannot be read.

Recommended solution. Figure 6.4 shows how intermediate signals may be used to resolve the component instantiation version of the inability to read an out port. Note how these intermediate signals are used as actual parameters in the port mapping construct and also how their respective updates are propagated to the appropriate out port via concurrent signal assignments.

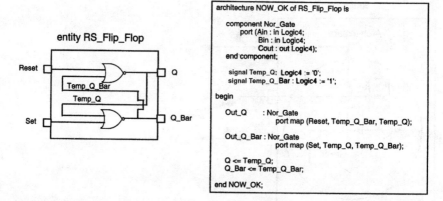

Figure 6.4 Use of intermediate signals to solve the component instantiation version of the VHDL restriction that an out port cannot be read.

Error. Subprogram conformance rules not honored.

For instance, suppose that a package specification contains the following subprogram declaration:

```
procedure Proc (A, B : INTEGER);
```

Consequently, this package's body must contain a procedure body for Proc or else there will be a compilation error. So far this requirement seems quite reasonable. Then the subprogram conformance rules take over and stipulate that the procedure Proc in the package's body must have the same lexical sequence of elements as given in the procedure's declaration. And so, as strange as it may seem, the following statements will all result in a compilation error:

```
procedure Proc (A: INTEGER; B: INTEGER) is          -- ERROR

procedure Proc (constant A, B : INTEGER) is         -- ERROR

procedure Proc (A, B : in INTEGER) is               -- ERROR

procedure Proc (constant A, B : in INTEGER) is      -- ERROR
```

There is only one correct choice for Proc's body header, and it is the one that exactly reproduces the original lexical sequence given in the procedure's declaration.

```
procedure Proc (A, B : INTEGER) is -- OK
```

An interesting (strange may be a better word for it) point about this subprogram conformance rule is that the above procedure headings for

Proc all provide the same information, even though the VHDL compiler will not view them as being equivalent: A subprogram parameter without a user-specified mode defaults to mode in; furthermore, a subprogram parameter of mode in but without a user-specified object class defaults to the class constant.

Incidentally, while on the topic of default subprogram parameter profiles, let me add that a procedure parameter of mode out or inout that is not assigned an object class by the VHDL modeler will be assumed to be of class variable.

Returning to the main theme of subprogram conformance errors it can be seen that in VHDL the editorial shortcut of cutting and pasting from a subprogram's declaration to its body has some theoretical merits as well.

Error. A subtype and its parent type were used to distinguish overloaded subprograms.

The following two declarations will result in a compilation error:

```
function To_Bit (Input : INTEGER; Bit_Length : NATURAL)
                              return BIT_VECTOR;

function To_Bit (Input : NATURAL; Bit_Length : NATURAL)
                              return BIT_VECTOR;
```

In general, the compiler must somehow be able to completely identify which overloaded function is being called. Since NATURAL is a subtype of INTEGER, the above two functions cannot be distinguished.

Recommended solution. The only solution to this problem is to not overload the function name To_Bit as shown earlier. Instead, two different names must be used as in:

```
function Integer_To_Bit (Input : INTEGER; Bit_Length : NATURAL)
                              return BIT_VECTOR;

function To_Bit (Input : NATURAL; Bit_Length : NATURAL)
                              return BIT_VECTOR;
```

Incidentally, the following criteria are used by compilers to distinguish between overloaded subprograms:

- Number of parameters.
- Base type of each parameter.
- The return type if the subprogram is a function.

Since this list does not include a parameter's class, the following two subprogram declarations cannot be distinguished by the compiler and hence an error will occur:

```
function To_Bit (constant Input : NATURAL; Bit_Length : NATURAL)
                                       return BIT_VECTOR;

function To_Bit (signal Input : NATURAL; Bit_Length : NATURAL)
                                       return BIT_VECTOR;
```

Error. A wait statement occurred inside a function.

But wait statements can appear in a procedure. This fact implies that you must be careful not to call a procedure having a wait statement from either a function (since functions cannot have any wait statement) or from a process having a static sensitivity list (process with static sensitivity list cannot contain an explicit wait statement).

Error. The concatenation operator & appeared on the left-hand side of an assignment.

Though the following code segment has good intentions, it will not compile since the concatenation operator & is a function and hence cannot be written to:

```
Scalar_Value & Vector_Value(1 to 7) <= Data;   -- ERROR
```

Recommended solution. Suppose that Data is an array having the range 0 to 7, then the aforementioned error may be corrected by explicitly writing out the intended two assignments that the concatenation operation hoped to abbreviate.

```
Scalar <= Data(0);

Vector_Value(1 to 7) <= Data(1 to 7);
```

Error. The actual parameter corresponding to a formal out subprogram parameter contained the & operator.

This is a variant of the previous error in which an & expression cannot be the target of an assignment.

Consider the following procedure declaration:

```
procedure Booth_Mult (L_V : in BIT_VECTOR;
                      R_V : in BIT_VECTOR;
                      Result : out BIT_VECTOR);
```

The & operator can only appear in those actual expressions that are associated with formal parameters having the in direction. This fact is expressed in the comments of the following erroneous call to this given procedure:

```
Booth_Mult (Value_A & Value_B,            -- OK
            Value_C & Value_B,            -- OK
            Product_High & Product_Low);  -- ERROR
```

Recommended solution. Call the procedure with a temporary array variable. When the procedure returns to the caller, you should then reference slices of this variable to make individual assignments to Product_High and Product_Low.

Error. The & operator was used in an expression that served as an actual parameter in a port map association.

On the surface this error seems similar to our previous two concatenation errors but, in actuality, the problem is a bit more involved. The VHDL syntax requires that all functions in a port map association have only one input parameter and that it be of the form F(x). These requirements are not supported in the following port map association:

```
port map (Value_A & Value_B,  -- ERROR because & function has two inputs
          Value_C & Value_D,  -- ERROR because & function has two inputs
          not Value_E);       -- ERROR since this usage of the logical
                              -- operator not does not satisfy the
                              -- requirement that functions in a port
                              -- map association be of the form F(X).
```

Recommended solution. Suppose that the signals Value_A, Value_B, Value_C, and Value_D are all 4 bits wide and that the original port declaration is:

```
port (Data_Out : out BIT_VECTOR(0 to 7);
      Data_In_1 : in BIT_VECTOR(0 to 7);
      Data_In_2 : in BIT_VECTOR(0 to 7));
```

Then slices of the formal parameters may be used as follows instead of the & operator:

```
port map (Data_Out(0 to 3)  => Value_A,
          Data_Out(4 to 7)  => Value_B,
          Data_In_1(0 to 3) => Value_C,
          Data_In_1(4 to 7) => Value_D,
          Data_In_2         => Value_E);
```

As for Value_E, instead of trying to negate it in the port map construct, it is better to merely accept it as is and then subsequently negate it inside the design entity that corresponds to this specific instantiation. If such a technique is not feasible, then you should use the concurrent signal assignment

```
Value_E_Not <= not Value_E;
```

In this case Value_E_Not should be referenced in the port map association instead of Value_E.

One final note regarding the previous usage of slices. VHDL requires these slices to be contiguous. Hence it is incorrect to write:

```
port map (Data_Out(0 to 3)   => Value_A,
          Data_In_1(4 to 7)  => Value_B,
          Data_In_1(0 to 3)  => Value_C,
          Data_Out(4 to 7)   => Value_D,
          Data_In_2          => Value_E);
```

Error. The prefix of a signal's attribute was not a static name.

An example of this annoying restriction is the erroneous usage of Signal_A(K)'EVENT in a for loop to iterate through the array Signal_A in order to determine which of its individual elements just had an event.

```
process (Signal_A)
begin
    for K in Signal_A'RANGE loop
        if Signal_A(K)'EVENT then  -- ERROR
            -- sequential code
```

Incidentally, testing for individual transactions via Signal_A(K)'ACTIVE is another coding scenario that is not accepted by VHDL.

Recommended solution. Determining which element just had an event can be achieved by manually keeping track of the previous value and then comparing it element-wise to the current updated array. Let's assume that Signal_A is 32 bits wide and was initialized to all '0'. Note well that for the following to work Signal_A and Previous_Value must both have the same initial value:

```
process (Signal_A)
    variable Previous_Value : BIT_VECTOR(31 downto 0);
begin
    for K in Signal_A'RANGE loop
        if Signal_A(K) /= Previous_Value(K) then  -- 'EVENT identified
            -- sequential code
        end if;
    end loop;
    Previous_Value := Signal_A;
    -- more sequential code
end process;
```

Unfortunately, this comparison trick does not port over to circumvent the Signal_A(K)'ACTIVE restriction. In this case "shadow registers" may be used to toggle an array subelement whenever its corresponding Signal_A subelement is scheduled to receive a value. Since toggling a bit creates an event, the aforementioned test can then be used to determine which bit toggled and, consequently, to deduce which Signal_A element had a transaction. A concrete example showing a real-world need for knowing the value of Signal_A(K)'ACTIVE is given in the Experiment segment of this book via Test_35b_1. Possible solutions to circumvent the resulting compilation error are then provided in full detail in Test_35j, Test_35k, and Test_40.

Error. Reserved words were referenced as enumeration literals.

Though you are never going to declare a signal with the name *begin*, this type of error might sneak up on you when you are creating an enumeration list of instructions. For example, since *AND* is a VHDL reserved word, the following enumeration type declaration is in error:

```
type Instructions is (ADD, LOAD, STORE, AND); -- ERROR
```

Recommended solution. In VHDL'87 the only solution is to modify the name somewhat as in:

```
type Instructions is (ADD, LOAD, STORE, CPU_AND); -- OK
```

A cleaner solution exists in VHDL'93 that applies the back slash notation to delimit previously nonallowable VHDL'87 identifiers (names). Hence the enumeration type declaration may now be written as:

```
type Instructions is (ADD, LOAD, STORE, \AND\); -- OK
```

While on the topic of legal identifiers let me point out the VHDL'87 and VHDL'93 rules. In VHDL'87 an identifier has to be a sequence of characters that satisfies the following rules:

- Every character is either a
  - Letter (upper or lower case)
  - Digit (0 to 9)
  - Underscore
- Adjacent underscores are not permitted.
- Last character cannot be an underscore.
- Blanks cannot be used.
- First character must be a letter.
- Characters cannot spell out a reserved word.

A key qualification has to be made with regards to the term *character*. In VHDL'87 the enumeration type CHARACTER consists of the 128-character set based on 7 bits. But in VHDL'93 CHARACTER has been expanded to 256 elements and is based on 8 bits. Identifiers without back slashes must still honor the above VHDL'87 rules. However, if back slashes are used, as previously shown, then the VHDL'87 rules no longer apply to those CHARACTERS that are enclosed by them.

Error. The case of an enumerated character literal was not honored. Consider the following enumeration type declaration.

```
type Logic4 is ('X', '0', '1', 'Z');
```

Because the upper case character literal 'X' was listed in the enumeration declaration, the following lower case assignment is in error, where Data_Logic4 is a variable of type Logic4:

```
Data_Logic4 := 'x'; -- ERROR
```

Recommended solution. The obvious solution is to use the correct upper case 'X'. While on the topic of case sensitivity recall that VHDL is case insensitive and so we have that LOAD and Load are the same. However, the VHDL'93 back slash notation introduced in the previous example is case sensitive. Consequently, \AND\ is not the same as \And\.

Error. An unconstrained array was referenced as an array's element.

```
type Illegal is array (NATURAL range 0 to 255)
                        of BIT_VECTOR; -- ERROR
```

In this example the problem is that BIT_VECTOR is an unconstrained array.

Recommended solution

```
type Acceptable is array (NATURAL range 0 to 255) of
                        BIT_VECTOR(0 to 7); -- OK
```

Incidentally, just for the record, let me point out that it is valid to have a two-dimensional array type where the ranges in both of the dimensions are unconstrained.

```
type Two_Dimensional is array (NATURAL range <>, NATURAL range <>)
                        of BIT; -- OK
```

This variation was shown because an array of an array and a two-dimensional array may both be used to model the same memory device.

Compilation Caveats

Error. Object declaration referenced an array slice of a data type that was declared as a constrained array:

```
type Byte_Type is BIT_VECTOR(31 downto 0);

variable Bad_Var : Byte_Type(7 downto 0);   -- ERROR
```

But since BIT_VECTOR is declared as an unconstrained array it is valid to declare

```
variable Good_Var : BIT_VECTOR(7 downto 0);  -- OK
```

Error. The reserved word *others* was used in the aggregate of an unconstrained array.

This error typically occurs in either subprograms or in formal port default assignments. The subprogram version of this error is illustrated in the Experiments segment of this book via Test_187 and Test_188. The port default version of this error is shown below:

```
library IEEE;
use IEEE.std_logic_1164.all;
entity Interface_N is
   port (Clock    : in  BIT;
         Dataout  : out std_logic_vector
                       := (others => 'Z'));    -- ERROR
end Interface_N;
```

Recommended solution

```
library IEEE;
use IEEE.std_logic_1164.all;
entity Interface_N is
   generic(Number_Of_Bits : NATURAL);
   port (Clock    : in  BIT;
         Dataout  : out std_logic_vector(Number_Of_Bits - 1 downto 0)
                       := (others => 'Z')       -- OK
        );
end Interface_N;
```

Error. Positional and named associations were mixed in an aggregate assignment to an array type object.

Actually, this is a very honest mistake since, as shown in the Excursion, port map associations permit the mixing of named and positional associations. And so one might be tempted to extrapolate this VHDL feature to the aggregate notation as follows, where Data has a range of 1 to 8:

```
Data := (5 => '0', 1 => '3', '0', others => '0');  -- ERROR
```

Recommended solution. The obvious solution is to never mix positional with named associations in an array aggregate notation.

Error. An element listed in an array aggregate did not have the same data type as the individual elements of the array.

Suppose the we have the following declarations:

```
variable A : BIT_VECTOR(1 to 8);
variable B : BIT_VECTOR(11 to 18);
```

Since the slice B(12 to 15) is not of type BIT, which is the base type of each BIT_VECTOR element, the following aggregate notation is in error:

```
A := ('1', '0', '1', '1', B(12 to 15));  -- ERROR
```

Recommended solution. The intended assignment can readily be achieved via a concatenation operation:

```
A := '1' & '0' & '1' & '1' & B(12 to 15);  -- OK
```

Incidentally, the following aggregate possibilities are all valid:

```
A := ('1', '0', '1', '1', B(12), B(13), B(14), B(15));  -- OK
A := (1 => '1', 3 => '1', 5 to 8 => B(14), others => B(12));  -- OK
A := (others => B(14));  -- OK
```

The first aggregate assignment is self-explanatory. The second aggregate assignment implies that A(1) and A(3) both get '1', A(5 to 8) each (!!) get the value B(14), and the remaining elements (A(2) and A(4)) not explicitly referenced are to get the value B(12).

Error. Aggregate notation used in an assignment did not account for all elements of the targeted array or record.

For example, the following aggregate notations are incorrect, where A is the array just declared in the previous example:

```
A := ('1', , , , , , , '0');  -- ERROR
A := (1 => '1', 8 => '0');  -- ERROR
```

The first example shows that VHDL does not allow blank spaces to be used as placeholders. The second example emphasizes that even with named association techniques all elements of the assignment's target must be accounted for in an aggregate notation.

Recommended solution. The aggregate notation should not be used to make selective assignments. Instead, the intended selective assignments should be explicitly written out as follows:

```
A(1) := '1';  -- OK
A(8) := '0';  -- OK
```

Error. A record field was declared as an unconstrained array.

First, a quick lesson on what a record is. Recall that an array is a collection of elements where each element is of the same type and each element can be selected via an index. A record, on the other hand, is a collection of elements, where these elements do not have to be of the same type. The various elements of a record are known as its fields, and they may be individually accessed via VHDL's dot notation.

The following declaration is an example of the error in which a record's field is an unconstrained array:

```
type Instruction is
record
    Opcode     : CPU_Mnemonics;
    Register_1 : BIT_VECTOR;     -- ERROR
    Register_2 : BIT_VECTOR;     -- ERROR
end record
```

Recommended solution. Constrained arrays should be used, as in:

```
type Instruction is
record
    Opcode     : CPU_Mnemonics;
    Register_1 : BIT_VECTOR(1 to 4);   -- OK
    Register_2 : BIT_VECTOR(1 to 4);   -- OK
end record
```

Incidentally, the aggregate notation also exists for records. Suppose that Current_Instruction is declared to be a variable of type Instruction. Then the following aggregate assignment can be made:

```
Current_Instruction := (LOAD, ('1','0','1','1'),
                              ('0','0','0','1'));  -- OK
```

However, analogous to a previously described error it would be incorrect to write:

```
Current_Instruction := (LOAD, , ('0','0','0','1'));  -- ERROR
```

The omission of an element to correspond with the field Register_1 implies that the modeler does not wish to modify this specific record field.

The solution in this record field omission problem mirrors the array solution given above. The only difference is that instead of using an index to select a specific element, as per an array technique, one has to now rely on the VHDL dot notation to select specific fields of a record. Here is how the dot notation works:

```
Current_Instruction.Opcode := LOAD;  -- OK

Current_Instruction.Register_2 := ('0', '0', '0', '1');  -- OK
```

As per this coding segment, the field Register_1 of the record variable Current_Instruction will not be modified.

Error. In a VHDL'87 compliant tool suite the hexadecimal, X, notation was applied to non-BIT_VECTOR data type object.

Suppose that the signal Data was declared as Logic4_Vector(31 downto 0). Then in VHDL'87 the following assignment would be incorrect since Data is not a BIT_VECTOR:

```
Data <= X"E1F0";  -- ERROR in VHDL'87
```

Fortunately, VHDL'93 very wisely allows this hexadecimal notation for any data type that contains the character literals '0' and '1'. Of course the expression X"01XZ" is not allowed. Analogous techniques may also be applied by the octal (O) and binary (B) notations in VHDL'93.

Error. Slice directions did not honor the direction that the object was originally declared with.

This error can very innocently come about as in the following sneaky attempt to reverse the bits of an array. Suppose that Data is declared as:

```
variable Data : BIT_VECTOR(0 to 7);
```

The following attempt at reversing Data's bits is in error because a descending slice is referenced that is contrary to the ascending direction that Data was declared with:

```
Data := Data(7 downto 0);  -- ERROR;
```

Recommended solution. Bit reversal can be achieved via:

```
Data := Data(7) & Data(6) & Data(5) & Data(4) & Data(3) & Data(2) &
        Data(1) & Data(0);
```

Clearly, this concatenation technique is inappropriate for a 64-bit array. Consequently, for longer arrays it is better to use a for loop as in

```
for K in 64 downto 0 loop
    Data(K) := Data(64 - K);
end loop;
```

- DECLARATIONS
type Memory is array (0 to 5, 0 to 3) of Logic4;
constant ROM_DATA : Memory := (('0', '0', '0', '0'),
         ('0', '1', '1', '0'),
         ('1', '1', '1', '0'),
         ('0', '1', '1', '0'),
         ('1', '1', '1', '0'),
         ('1', '1', '0', '0'));
variable Data_Logic4 : Logic4;
variable Data_Logic4_Vector : Logic4_Vector(0 to 3);

- USAGE
Data_Logic4 := ROM_DATA(4, 2);

- CAVEAT
Data_Logic4_Vector := ROM_DATA(4, 0 to 3); -- *** ERROR ***

Figure 6.5 Accessing a two-dimensional array.

Error. A slice technique was applied to a multidimensional array.

This error typically comes about when one tries to access a complete row of a multidimensional array via a range specification. This erroneous attempt is depicted in Fig. 6.5.

Recommended solution. The best solution is to view the two-dimensional array as an array where each of its elements is also an array. This viewpoint is shown in Fig. 6.6. By the way, note from Figs. 6.5 and 6.6 how

- DECLARATIONS
  type Memory is array (0 to 5) of Logic4_Vector(0 to 3);
  constant ROM_DATA : Memory := (('0', '0', '0', '0'),
           ('0', '1', '1', '0'),
           ('1', '1', '1', '0'),
           ('0', '1', '1', '0'),
           ('1', '1', '1', '0'),
           ('1', '1', '0', '0'));
  variable Data_Logic4 : Logic4;
  variable Data_Logic4_Vector : Logic4_Vector(0 to 3);

- USAGES
  Data_Logic4 := ROM_DATA(4)(2);
  Data_Logic4_Vector := ROM_DATA(4);

Figure 6.6 Accessing an array element of an array.

the notation used to access a specific element must honor the data type that is being used: array of arrays or a two-dimensional array.

However, if the data type must, for some reason, be defined as a two-dimensional array, then a for loop may be used to collect each element of the row as in:

```
for K in 0 to 3 loop
    Data_Logic4_Vector := ROM_DATA(4,K);
end loop;
```

Error. Nonassociative logical operators were not parenthesized.

This rule should be interpreted as a VHDL safety net deliberately designed to help you since the computational order used will, in general, yield different results. The following two assignments are both erroneous:

```
W := A nand B nand C;  -- ERROR

V := A and B or C;  -- ERROR
```

By the way, this error may also pop up in any construct that does a BOOLEAN-related test. Here are two such examples:

```
if (A = ENABLED) and (B = TEMPERATURE_DANGER) or
               (C = PRESSURE_DANGER) then -- ERROR
-- sequential code


assert (A /= ENABLED) or (B /= TEMPERATURE_DANGER) and
       (C /= PRESSURE_DANGER)                       -- ERROR
```

Recommended solution. Parentheses should be appropriately placed that capture the intent of your design.

```
W := (A nand B) nand C;  -- OK

V := A and (B or C);  -- OK
```

While on the topic of parentheses, let me recommend that you always use them instead of relying on the VHDL operator precedence rules. The reason why I say this is that you have much better things to do with your life than to go around memorizing precedence tables. And besides, whenever memorization is relied on, there always is an element of risk. So why take chances? Use parentheses to explicitly specify your intended order of operations. The resulting computation will not only be more readable, but it will also work exactly the way that you want it to.

Error. A function declaration did not return a type mark. Recall from the Excursion that the term *type mark* refers to the name of a data type or subtype.

Since the slice BIT_VECTOR(31 downto 0) is not the name of a type or a subtype, the following function specification will not compile:

```
function Bad_Spec (Input : INTEGER)
                    return BIT_VECTOR(31 downto 0); -- ERROR
```

Recommended solution. Ideally, when writing utility functions that manipulate arrays either as input or returned output, you should strive to use unconstrained arrays. This way the function is elastic and can adjust to any input or output array length. However, if you do have a special need to return a constrained array, then here is how to do it:

```
subtype Word_Subtype is BIT_VECTOR(31 downto 0);

function Needed_Spec (Input : INTEGER) return Word_Subtype; -- OK
```

Error. In a subprogram call the actual associated with a formal parameter of class variable was not a variable.

For example, the TEXTIO read procedure cannot be called with a signal parameter since its formal parameter is declared to be of class variable. Hence, signals cannot be directly read via this procedure. A concrete example of this error is as follows, where Signal_A is a signal-classed object.

```
read(Buffer_Line, Signal_A); -- ERROR
```

Recommended solution. The TEXTIO read procedure should be called with a variable, and then this variable should be assigned to the desired signal.

```
read(Buffer_Line, Variable_A);

Signal_A <= Variable_A;
```

Incidentally, while on this subject, let me point out the other variations on this subprogram parameter class theme. To begin with, there are four classes in VHDL: constant, signal, variable, and file. We just saw that an actual parameter associated with a formal parameter of class variable must be a variable. Likewise, an actual associated with a formal parameter of class signal must also be a signal. But an actual associated with a formal of class constant only has to be an expression. In particular, a formal of class constant may be associated with an actual that is either a signal or a variable. The key point is that the actual will be treated as a constant inside the function's body and cannot be written to. As for formals of class file, they can only be associated with actuals of class file.

Error. The actual signal associated with a subprogram's signal parameter was the K'th element of a signal array where K is dynamically determined during the simulation.

Consider the function rising_edge from the IEEE standard std_logic_1164 package. It is declared as

```
function rising_edge (signal s : std_ulogic) return BOOLEAN;
```

From the previous example we already know that the associated actual must be of class signal. Here we learn that the following for loop, though well intended, will result in a error since the specific element's signal name Signal_A(K) is dynamically determined during the simulation:

```
            for K in Signal_A'RANGE loop

                if rising_edge(Signal_A(K)) then -- ERROR
```

Recommended solution. This error is closely related to our previous error of referencing Signal_A(K)'EVENT. Therefore, it should come as no surprise that the solution is very similar. Instead of calling the rising_edge function, simply compare Signal_A element by element to its previous value, all along checking if there was a change from '0' to '1'. If such a '0' to '1' transition did occur, then we have a rising edge on that specific subelement of the array.

Another solution is to write a new rising-edge checking function that will have two inputs instead of just one: Signal_A and its previous value. This new function will then contain the for loop mentioned in the previous alternate solution.

Error. Report clause of an assert statement spanned more than one physical line.

```
assert (S = '0' or R = '0')
   report "S and R both equal to 1
          which is contrary to RS flip-flop operation" -- ERROR
   severity ERROR;
```

Recommended solution. The concatenation operator may be used to join a long line as in

```
   assert (S = '0' or R = '0')
      report "S and R both equal to 1" &
             "which is contrary to RS flip-flop operation" -- OK
      severity ERROR;
```

However, this solution has a slight drawback in that the modeler has no control over how this long sentence will be wrapped around on the display screen. Control may be achieved by using the CHARACTER enumeration literals CR (Carriage Return) and LF (Line Feed) to explicitly define the boundaries for two consecutively written lines.

```
assert (S = '0' or R = '0')
   report "S and R both equal to 1" & CR & LF &
          "which is contrary to RS flip-flop operation"
   severity ERROR;
```

Unfortunately, there is a slight problem with this solution in that it might not be portable across different operating systems. On some operating systems this given solution will work fine, whereas on others the CR & LF combination might create havoc with your screen. In that case the LF CHARACTER enumeration literal should not be used.

Error. An entity declaration was preceded by a data-type declaration.

Sometimes, when setting up a quick experiment, a user-defined data type might be required for an entity's port declaration. One might then be tempted to write the following erroneous code:

```
type Logic4 is ('X', '0', '1', 'Z');  -- ERROR
entity Interface is
   port (Ain : in Logic4);
end Interface;
```

Recommended solution. Since connected ports must have the same data type, it follows that the common data type should be shareable among those design entities in which these ports are declared. Packages are the only vehicle for broadcasting a collection of data types to different design entities. Consequently, the solution is to define Logic4 inside a package and then to make this package visible (available) via a use clause.

```
library MVL_Packages_Lib;
use MVL_Packages_Lib.MVL4_Pkg.all;
entity Interface is
   port (Ain : in Logic4);
end Interface;
```

Chapter 7
Simulation Caveats

The purpose of this chapter is to enhance your VHDL problem-solving skills by exposing and solving a broad spectrum of VHDL coding scenarios that compiled but did not simulate as expected.

Problem. Misconception regarding the sensitivity list membership rules.

Since the statement wait until Enable = '1' has the signal Enable in its sensitivity list, you might write the following code segment thinking that it too will have Enable in its sensitivity list.

```
wait on A, B until Enable = '1';
```

Fact. When wait on is used the until clause does not contribute any signals toward the sensitivity list. Consequently, in this given example Enable is not in the sensitivity list. To include Enable you would have to write

```
wait on A, B, Enable until Enable = '1';
```

This coding scenario is explored in Test_5 of the Experiments.

Problem. VHDL'93 changes to the sensitivity list membership rules are not upward compatible.

In VHDL'87 only Signal_A would be in the sensitivity list of the following wait statement because Signal_B'LAST_VALUE is not a signal:

```
wait until (Signal_A = '1') and (Signal_B'LAST_VALUE = '0');
```

However, this code behaves differently in a VHDL'93 compliant environment. Consequently, any VHDL model containing such code is not upward compatible with VHDL'93. If the modeler is not aware of the new VHDL'93 rules, then VHDL models will be written that will not behave as intended.

Fact. In VHDL'87 the given wait until statement is equivalent to:

```
wait on Signal_A until
        (Signal_A = '1') and (Signal_B'LAST_VALUE = '0');
```

However, in VHDL'93 the given wait statement is equivalent to:

```
wait on Signal_A, Signal_B until
        (Signal_A = '1') and (Signal_B'LAST_VALUE = '0');
```

This difference is due to the new VHDL'93 rule that in the absence of a wait on construct all signals appearing in the until clause will be placed into the sensitivity list even if those signals serve only as prefixes for signal attributes that are themselves not signals. But if the attribute is of the class signal, then both VHDL versions concur. For instance, both Signal_A and Signal_B'STABLE are in the sensitivity list associated with the following wait statement:

```
wait until (Signal_A = '1') and (Signal_B'STABLE = FALSE);
```

But keep in mind that if Signal_B'EVENT would have been used instead then, in VHDL'93, Signal_A and Signal_B would both be in the sensitivity list, whereas in VHDL'87 only Signal_A would be. This discrepancy can lead to models that are not upward compatible.

Problem. Misapplication of the relational operators between arrays.

Suppose that Datain is declared as BIT_VECTOR(1 to 4) and that it is accidentally compared to a three-element array instead of to a four-element array as intended.

```
            if Datain = "010" then
```

A length mismatch during an array comparison is not viewed as an error. Hence, the VHDL compiler will not flag the previous comparison as an error. Instead, the resultant model will not simulate as expected, and it might even be somewhat tricky to track down this unintentional mistake.

Facts. Array comparisons are viewed as string comparisons and are made element-wise beginning from the left. Instead of a lengthy discourse of the pertinent VHDL rules it is more beneficial to see some concrete examples. Keep in mind that for BIT we have that '0' is less than '1' since '0' occurs to the left of '1' in the enumeration list that defines the data type BIT. BIT '0' is less than '1' because the left-to-right physical order in which the elements of an enumeration type are listed defines an ascending order relationship for these elements.

Here are some examples of BIT_VECTOR comparisons which, as stated earlier, will rely on a left-to-right comparison of its individual BITs:

```
"010101" = "010101"

"010101" /= "010100"

"010101" /= "01010"

"010101" < "0101010"

"010101" < "110101"

"010101" < "100"

"010101" < "1"
```

Observe that the lengths have to match only when the equality relation is being checked. These comparisons show that you have to be very careful when comparing two BIT_VECTORs such as "1111" to "0011". As per the previous string-motivated comparisons "1111" is greater than "0011," which is correct when both these bit streams are interpreted as the magnitude numbers 15 and 3, respectively. But what if "1111" is to be interpreted as a twos complement 4-bit number? In this case the string-motivated relation is wrong since -1 (= "1111") is not greater than 3 ("0011"). Most vendors supply an auxiliary package containing overloaded relational operators that allow you to correctly execute such signed comparisons.

By the way, just for the record, it is considered an error if the source and destination of an array assignment do not have the same length.

Problem. An equality check was made between two real numbers.

Different CPUs may have different internal representations for floating point numbers. Hence, the following equality check is not portable since they might behave differently on different hardware platforms:

```
if A_Real = 3.14 then -- Not portable.
```

Recommended solution. Comparison should be made within an epsilon distance as shown by the following example:

```
if abs(A_Real - 3.14) < ALLOWED_TOLERANCE then
```

Problem. Incomplete conditional test (wait version).

Simulation models must take into account the full range of an object's data type. Suppose that the signal Enable is declared to be of type std_ulogic. Then the following wait condition will erroneously interpret a transition from 'Z' to '1' as a leading edge:

```
wait until Enable = '1';
```

Recommended solution

```
wait until (Enable = 1) and (Enable'LAST_VALUE = '0');
```

Problem. The ramifications of the initialization phase were not considered.

Suppose that the BIT type signal Clock has an initial value of '1'. Then the following process is an erroneous model of a leading edge-triggered flip-flop. The problem is that the output Q will be updated 1 delta time after the initialization phase (implicit wait statement is at the bottom of the process Erroneous_FF since it has a static sensitivity list) even though the first leading edge has not occurred yet.

```
Erroneous_FF:
process(Clock)
begin
    if Clock = '1' then
        Q <= D;
    end if;
end process Erroneous_FF;
```

Recommended solution

```
Technique_1:
process(Clock)
begin
    if Clock = '1' and Clock'EVENT then
        Q <= D;
    end if;
end process Technique_1;
```

Technique_1 works by blocking the entrance into the if statement since Clock'EVENT is always FALSE during the initialization phase. So we have solved our initialization ramification problem but, while doing so, have introduced an inefficiency. When an event on Clock occurs, then this process will loop from its bottom back to the top and execute the if statement. So, excluding the initialization phase, every time this if statement's conditions are tested the attribute Clock'EVENT will always be TRUE. Hence the test to determine the value of Clock'EVENT is redundant. A more efficient way to write this VHDL model is to use our old friend wait until Clock = '1' as follows:

```
Technique_2:
process
begin
    wait until Clock = '1';
    Q <= D;
end process Technique_2;
```

Problem. Incomplete conditional test (if...then...end if version).

Suppose that Clock is of type std_ulogic. Then the following if statement will erroneously interpret a transition from 'Z' to '1' as a leading edge:

```
if Clock = '1' and Clock'EVENT then
    Q <= D;
end if;
```

Recommended solution. The attribute 'LAST_VALUE should be used as follows:

```
if Clock = '1' and Clock'LAST_VALUE = '0' and Clock'EVENT then
    Q <= D;
end if;
```

Problem. Incorrect usage of the attribute 'LAST_EVENT.

The following if statement is attempting to determine whether or not Signal_A has timed out, where NOW is the VHDL predefined function that provides the current absolute simulation time:

```
if (NOW - Signal_A'LAST_EVENT) >= TIME_OUT then
```

The subtraction of Signal_A'LAST_EVENT from the current absolute simulation time would be correct if the Signal_A'LAST_EVENT would be the absolute time at which Signal_A last had an event. But it is not. Recall that Signal_A'LAST_EVENT is the amount of time that has elapsed since Signal_A last had an event. Consequently, this if statement is incorrect.

Recommended solution. Since Signal_A'LAST_EVENT already represents the desired duration, it is sufficient to write:

```
if Signal_A'LAST_EVENT >= TIME_OUT then
```

But even this if statement must be thoroughly tested to determine whether Signal_A'LAST_EVENT or Signal_A'DELAYED'LAST_EVENT is to be used.

Problem. Unexpected behavior of the conditional concurrent signal assignment.

Consider the following conditional concurrent signal assignment, where Value_1, Value_2, Value_3, Cond_1, and Cond_2 are all signals:

```
Destination <= Value_1 after Delay_1 when Cond_1 else
               Value_2 after Delay_2 when Cond_2 else
               Value_3 after Delay_3;
```

To best understand the behavior of this conditional signal assignment, you should examine its equivalent process shown that follows:

```
process(Value_1,Value_2,Value_3,Cond_1,Cond_2)
begin
   if Cond_1 = TRUE then
      Destination <= Value_1 after Delay_1;
   else
      if Cond_2 = TRUE then
         Destination <= Value_2 after Delay_2;
      else
         Destination <= Value_3 after Delay_3;
      end if;
   end if;
end process;
```

Inspection of this process shows that the lexical order used in the conditional signal assignment statement defines a priority scheme that will determine how Destination is updated. This is where the problem might creep in. Suppose that Value_2 has an event and Cond_1 is currently TRUE. Following the logic flow illustrated by the equivalent process we see that Value_1 will be assigned to Destination instead of Value_2. If this behavior is what you want, then all is well. However if you really intended Value_2 to be assigned when it has an event and its corresponding Cond_2 is TRUE, then your model will not behave as expected.

Recommended solution. Ideally, you should replace this conditional concurrent signal assignment with a process that explicitly captures the logic and intent of your modeled device. However, if you insist on using a conditional concurrent signal assignment, then you must qualify the conditions to gain a finer granularity of control.

```
Destination <= Value_1 after Delay_1 when (Cond_1 and
                  Value_1'EVENT) else
               Value_2 after Delay_2 when (Cond_2 and
                  Value_2'EVENT) else
               Value_3 after Delay_3;
```

Problem. Unexpected behavior due to the else clause of a conditional concurrent signal assignment.

This potential problem is illustrated in Test_35a of the Experiments segment of this book.

Recommended solution. Both VHDL'87 and VHDL'93 solutions are discussed in Test_39 of the Experiments.

Problem. Incorrect usage of 'EVENT in a GUARD expression to model an edge-triggering device.

This problem has already been explored during our Excursion. Also Test_16 of the Experiments will give you a hands-on feel for this erroneous usage of the nonsignal attribute 'EVENT.

Recommended solution. As pointed out in the Excursion and Test_16 the attribute 'STABLE should be used to model edge-triggering devices since this attribute is a signal and hence may serve as a member of a sensitivity list.

Problem. Index range mismatch between actual input parameters when subprogram is called.

An index range mismatch might result in a simulation run time error as highlighted by the coding scenario in Test_28 of the Experiments.

Recommended solution. Test_28a illustrates an application of the alias construct to overcome this index range mismatch problem. Examination of Test_28 and Test_28a will give you a better understanding of why the alias construct is used so often in subprograms.

Problem. In a VHDL'87 environment the concatenation operator in a subprogram's actual parameter expression might result in a simulation run time error.

This problem is unique to VHDL'87 because of its poor definition for the resulting range bounds of concatenated arrays. The VHDL'87 definition might result in an array with negative indices even though the indices are declared to be of type NATURAL. Test_27d illustrates this problem. Here again, the solution is to use an alias construct to ensure that the resultant indices for the concatenated arrays are valid.

Fortunately, this problem does not arise in VHDL'93 which has a new and much improved definition for the direction and range bounds of a concatenated array. In VHDL'93 the leftmost range bound of a concatenation result will be base_type'LEFT, where base_type is the base data type of the indices defined for the individual arrays being concatenated. Furthermore, the direction of a concatenation result will follow the direction of this base_type. Consequently, the concatenation of arrays that are indexed by NATURAL will result in an array with an ascending index range that begins at 0. Negative indices will not occur.

Problem. A function potentially might not execute because it is located on the wrong side of a short circuit operation.

Consider the following if statement:

```
if (W = '1') and Error_Check(Y) then
```

Suppose that the function Error_Check uses an assert statement to inform the modeler that an error has occurred. If W does not equal '1' then, because of the short circuit property of the and operator, the function Error_Check will not be called. And consequently, a serious design flaw might not be reported.

Recommended solution. Any function containing an assert statement should be on the left side of a short circuit operator. The topic of short circuit operators will be described in more detail later in the chapter on Model Efficiency.

```
if Error_Check(Y) and (W = '1') then
```

Problem. There is a run time error due to an inappropriate slice referencing of a function's returned value.

This problem is illustrated in Test_105 of the Experiments.

Recommended solution. Provided in Test_105.

Problem. Unintended report message generated by concurrent assert statement.

Recall that a concurrent assert statement is an assert statement that physically appears in a VHDL concurrent region. To best understand its behavior, one must, as always, look at the equivalent process.

Consider the following concurrent assert statement:

```
assert Signal_A /= 'X'
    report "Value of Signal_A is currently unknown."
    severity ERROR;
```

The equivalent process is as follows:

```
process
begin
    assert Signal_A /= 'X'
        report "Value of Signal_A is currently unknown."
        severity ERROR;
    wait on Signal_A;
end process;
```

This equivalent process very quickly identifies a potential problem that you might have when working with the concurrent assert statement. Since the wait statement is at the bottom of the process, the assert statement will be executed during the initialization phase. Suppose that Signal_A was initialized to 'X'. Then a report message will be generated onto the screen during the initialization phase. If your intentions were to use this assert construct to identify the situation when Signal_A becomes 'X', then these initialization phase messages are inappropriate. In fact, you might even find these messages very distracting. And even more annoying is the fact that the simulation will probably halt due to the severity level associated with this assert statement.

Recommended solution. The trick is to stop the assert statement from executing during the initialization phase. To do so you must replace the concurrent assert statement with the following process that blocks the assert's execution during the initialization phase by placing the wait condition at the top of the process instead of at the bottom.

```
process
begin
   wait on Signal_A;
   assert Signal_A /= 'X'
          report "Value of Signal_A is currently unknown."
          severity ERROR;
end process;
```

While on the subject of concurrent assert statements, let me point out a practical location where they may be placed. Suppose that an entity declaration is associated with several architecture bodies. Instead of writing the same concurrent assert statement into each architecture, you may alternatively place it into the entity statement part as in the following (note well the necessary introduction of the reserved word *begin*). Then the given assert statement will be applicable to all architecture bodies associated with this entity.

```
entity RS_Flip_Flop is
    port(Set : in  BIT; Reset : in BIT;
         Q   : out Bit; Q_Bar : out BIT);
begin
        assert (Set = '0') or (Reset = '0')
           report "Set and Reset simultaneously 1."
           severity ERROR;
end RS_Flip_Flop;
```

Now for some theoretics. The only reason why the assert statement can be placed inside an entity statement region is that an assert statement does not schedule any assignments to signals. It merely monitors them. Signal assignments are forbidden in the entity statement region. Concurrent VHDL constructs that do not make assignments to signals are said to be passive constructs. In particular, a process that does not make any signal assignments is called a passive process. Therefore, it follows that a process can only be placed inside an entity statement region if it is a passive process.

Problem. Unexpected values in a postponed process.

Before discussing this problem let me first introduce you to the VHDL'93 construct known as a postponed process. As we saw in our Excursion it is possible to advance into successive delta time units (up the vertical axis) while still maintaining the same absolute simulation time. By declaring a process to be postponed (technique shown in the example that follows) we are ensuring that it will be executed only during that last delta time unit just before there is an advancement to the next scheduled absolute time unit. The implication of this requirement is that a postponed process cannot schedule any activity to occur during the next delta time. Consequently, a postponed process cannot contain any of the following two statements:

```
A <= B; -- NOT ALLOWED IN POSTPONED PROCESSES

wait for 0 ns; -- NOT ALLOWED IN POSTPONED PROCESSES
```

It is anticipated that postponed processes will bridge the VHDL digital, event-driven simulation world to the analog simulation world when VHDL officially expands into this continuum domain. Postponed processes will help to identify digital steady states at which point the analog algorithms may be executed. An analog extension will probably be included in the next IEEE update of VHDL which should occur by 1998.

Let's now return to our given problem. Consider the following coding scenario:

```
postponed process  -- Note usage of VHDL'93 reserved word postponed.
begin
    wait on Signal_A, Signal_B until Enable = '1';
    if Signal_A'EVENT then
        -- sequential code
    else
        if Signal_B'EVENT then
            -- sequential code
        end if;
    end if;
    -- sequential code
end postponed process;
```

Suppose that Signal_A just had an event and Enable is currently equal to '1'. Then this postponed process will be reactivated and placed on the queue of pending postponed processes. When the simulation engine determines that it is currently executing the last delta before the advancement to the next absolute time, then the pending postponed processes will be executed one by one but in no particular order. However, in the mean time, any number of delta time units may have already transpired during which either Signal_A, Signal_B, or Enable may have been modified. Consequently, when this postponed process is finally being executed, it is totally pointless to assume that the current values of Signal_A, Signal_B, Enable, Signal_A'EVENT, Signal_B' EVENT, or Enable'EVENT are the same as when this process was initially placed onto the pending postponed process queue. Consequently, your postponed process should not rely on the current values of any of these potentially volatile objects.

By the way, VHDL'93 also allows concurrent asserts, concurrent procedures, and any form of concurrent signal assignments to be postponed as well.

Recommended solution. Postponed processes should be used with great care and thoroughly tested.

Problem. Erroneous application of variables.

When working with variables, you must always keep in mind that their assignments occur immediately. This fact is crucial when using variables; otherwise, your model may not behave as expected. Section 2.13 of the Excursion exposed this potential problem.

Problem. Erroneous application of signals.

When working with signals, you must always keep in mind that signals are never updated immediately. Moreover, in the absence of an after clause the signal will be updated 1 delta time later. For instance, consider the following sequence of assignments:

```
Signal_A <= Signal_A(1 to 7) & Signal_A(0);
Var_B := Signal_A;
```

Because Signal_A will only be updated 1 delta delay later, the assignment to Var_B will use the old, unshifted value of Signal_A. If this usage of Signal_A is not what you want, then it follows that this model will not behave as expected.

Recommended solution. If you intended to use the updated value of Signal_A, then you have several options available to you: You can replace Signal_A with a variable (preferred solution whenever feasible); you can write wait on Signal_A'TRANSACTION between the two assignments; or, you can write the infamous wait for 0 ns between the two assignments.

By the way, while on the topic of shifting, let me point out that VHDL'93 has introduced six new shift operators: sll (shift left logical), srl (shift right logical), sla (shift left arithmetic), sra (shift right arithmetic), rol (roll left), ror (roll right). These new operators can shift any one-dimensional array type whose element type is BIT or BOOLEAN. Current MVLN packages, such as the IEEE std_logic_1164 package, should be updated to overload these shift operators so that they can apply to one-dimensional arrays of the respective MVLN package's base type. In our particular example assume that Signal_A was declared as BIT_VECTOR(0 to 7). Then the above concatenation technique may be replaced by rol as follows:

```
Signal_A <= Signal_A rol 1;
wait on Signal_A'TRANSACTION;
Var_B := Signal_A;
```

Problem. Erroneous usage of global variables.

Global variables may be declared in the architecture declarative part, entity declarative part, package declaration, and package body using the reserved word *shared*, as in:

```
shared variable Global_A : INTEGER;
```

As pointed out in the Excursion the problem with global variables is that they are updated immediately in our concurrent VHDL environment. Consequently, a model that uses global variables is nondeterministic and cannot be considered portable. Nonetheless, there is still a need for global variables in the area of performance modeling as discussed in the Excursion (Sec. 2.4). And as always, human ingenuity will find many other innovative and worthwhile applications for global variables. However, in your day-to-day VHDL hardware modeling activities you should, in general, not use global variables. But if you insist on working with global variables, then do so very carefully and thoroughly test out your model.

Problem. Incorrect waveform generation via two consecutive sequential signal assignment statements.

Suppose that your intentions are to currently assign '1' to Dataout and then '0' 15 ns later. Because of the VHDL inertial signal update algorithms, the following SEQUENTIAL code segment will not give you what you want:

```
Dataout <= '1';

Dataout <= '0' after 15 ns;
```

The net result of these two scheduled assignments is that Dataout will be assigned '0' 15 ns from now. The value '1' will be preempted from

Dataout's projected driver as per the VHDL algorithm that was described in Chap. 3.

Recommended solution. There are two options that you can use. You may write:

```
Dataout <= '1';

Dataout <= transport '0' after 15 ns;
```

Or, alternatively, you may use the following waveform assignment construct which, incidentally, is equivalent to the above solution:

```
Dataout <= '1', '0' after 15 ns;
```

Problem. Incorrect waveform generation in a for loop.

Suppose that your intentions are to create a scheduled waveform by iteratively reading the data from an input file. The overall objective is to apply this scheduled waveform as the stimulus for your device under test in a test bench environment. For the sake of argument, assume that Data is of type INTEGER and that the Data-related values in the input file are all different. Then, because of the effects of the VHDL inertial update algorithm, only the final read in value will be assigned to Test_Vector in the following for loop:

```
for K in 1 to NUMBER_OF_WAVEFORM_DATA loop
    readline (Waveform_Data, Buffer_Line);
    read (Buffer_Line, Data);
    read (Buffer_Line, Update_Time);
    Test_Vector <= Data after Update_Time;
end loop;
```

Recommended solution. The intended waveform can be achieved by using a transport delay updating scheme. The assignment to Test_Vector should be:

```
Test_Vector <= transport Data after Update_Time;
```

Hence, you must always be aware of the updating differences between VHDL's inertial and transport-delay models.

Problem. Same buffer line (pointer) used for reading from one file and then writing into another.

The unexpected results due to this duplicate usage of a buffer line are exposed in Test_172 of the Experiments.

Recommended solution. The same buffer line (pointer) should not be used for reading from one file and then later writing into another.

Problem. In VHDL'87 writing to an output file is not portable.

VHDL'87 does not address the issue of what is to be done when a model wishes to write to a file and that file already exists. Should the new data merely be appended to the already existing file or should the old file be deleted so that the data may be written into a new, empty file? Some VHDL'87 tools append to the existing file while others delete the old file in order to begin anew. This is a serious problem especially if you wish to postprocess the captured data at both your site and the customers' sites as well.

Recommended solution. More than likely you will wish to delete the old file. Consequently, in VHDL'87 you should write a script file that, if necessary, will automatically delete the old files just before you begin another simulation. This script file should then be included as a deliverable item to your VHDL customers.

Fortunately, with VHDL'93 you do not have to go to such lengths. The VHDL'93 update has introduced syntax for file declarations that incorporate your appending intentions right into its syntax. For instance, if you do not wish to append to a preexisting file, then your file declaration should be similar to the following:

```
file Captured_Data : TEXT open WRITE_MODE is "simulation_data";
```

WRITE_MODE is a member of the new VHDL'93 enumeration literal FILE_OPEN_KIND, which is declared in the updated STANDARD package as follows:

```
type FILE_OPEN_KIND is (READ_MODE, WRITE_MODE, APPEND_MODE);
```

The purpose of the other members of this enumeration list are self-explanatory, and READ_MODE is assumed when no FILE_OPEN_KIND is given as in

```
file Data_File : TEXT is "cpu_data";
```

Problem. Full path name specified in file declaration.

Suppose that the following file declaration is made:

```
file Input_Data : TEXT open READ_MODE is
                    "/usr/home/projects/cpu/test_vectors";
```

When this model is delivered to a customer, it is highly unlikely that the subdirectory /usr/home/projects/cpu will exist at the delivery site. Consequently, an elaboration phase error will occur when the VHDL tool suite attempts to open the file *test_vectors* in this nonexistent subdirectory.

Recommended solution. Full path names should not be used. This recommendation implies that the files referenced by your model should be in the subdirectory from which the simulation tool suite was launched.

```
file Input_Data : TEXT open READ_MODE is "test_vectors"
```

Problem. Port artificially declared to be of mode inout just so that the port's name can occur on both the left-hand and the right-hand side of a signal assignment.

Test_195 of the Experiments exposes the potential hazards of this artificial application of the mode inout and provides some viable VHDL'87 and VHDL'93 alternatives.

Chapter 8

Model Efficiency

The purpose of this chapter is to expose inefficient coding scenarios that will have a detrimental effect on the overall simulation performance of your VHDL models. Each inefficient coding style is countered by a recommended optimization technique.

8.1 The Efficiency War Room

Designing efficient VHDL models is a two-pronged attack. You must simultaneously assault both the VHDL language syntax and also the program coding style.

By attacking the VHDL language syntax I am inferring that you should strive to use optimal VHDL constructs. For instance, whenever possible use variables instead of signals since signals have a higher operational overhead. Another point to keep in mind is that blocks and conditional concurrent signal assignments must monitor the numerous signals that are in their sensitivity lists. It is often more efficient to write a process that is waiting for events on a smaller but judiciously selected set of signals while still capturing the essence of the intended behavior. Less signals in a sensitivity list means less signals to be monitored. And as a result your simulation will run faster.

When attacking the program coding style, you should always remember that, in the final analysis, VHDL is a software medium. Hence, the full spectrum of programming optimization tricks and techniques are available to you. Moreover, the high-order language capabilities of VHDL may also be exploited to design efficient models.

8.2 Examples

Inefficiency. 'STABLE should be avoided whenever possible since this signal attribute is also a signal, and the updating management of signals has a bigger overhead than that of nonsignals. Typically, 'STABLE is applied in the following inefficient manner:

```
if Clock = '1' and (not Clock'STABLE) then
```

Recommended efficient alternative

```
if Clock = '1' and Clock'EVENT then
```

As a rule of thumb 'EVENT should be used in any sequential region of code instead of 'STABLE.

Inefficiency. Unless your VHDL compiler is clever enough to convert 2**N into an appropriate shift left operation, you should avoid this CPU intensive computation.

2**N is used most often when converting a BIT_VECTOR into a NATURAL or INTEGER number. The following subprogram code fragment shows an application of this inefficient computation, where Input_Alias is an alias for the function's Input parameter and has the range Input'LENGTH—1 downto 0:

```
for N in 0 to Input_Alias'LEFT loop
    Converted_Value := Converted_Value + 2**N;
end loop;
```

Recommended efficient alternative

```
for N in Input_Alias'LEFT downto 0 loop
    Converted_Value := Converted_Value +
                       Converted_Value +
                       Bit'POS(Input_Alias(N));
end loop;
```

Note that this solution added Converted_Value to itself instead of multiplying it by 2. This autoaddition strategy was selected just in case the compiler does not realize that multiplication by 2 is equivalent to a 1-bit shift to the left. Whereas multiplication requires 10 to 20 CPU cycles, the shift operation can be accomplished in one CPU cycle.

Another key operational technique to observe is that aliasing is used as indicated by my naming convention, Input_Alias. By doing so I was able to have control over the range direction of the value being converted. Otherwise, I would first have to determine whether the input

BIT_VECTOR array has an ascending or descending range. Then two for loops would have to be written (one for each direction) and conditionally entered via an if statement. Such an approach would be inefficient in both memory consumption (area must be set aside to accommodate both of these for loops) and time utilization (conditional branching must be executed to determine which of these two for loops is to be implemented).

By the way, did you notice that the loop range of this efficient solution is the reverse of the loop range from the above inefficient technique? Subtle patterns like this always fascinate and amuse me.

Inefficiency. Unless your VHDL compiler is clever enough to convert a division by two into a shift right operation, you should avoid this CPU intensive computation.

Division by 2 is used most often when converting a Natural data type object into a BIT_VECTOR. The following subprogram code fragment shows an example of this inefficient computation. The key elements of this code fragment are the subprogram variables Input_Local, Converted_Value, and the subprogram constant Bit_Width_M1. Input_Local is initially set to the NATURAL number that is to be converted into a BIT_VECTOR. Converted_Value is initialized to all '0's. The subprogram constant Bit_Width_M1 is set to the user-specified bit width minus one.

```
for N in 0 to BIT_WIDTH_M1 loop
    if (Input_Local mod 2) = 1 then
        Converted_Value(N) := '1';
    end if;
    Input_Local := Input_Local / 2;
end loop;
```

In addition to the division by 2 this conversion algorithm also applies the mod operator to determine whether the result after division is an odd or even number. If Input_Local mod 2 is derived by an examination of the remainder after a division operation, then we have an additional inefficiency to contend with. A clever implementation alternative to a mod 2 computation is to examine the least significant bit (LSB) of the register holding the binary equivalent of Input_Local (0 implies even and 1 implies odd). Here again, the compiler must be clever enough to implement the mod 2 operation via a simple LSB examination.

Recommended efficient alternative. Both the division and mod operations can be eliminated by using a table lookup technique. In this efficient alternative the constant lookup table N_To_2 will be referenced. As the name suggests, this one-dimensional constant array will consist of the values $N^{**}2$ for $N = 0, 1, 2,\ldots 30$.

```
type Natural_Array_Type is array (NATURAL range <>) of NATURAL;
constant N_To_2 : Natural_Array_Type(0 to 30) :=
   (1, 2, 4, 8, 16, 32, 64, 128, 256, 512, 1024, 2048, 4096,
    8192, 16384, 32768, 65536, 131072, 262144, 524288, 1048576,
    2097152, 4194304, 8388608, 16777216, 33554432, 67108864,
    134217728, 268435456, 536870912, 1073741824);
```

An efficient conversion algorithm can now be implemented as follows:

```
for I in BIT_WIDTH_M1 downto 0 loop
   N_To_2_Value := N_To_2(I);
   if Input_Local >= N_To_2_Value then
      Converted_Value(I) := '1';
      Input_Local := Input_Local - N_To_2_Value;
   end if;
end loop;
```

Observe how, in essence, division has been replaced by subtraction which requires less time (1 CPU cycle versus 20 to 30) to execute. Also note the usage of the temporary variable N_To_2_Value so that the table lookup will only have to be done once per loop iteration. And did you notice that here again the loop range of the efficient solution is the reverse of the loop range from the above inefficient technique? As I said before, such subtle patterns always fascinate and amuse me.

But there is one minor drawback to this efficient solution that you should be aware of. The lookup table N_To_2 has only 31 elements, and so it assumes that X"7FFFFFFF" is the largest NATURAL number when written in hexadecimal notation. Currently, this is a true statement since all the vendors abide with the minimum VHDL requirement that the data-type INTEGER be represented at least as a 32-bit twos complement number. If, sometime in the future, the vendors will upgrade INTEGER to a 64-bit twos complement number then you will have to enlarge N_To_2 accordingly.

Inefficiency. VHDL simulation engine should not conduct unnecessary data validation tests.

Suppose that a range constraint is levied onto a variable as follows:

```
variable Count_7 : INTEGER range 0 to 7;
```

Such range constraints are highly recommended for synthesis modeling since it communicates to the synthesis tool the number of bits to be used when converting an abstract software object into a concrete hardware element. Unfortunately, from the simulation perspective, it may introduce redundant type checking. Consider the following code segment:

```
if Count_7 = 7 then
   Count_7 := 0;
else
   Count_7 := Count_7 + 1;
end if;
```

This if statement guarantees that Count_7 will always be within the range boundaries defined for it. Nonetheless, every time an assignment will be made to Count_7 the simulation engine will first check to confirm that the assigned value is within the stated bounds. Clearly, this test by the simulation engine is redundant.

Recommended efficient alternative. Data-type range constraints should be avoided unless the VHDL model is targeted for synthesis.

Inefficiency. Fixed values should not be recomputed.

Suppose that PULSE_WIDTH is a generic parameter. Then the product 2*PULSE_WIDTH will be a fixed value for the duration of the simulation run. Hence, the following wait statement will needlessly recompute this fixed value on each and every iteration of its encapsulating process:

```
wait for 2*PULSE_WIDTH;
```

Recommended efficient alternative. The given product should be computed only once during the elaboration phase and then assigned to an object of class constant.

```
constant TWICE_PULSE_WIDTH : TIME := 2*PULSE_WIDTH;
```

The wait statement should then use this constant value as follows:

```
wait for TWICE_PULSE_WIDTH;
```

Inefficiency. Unnecessary calls to the same function.

Consider the following assert condition where NOW is the predefined VHDL function that returns the current absolute simulation time:

```
assert (NOW = 0 ns) or (NOW - R_Last_Event) > SPIKE_WIDTH;
```

The inefficiency here is that the function NOW will be called and executed twice, despite the fact that the returned values will be the same.

Recommended efficient alternative. Declare an intermediate variable, assign NOW's returned value to it, and then use this variable instead of repeatedly calling NOW:

```
Now_Value := NOW;

assert (Now_Value = 0 ns) or (Now_Value - R_Last_Event)
                                        > SPIKE_WIDTH;
```

Inefficiency. Redundant determination of 'EVENT.

```
            wait until Clock = '1' and Clock'EVENT;
```

As pointed out in the Excursion the condition in the until clause will only be tested whenever any of the signals appearing in this clause has an event. So in this particular example the condition will be checked only when Clock has an event. But if Clock has an event then Clock-'EVENT will automatically be TRUE. Consequently, the test Clock-'EVENT is redundant and hence inefficient.

Recommended efficient alternative. Eliminate the 'EVENT test since it is not needed at all. The following wait statement should be used instead:

```
                   wait until Clock = '1';
```

Inefficiency. Usage of nested if...then...else statements to derive fixed values.

At the assembly language level an if statement translates into a conditional branch instruction. Unnecessary usage of such instructions should be avoided whenever possible since they have a very detrimental effect on the host CPU's pipeline flow.

Recommended efficient alternative. The various fixed values should be saved into a constant lookup table that can be referenced as needed. This technique has already been discussed during the Excursion part of this book.

Inefficiency. Unnecessary regeneration of a multidimensional constant lookup table.

This inefficiency is illustrated in Fig. 8.1. Since the constant OR_TABLE is declared in the function's declarative part, it will be written onto the host computer's stack structure every time the function "or" is called. This regeneration of the constant lookup table consumes precious CPU time. By the way, it should be well noted that it is indeed acceptable to declare a few scalar (single-element) constants inside a subprogram's declarative part. The inefficiency in Fig. 8.1 comes about only

```
type Logic4 is ('X', '0', '1', 'Z');
type Two_D_Table_Type is array (Logic4, Logic4) of Logic4;
........................
function "or" (L, R: Logic4) return Logic4 is
  constant Or_Table : Two_D_Table_Type :=
          -- 'X' '0' '1' 'Z'
          (('X', 'X', '1', 'X'),   -- 'X'
           ('X', '0', '1', 'X'),   -- '0'
           ('1', '1', '1', '1'),   -- '1'
           ('X', 'X', '1', 'X'));  -- 'Z'
begin
  return Or_Table(L,R);
end "or";
```

Figure 8.1 Inefficient location for declaration of multidimensional array.

because of the large number of write operations that are required to copy the numerous elements of the multidimensional array onto the stack.

Recommended efficient alternative. The constant lookup table should be placed outside of the function declarative region and into the general package body (or specification) area. In this way the constant table will be stored in the computer's memory region known as the HEAP. This global location will allow the constant table to always be accessible. Consequently, there will no longer be a need to regenerate the lookup table onto the CPU stack during each function call. This efficient alternative is illustrated in Fig. 8.2.

Inefficiency. Recomputation of the value 'LENGTH and the repeated calls to the scalar version of a function by its array counterpart.

```
constant Or_Table : Two_D_Table_Type :=
        (('X', 'X', '1', 'X'),
         ('X', '0', '1', 'X'),
         ('1', '1', '1', '1'),
         ('X', 'X', '1', 'X'));

function "or" (L, R: Logic4) return Logic4 is
begin
  return Or_Table(L,R);
end "or";
```

Figure 8.2 Efficient location for declaration of multidimensional array.

254   VHDL Techniques and Recommendations

```
function "or" (L, R: Logic4_Vector) return Logic4_Vector is
  alias L_Alias : Logic4_Vector(L'LENGTH downto 1) is L;
  alias R_Alias : Logic4_Vector(R'LENGTH downto 1) is R;
  variable Result := Logic4_Vector(L'LENGTH downto 1);
begin
  assert L'LENGTH = R'LENGTH
    report "Length mismatch during overloading or"
    severity ERROR;
  for I in L_Alias'RANGE loop
    Result(I) := L_Alias(I) or R_Alias(I);
  and loop;
  return Result;
end "or";
```

Figure 8.3   Inefficient array version of a scalar function.

Both these inefficiencies are depicted in Fig. 8.3. When this subprogram is called, the simulator may recompute L'LENGTH every time it encounters this expression. (The previous chapter contained a counterexample that proved why the alias construct is required in the first place.) The second listed inefficiency occurs in the for loop where the scalar version of the overloaded "or" function is iteratively called for each element of the function's input array. The subprogram parameter passing and context switching mechanisms will incur a performance penalty, which in this case is easy to avoid.

Recommended efficient alternative.   The various 'LENGTH values should be stored into constant objects that can later be referenced as needed. Also, instead of calling the scalar version of this function for each element of the array, the constant lookup table should be directly referenced to determine the desired scalar value. There is one additional optimization that should be considered. Note that the lower bound of the for loop is now 0 instead of a 1. This slight modification will save one to two assembly language instructions by a direct comparison to the host computer's register R0 which, in general, is hard-wired to 0. Otherwise, a lower loop bound of 1 would mean that this value would first have to be loaded into a register and then compared to that register containing the current loop iterator value. All these optimizing alternatives are illustrated in Fig. 8.4.

Inefficiency.   Unnecessary calls to the same function.

Consider the following two consecutive assignments:

```
Sign_Bit   := ALU_Result(A_Bus, B_Bus)(31);

Lower_Byte := ALU_Result(A_Bus, B_Bus)(7 downto 0);
```

```
function "or" (L, R: Logic4_Vector) return Logic4_Vector is
   constant L_Length_M1 : NATURAL := L'LENGTH - 1;
   constant R_Length_M1 : NATURAL := R'LENGTH - 1;
   alias L_Alias : Logic4_Vector(L_Length_M1 downto 0) is L;
   alias R_Alias : Logic4_Vector(R_Length_M1 downto 0) is R;
   variable Result := Logic4_Vector(L_Length_M1 downto 0);
begin
   assert L_Length_M1 = R_Length_M1
         report "Length mismatch during overloading or"
         severity ERROR;
   for K in L_Length_M1 downto 0 loop --
      Result(K) := Or_Table(L_Alias(K),R_Alias(K));
   end loop;
   return Result;
end "or";
```

Figure 8.4 Efficient array version of a scalar function.

The function ALU_Result will return exactly the same value on both calls. Hence, during the second call ALU_Result is needlessly recomputing and deriving the same value.

Recommended efficient alternative. An intermediate variable should be used to capture the current returned value of the function ALU_Result. The appropriate slices of this variable may then be referenced as required.

```
Returned_Value := ALU_Result(A_Bus, B_Bus);

Sign_Bit      := Returned_Value(31);

Lower_Byte    := Returned_Value(7 downto 0);
```

By the way, as a friendly reminder recall from the Simulation Caveats chapter that there are many traps that one can fall into when referencing a slice of a function's return value.

Inefficiency. Inefficient instruction decoding.

An example of this inefficiency is shown in Fig. 8.5. The array Instruct_Field_Bits is compared element by element to "000000", "000001", "000010", ... until a match is found. As the array length increases, this inefficiency gets worse and worse.

Recommended efficient alternative. The case comparisons should be done via enumeration literals. The case statement can then be implemented down at the assembly language level as a jump table. Though it is true that the original bit stream must first be converted into the appropri-

```
case Instruct_Field_Bits is
    when "000000" =>
        ..........
    when "000001" =>
        ..........
    when "000010" =>
        ..........
end case;
```

Figure 8.5 Inefficient instruction decoding.

ate enumeration literal, this conversion is just a one-shot deal and may be done very efficiently. The actual technique is shown in the Type Conversions chapter (Chap. 9). The recommended efficient case comparison for arrays is shown in Fig. 8.6. Note that this technique also improves the readability of the model.

Inefficiency   Periodic waveform not generated efficiently.

Consider the following two statements that together will generate a periodic waveform, where PULSE_WIDTH is a generic parameter:

```
wait for PULSE_WIDTH;

Clock <= not Clock'DRIVING_VALUE;
```

The inefficiency of this code segment is that the encapsulating process will be reactivated on each of Clock's two edges. The operat-

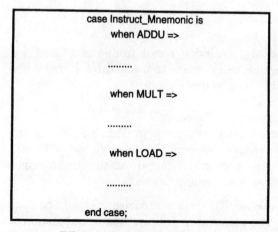

```
case Instruct_Mnemonic is
    when ADDU =>

        ..........

    when MULT =>

        ..........

    when LOAD =>

        ..........
end case;
```

Figure 8.6 Efficient instruction decoding.

ing overhead required for process reactivation should be avoided whenever possible.

Recommended efficient alternative. The following coding scenario will reactivate the encapsulating process only once per Clock's period, where PULSE_WIDTH is a generic parameter and PERIOD is a constant equal to twice the PULSE_WIDTH:

```
Clock <= '1' after PULSE_WIDTH, '0' after PERIOD;

wait for PERIOD;
```

Inefficiency. Common sensitivity lists are not exploited.

Consider the following three concurrent signal assignments:

```
A <= B + C;
W <= B - C;
H <= B and C;
```

In essence, we have three processes here, each of which must be reactivated whenever the signal B or signal C has an event. Incidentally, there are many variations of this inefficiency, and so you should always be on the lookout for them.

Recommended efficient alternative. All three concurrent processes should be combined into a single process so that only one process will have to be reactivated and resuspended instead of three.

```
process(B,C)
begin
   A <= B + C;
   W <= B - C;
   H <= B and C;
end process;
```

Inefficiency. Process reactivated unnecessarily.

Consider the following model for a level-sensitive device:

0 ns;

Suppose that Datain would have several events while Enable is equal to '0'. This process would then be reactivated and resuspended without any scheduled update to Destination. Consequently, the effort spent to reactivate and resuspend this process would be a waste of precious computer time.

Recommended efficient alternative. The given inefficient process should be rewritten as follows:

```
process
begin
    if Enable /= '1' then
        wait on Enable;
    end if;
    Destination <= Datain after 10 ns;
    wait on Datain;
end process;
```

In this model the process will only be awakened when it is valid to schedule the signal assignment to Destination.

Inefficiency. Conditional concurrent signal assignment may cause unnecessary monitoring of signals and nonproductive reactivations.

This inefficiency is a variation of the preceding one that you have just read about. Consider the following conditional concurrent signal assignment that is modeling a latchlike behavior:

```
Destination <= Datain after 10 ns
                    when Enable = '1' else
               Destination;
```

This concurrent signal assignment is sensitive to Datain, Enable, and Destination. In VHDL'93 the else clause may be eliminated, but that will still leave us with the following inefficiency. Suppose that Datain has an event while Enable is currently equal to '0'. Under this condition Datain will not be assigned to Destination. But the event on Datain will cause this conditional concurrent signal assignment to be reactivated. Valuable simulation time will be consumed by this unnecessary reawakening and immediate resuspension.

Recommended efficient alternative. Write a process that models the intent of the given concurrent signal assignment but does so in an efficient manner.

```
process
begin
    if Enable /= '1' then
        wait on Enable;
    end if;
    Destination <= Datain after 10 ns;
    wait on Datain;
end process;
```

The similarity between this solution and the one in the previous example is due to the theoretical equivalences between processes and conditional concurrent signal assignments. Take away the else clause and these last two examples are basically saying the same thing.

Inefficiency. Redundant signal transactions due to the else clause occurring in a conditional signal assignment.

Consider the following conditional concurrent signal assignment:

```
X <= X_Proc_1 when (not X_Proc_1'QUIET) else
     X_Proc_2 when (not X_Proc_2'QUIET) else
     X;
```

Whenever the assignment of X_Proc_1 or X_Proc_2 results in an event on X, then 1 delta time later X will be assigned to itself. This autoassignment is not only a waste of time, but it also could lead to potential problems elsewhere in your model (see Test_35a in the Experiments segment of this book).

Recommended efficient alternative. Test_39 in the Experiments illustrates a VHDL'87 workaround that replaces the given conditional concurrent signal assignment with a process that deliberately avoids this redundant transaction inefficiency. Fortunately, VHDL'93 allows you to completely bypass this problem while still working within the framework of a conditional concurrent signal assignment. In VHDL'93 the else clause may be either omitted or you may use the new reserved word *unaffected* (behaves exactly as it sounds) as follows:

```
X <= X_Proc_1 when (not X_Proc_1'QUIET) else
     X_Proc_2 when (not X_Proc_2'QUIET) else
     unaffected;
```

Inefficiency. Multiple processes all waiting for the same edge of a periodic signal.

Suppose that an architecture body contains numerous processes that are all waiting for the leading edge of the periodic signal Clock. Typically, they would all contain the following standard edge detection statement:

```
wait until Clock = '1';
```

Consequently, all these processes will be reactivated whenever Clock has an event. If Clock has a 50 percent duty cycle, then its leading edge will only occur 50 percent of the time. Hence, all these processes will only continue executing 50 percent of the time. For the other 50 percent they will just be resuspended because the leading edge did not occur. This is yet another case when valuable simulation time will be consumed by unnecessary reawakenings and immediate resuspensions.

Recommended efficient alternative. Write a new process that toggles a bit-valued signal on the leading edge of Clock.

```
wait until Clock = '1';

Toggle <= not Toggle;
```

The other processes should replace their wait until Clock = '1' statement with the following:

```
wait on Toggle;
```

The beauty of this approach is that, except for the newly written process, all the other, original processes will now only be reactivated on Clock's leading edge. Clearly, the more processes we have the greater are the savings that can be earned via this optimization technique.

Inefficiency. Infrequent events tested on a periodic basis.
Consider the following code segment:

```
wait until Clock = '1';

if Busy = SET then
```

Suppose that Busy is typically not equal to SET. Then the encapsulating process will needlessly be reactivated and resuspended without any action taken since Busy is not equal to SET.

Recommended efficient alternative. Instead of waiting for Clock's leading edge you should focus on the signal Busy.

```
wait until Busy = SET;

wait until Clock = '1';
```

Observe that when Busy goes to SET, we must then synchronize to the next leading edge of Clock. A simple sender or receiver timing diagram will convince you that this synchronization must take place or else the receiver will prematurely identify Busy's SET value.

But there still is a slight problem with this proposed efficient alternative. Suppose that Busy going to SET was just a glitch and that by the time this model synchronizes to Clock's next leading edge Busy no longer has the value SET. The following UART inspired solution will correctly handle this glitch-related problem:

```
Busy_Loop:
loop
wait until Busy = SET;
wait until Clock = '1';
if Busy = SET then
    exit Busy_Loop;
end if;
end loop Busy_Loop;
```

Inefficiency. Inefficient usage of short circuit operation.
The VHDL predefined logical operators AND, OR, NAND, and NOR are short circuit computations. This means that the right operand is

evaluated only if the value of the left operand does not determine the value of the overall logical expression. For instance, consider the following if statement:

```
if (A = '1') and (B = '0') then
```

Suppose that in the given expression (A = '1') is FALSE. Therefore the compound expression ((A = '1') and (B = '0')) will be FALSE irrespective of whether (B = '0') is TRUE or FALSE. Therefore, there really is no need to continue checking the relational status of (B = '0') and, indeed, the earlier stated short circuit rule ensures that this redundant test will not be done.

Based on this illustration of the short circuit rule, it immediately follows that an inefficiency will occur in the and's case if the more frequently TRUE expression is on the left side of the and operator. Since a TRUE left operand does not determine a compound and statement, it follows that the expression on the right will have to also be computed as well.

Recommended efficient alternative. When working with the and operator, you should place the expression more likely to be FALSE on this operator's left side. This way it is more probable that you will not have to continue to determine the BOOLEAN value of the expression on the right-hand side of the and operator.

By the way, when working with the or operator, it is more efficient to place the expression more likely to be TRUE on the left side of this operator since this BOOLEAN value will fully determine the overall value of the compound expression involving the or operator.

Inefficiency. Demorgan's law not applied in an assert condition.

The tricky part of working with the assert construct is that you have to think backwards when deriving the assert condition. The report string will only be displayed on the screen if the assert condition is FALSE. What is desirable for a correct hardware implementation is totally the opposite from the criteria to be specified in the assert condition. And so, typically, VHDL users will first derive the hardware rejection condition and then negate this expression to acquire the appropriate assert condition. Unfortunately, this negation operation introduces another BOOLEAN operation that has to be executed. Consider the following assert construct:

```
assert not((S = '1') and (R = '1'))
    report "SET and RESET erroneously both 1 in RS flip-flop"
    severity ERROR;
```

Recommended efficient alternative. Apply Demorgan's law to eliminate the outermost not operator.

```
assert (S = '0') or (R = '0')
    report "SET and RESET erroneously both 1 in RS flip-flop"
    severity ERROR;
```

Inefficiency. Assert condition always FALSE.

This VHDL'87 inefficiency comes about when the assert statement is used to unconditionally force messages onto the display screen.

```
assert FALSE
    report "Executed Line 55 in the process Transfer_IO"
    severity NOTE;
```

There is no other alternative in VHDL'87 because this version insists on the reserved word *assert*.

Recommended efficient alternative. You should exercise the VHDL'93 option to omit the reserved word *assert* and instead write only a report statement:

```
report "Executed Line 55 in the process Transfer_IO"
    severity NOTE;
```

In fact, the severity clause may be omitted altogether, and you may just write the following report statement (note that a semicolon is now required after the report clause):

```
report "Executed Line 55 in the process Transfer_IO";
```

When both the assert condition and the severity level are omitted, as in this example, then NOTE will be inferred as the report statement's severity level.

By the way, when the reserved word *assert* is used, the resulting statement is formally referred to as an assert statement to differentiate it from a report statement. An assert statement that omits the severity clause will infer an ERROR-valued severity level.

Chapter 9

Type Conversion Tricks and Methodologies

This section will highlight many of the techniques available to convert between data types. These conversion techniques utilize many of VHDL's advanced programming language capabilities. The aim of this chapter is to show that the topic of type conversions is both an art and a science.

The class and data type of each object used in this section will be reflected in their respective names. Consequently, it will not be necessary to provide a formal declaration statement for them.

Conversions via 'POS AND 'VAL. When the data types being converted belong to enumeration lists that have the same number of elements, then the attributes 'POS and 'VAL may be applied. A conversion between BIT and BOOLEAN may be achieved via

```
Boolean_Signal <= BOOLEAN'VAL(BIT'POS(Bit_Signal));

Bit_Signal <= BIT'VAL(BOOLEAN'POS(Boolean_Signal));
```

In this example the BIT and BOOLEAN enumeration literals lined up perfectly. But what if the number of listed items are different? And what if there is a positional skew in the respective enumeration lists? Recall that std_ulogic is defined as:

```
type std_ulogic is ('U', 'X', '0', '1', 'Z', 'W', 'L', 'H', '-');
```

To convert a BIT object into a std_ulogic object, you have to take into account that the std_ulogic element '0' is in position 2, whereas the BIT element '0' is in position 0. This difference must be used as a positional offset as follows:

```
Std_Ulogic_Signal <= std_ulogic'VAL(BIT'POS(Bit_Signal) + 2);
```

Another interesting application of 'POS and 'VAL is to convert from lower to upper case. The need to do case conversions may come about as follows. Suppose that the test vectors for your model contain CHARACTERs that must be read in and interpreted (decoded). To relieve the developer of these test vectors from having to remember whether to use lower or upper case characters you can allow them to use any case that they wish. Once these characters are read into your model you can then, if necessary, convert them into an upper case character. This way all the decoding activities of your VHDL program need only be concerned with upper case characters. The interesting point about this conversion algorithm is that it is an adaptation of a C programming trick into the VHDL domain. First a test is made to determine whether or not Char_Read is a lower case character. Next 'POS and 'VAL are used in conjunction with the special number 32. The significance of this magic number 32 is that it is the difference between the positional locations of upper case 'A' and lower case 'a'. By subtracting 32 you are linearly repositioning any lower case CHARACTER onto its upper case counterpart.

```
if (Char_Read >= 'a') and (Char_Read <= 'z') then
    Char_Read := CHARACTER'VAL(CHARACTER'POS(CHAR_READ) - 32);
end if;
```

It is worthwhile to note that 'POS and 'VAL techniques cannot be used to convert from std_ulogic to BIT since there are more std_ulogic elements (9) than BIT elements (2). Such a data-type conversion requires an altogether different approach that is shown next.

Conversions via table lookup. This conversion method requires that an array type declaration first be made. In our particular case since we are converting from std_ulogic to BIT an array data type must be declared having std_ulogic indices and BIT type elements:

```
type Std_Ulogic_To_Bit_Template is array (std_ulogic) of BIT;
```

In the Excursion (Sec. 2.27) you already saw how easy it is to work with an array that is indexed by elements from an enumeration data type.
Next a constant lookup table must be declared:

```
constant STD_ULOGIC_TO_BIT_TABLE : Std_Ulogic_To_Bit_Template
              := ('0', '0', '0', '1', '1', '0', '0', '1', '0');
              -- ('U', 'X', '0', '1', 'Z', 'W', 'L', 'H', '-')
```

The commented line displays how the nine elements of std_ulogic are to be mapped onto the two elements of BIT. The following is an example illustrating the application of this constant array to achieve our desired conversion:

```
Bit_Signal <= STD_ULOGIC_TO_BIT_TABLE(Std_Ulogic_Signal);
```

An analogous table lookup technique may also be used to convert from BIT to std_ulogic instead of relying on 'POS and 'VAL as we did earlier. The pertinent declarations for this method are:

```
type Bit_To_Std_Ulogic_Template is array (BIT) of std_ulogic;
constant BIT_TO_STD_ULOGIC_TABLE : Bit_To_Std_Ulogic_Template
                := ('0', '1');
```

The desired conversion can then be achieved via:

```
Std_Ulogic_Signal <= BIT_TO_STD_ULOGIC_TABLE(Bit_Signal);
```

Similarly, conversions from std_ulogic to CHARACTER can also be accomplished with a table lookup technique. The required declarations are:

```
type Std_Ulogic_To_Char_Template is array (std_ulogic) of CHARACTER;
constant STD_ULOGIC_TO_CHAR_TABLE : Std_Ulogic_To_Char_Template
                := ('U', 'X', '0', '1', 'Z', 'W', 'L', 'H', '-');
```

An example of this conversion is:

```
Char_Var := STD_ULOGIC_TO_CHAR_TABLE(Std_Ulogic_Var);
```

It also seems tempting to apply table lookup techniques to convert from CHARACTER to std_ulogic. However, such a technique is not practical because of the numerous character elements that would have to be listed in the constant array (128 in VHDL'87 and 256 in VHDL'93). Consequently, we are forced to search for a better solution.

Conversions via functions. Figure 9.1 illustrates how a function may be used to convert from CHARACTER to std_ulogic. The nonappropriate CHARACTERs are collectively handled by the when others arm of the case statement. An application of this conversion function is as follows:

```
Std_Ulogic_Var := To_Std_Ulogic (Char_Var);
```

Functions may also be used to convert between NATURAL and BIT_VECTOR data types. Excerpts of these functions were shown earlier in the Model Efficiency chapter. Conversions from BIT_VECTOR to NATURAL via the function To_Natural may also be applied toward an effi-

```
function To_Std_Ulogic (S : CHARACTER) return std_ulogic is
    variable Converted : std_ulogic;
begin
    case S is
        when 'U' => Converted := 'U';
        when 'X' => Converted := 'X';
        when '0' => Converted := '0';
        when '1' => Converted := '1';
        when 'Z' => Converted := 'Z';
        when 'W' => Converted := 'W';
        when 'L' => Converted := 'L';
        when 'H' => Converted := 'H';
        when '-' => Converted := '-';
        when others =>
            assert FALSE
                report "Unexpected character"
                severity ERROR;
            Converted := 'X';
    end case;
    return Converted;
end To_Std_Ulogic;
```

Figure 9.1 Subprogram body to convert from CHARACTER to std_ulogic.

cient instruction-decoding algorithm. The set of instruction mnemonics should first be listed in an enumeration data type. The key point is that each instruction's location in the list must be such that its 'POS value is the decimal equivalent of its bit pattern representation. For instance, suppose that the mnemonic ADD is associated with the bit pattern "00", which is equivalent to decimal zero. Then ADD would be the leftmost element of the instruction enumeration data type since this position has a 'POS value of 0. So, let's see all of this in action. First, the type declaration for all the available instructions must be declared as follows:

```
type Instruction_Repertoire is (ADD, SUBT, LOAD, STORE);
```

Suppose that the CPU's op-code is in bits 0 to 5 of the instruction register. The conversion of this bit-valued op-code into an instruction mnemonic can be achieved via:

```
Instruction_Mnemonic := Instruction_Repertoire'VAL(
                    To_Natural (Instruction_Register(0 to 5)));
```

The specific instruction corresponding to the current value of Instruction_Mnemonic can then be executed via the case statement technique that was shown in Chap. 8, Model Efficiency.

Conversions via 'IMAGE and 'VALUE. As shown in Chap. 4 these attributes may be applied to convert any scalar (single element) into a STRING data type and vice versa. Consequently, you should no longer have to apply the above CHARACTER-related conversions to read and/or write those enumeration literals that you have defined in your models. The only caveat here is the key word *scalar*. It is valid to reference std_ulogic'IMAGE(Std_Ulogic_Var), but it is illegal to write std_ulogic_vector'IMAGE(Std_Ulogic_Vector_Var) since std_ulogic_vector is not a scalar type or subtype. Instead, a for loop must be used to individually convert each std_ulogic typed element of the array Std_Ulogic_Vector_Var into a single-string element. Analogously, a for loop must also be applied when converting a STRING-typed object into an array-typed object such as std_ulogic_vector.

Conversions between closely related types. Let's ease into this subject by first converting between INTEGERs and REALs. VHDL views these two data types as being closely related. Consequently, it is possible to convert from INTEGER to REAL by merely parenthesizing an INTEGER data-type object and then prefixing the result with the type mark REAL. Here is how it looks in practice:

```
Real_Var := REAL(Integer_Var);
```

A conversion in the opposite direction can be similarly implemented as follows:

```
Integer_Var := INTEGER(Real_Var);
```

However, there is a slight portability problem here. Suppose that the number being converted is 27.5. VHDL does not specify whether this number should be rounded up to the INTEGER 28 or truncated to the INTEGER 27. VHDL's acceptance of both these values is intended to honor the differences in CPU representations of floating point numbers. Consequently, different platforms might yield different conversion results.

Let's ease into another example of the conversion between closely related types by first looking at the implicit conversion between Logic4 and LX01, which are defined as:

```
type Logic4 is ('X', '0', '1', 'Z');

subtype LX01 is Logic4 range 'X' to '1';
```

Since one is a subtype of the other, a formal conversion mechanism is not required. You can merely write:

```
Logic4_Var := LX01_Var;

LX01_Var := Logic4_Var;
```

However, since LX01 is only a proper subset of Logic4, it follows that the VHDL simulation engine will conduct a run time check to guarantee that any value assigned to LX01_Var lies within the range bounds defined for LX01. If Logic4_Var is within these range bounds, then all is well, and the assignment will be made to LX01_Var. However, if Logic4_Var is equal to 'Z', which is outside of LX01's defined range bounds, then the assignment to LX01_Var will not be made. In fact, the simulation will halt and a constraint error message will be displayed. Clearly, there should not be any range checking when assigning LX01_Var to Logic4 since every element of LX01 is already known to be an element of Logic4. In a nutshell, the key point of this example is that when subtypes are involved then, depending on the situation, range checking might be executed by the VHDL simulation engine.

Now, for comparative purposes, let's look into implicit conversions between std_ulogic and std_logic. Recall that std_logic is defined in the IEEE standard std_logic_1164 package as:

```
subtype std_logic is resolved std_ulogic;
```

Note well that std_logic is not a subtype in the same spirit as LX01 is. Yes, it is true that std_logic is declared as a subtype of std_ulogic. But std_logic is only a subtype by virtue of its association with the resolution function called resolved. It is not a subtype in the sense of a proper subset of a parent data type as was the case for LX01 and Logic4. Every element of std_ulogic is also an element of std_logic and vice versa. They are one and the same except that std_logic is associated with a resolution function, whereas std_ulogic is not. Furthermore, since std_ulogic and std_logic consist of the same elements, there is no need for the tool suite to do any type checking when an assignment is being made between objects that belong to these two data types. The following are two assignments between objects of type std_ulogic and std_logic:

```
Std_Logic_Signal <= Std_Ulogic_Signal;

Std_Ulogic_Signal <= Std_Logic_Signal;
```

With all this valuable information, let's now investigate conversions between objects of type std_ulogic_vector and std_logic_vector. Here are their definitions again for quick reference:

```
type std_ulogic_vector is array (NATURAL range <>) of std_ulogic;

type std_logic_vector is array (NATURAL range <>) of std_logic;
```

The key point to observe is that std_logic_vector is a type in its own right. Std_logic_vector is not a subtype of std_ulogic_vector. Consequently, the following assignments are incorrect:

```
Std_Logic_Vector_Signal <= Std_Ulogic_Vector_Signal; -- ERROR

Std_Ulogic_Vector_Signal := Std_Logic_Vector_Signal; -- ERROR
```

By the way, there is a legitimate need for such an assignment. Internal to your modeled device you might be working with objects of type std_ulogic_vector. But suppose that an output array is to be connected to a tristated bus having multiple sources. Then each element of this output array must be associated with the resolution function called resolved that is provided in the std_logic_1164 package. Hence, this output array must be of type std_logic_vector, since each element of this array type is associated with the resolution function resolved. And so you must somehow convert from std_ulogic_vector to std_logic_vector. As per the aforementioned described techniques you already can come up with two conversion possibilities: lookup tables or conversion functions. However, a lookup table approach is unfeasible because one fixed table cannot handle the totality of the infinite array lengths that are possible. Then what about a function conversion technique? In fact, the IEEE std_logic_1164 package supplies you with functions for these conversions, one of which follows (written exactly as it appears in the std_logic_1164 package):

```
FUNCTION To_StdLogicVector (s : std_ulogic_vector)
         RETURN std_logic_vector is
   ALIAS sv : std_ulogic_vector (s'LENGTH - 1 downto 0) is s;
   VARIABLE result : std_logic_vector ( S'LENGTH - 1 downto 0);
BEGIN
   FOR i in result'RANGE LOOP
      result(i) := sv(i);
   END LOOP;
   RETURN result;
END;
```

A call to this conversion function may be invoked as follows:

```
Std_Logic_Vector_Signal <= To_StdLogicVector(Std_Ulogic_Vector_Var);
```

There are numerous objections that I have with this function, the least serious of which is the name To_StdLogicVector. Yes, everyone is allowed to have their own personal VHDL style guide as long as there is an effort to maintain some degree of consistency. As it stands, the identifier To_StdLogicVector is not only written inconsistently, it is also an eyesore. It would have been more consistent and readable to have used To_Std_Logic_Vector. I also dislike uppercase reserved words and lowercase loop iterators. Enough said about trivial matters. Let's move on to some meatier issues. This proposed functional approach is inefficient. First, there is the overhead of making a function call. And second, there is the iterative assignment of sv(i) to result(i). If this function is called with a 64-bit array, then this assignment will require 64 reads and writes. Precious CPU time will be required to make all these assignments. So let's look for a more efficient way to do this conversion.

Recall that every element of std_ulogic_vector is of type std_ulogic and every element of std_logic_vector is of type std_logic. Hence, each element of std_logic_vector is a subtype of each element of std_ulogic_vector. And so, as per the VHDL rules, these two array types are closely related. Therefore, conversions between std_ulogic_vector and std_logic_vector may be done very simply by using the parenthesizing conversion technique that was shown earlier. Here is how these conversions may be implemented:

```
Std_Logic_Vector_Signal <= std_logic_vector(Std_Ulogic_Vector_Var);

Std_Ulogic_Vector_Var := std_ulogic_vector(Std_Logic_Vector_Signal);
```

Based on our earlier discussion regarding assignments between single elements of type std_logic and std_ulogic, you now know that range checking is not required for assignments between these data types. Consequently, range checking of the individual elements should not be done when converting between std_ulogic_vector and std_logic_vector. Can you think of any other tests that must be done when converting between std_ulogic_vector and std_logic_vector? Neither can I. Consequently, in this particular case of conversions between closely related types the only purpose of the given parenthesizing technique is to acquiesce the compiler. Nothing extra should be done. But if range checking is being done during simulation, then here is a golden opportunity for a vendor to improve the run time efficiency of his/her VHDL product. Clearly, the closely related conversion method easily beats the function technique in that all-important efficiency race.

Conversion using the division operator. In case you ever have a need to convert from type TIME to a NATURAL number, you can use the following technique:

```
Natural_Var := Time_Var / 1 ns;
```

If Time_Var currently is 20 ns, then the result of this computation will be the number 20.

Incidentally, the following method using 'POS would have converted 20 ns into the number 20*(10**6), which might not be appropriate for your intended algorithm:

```
Natural_Var := TIME'POS(Time_Var);
```

Conversion using the multiplication operator. A conversion from NATURAL to TIME can easily be accomplished via:

```
Time_Var := Natural_Var * 1 ns;
```

Conversion into the appropriate STD_LOGIC strength. First, let's see how the need for such a conversion comes about. Suppose that internal to your model you are manipulating a BIT_VECTOR object. After completing a certain algorithm you now wish to place the BIT_VECTOR result onto a modeled wired-or bus. If you would be working with the MVL4_Pkg, then the required conversion would be analogous to the technique used in the Excursion to create the look and feel of a tristated bus (refer to Fig. 2.70 of Sec. 2.29). First, you would apply the overloaded conversion function To_Logic4 to convert your BIT_VECTOR object into Logic4_Vector and then continue with a conversion to Wired_Or_RS_Vector using closely related array techniques. This converted result may now be assigned to an out port, which is also of type Wired_Or_Rs_Vector. Since each element of the data type Wired_Or_RS_Vector is associated with the resolution function Wired_Or_RF, the net effect will be that this out port will have the look and feel of being connected to a wired-or bus. All of this should be a review from the Excursion part of this book.

Now, suppose that the IEEE std_logic_1164 package environment is to be used. Instead of three resolution functions, as in the case of our MVL4_Pkg, this IEEE standard data type package has only one. Different bussing schemes are now modeled by assigning the appropriate strength onto a resolved signal. The manner in which this strength value interacts with the other strength values, as defined by the sole resolution function, will determine the actual bussing scheme that is being modeled. For instance, the assignment of a '0' to a signal associated with the resolution function resolved will create the look and feel of a '0' being placed onto a tristate bus. However, to create the effect of sending a '0' onto a modeled wired-or bus, you must transmit the weak zero 'L' instead of the strong zero '0'. The terms weak and strong have nothing to do with voltage strengths. Rather, they have to do with their respective dominating characteristics as defined by the resolution function called resolved. For instance, if the two sources driving the same signal are an 'L' and a '1', then their effective resolved value will be a '1'. On the other hand, a '0' combined with a '1' will result in an 'X'. In the former case '1'

```vhdl
function Drive_Wired_Or (V : BIT_VECTOR) return std_logic_vector is
   variable Result : std_logic_vector(V'RANGE);
begin
   for K in V'RANGE loop
      case V(K) is
         when '0' => Result(K) := 'L';
         when '1' => Result(K) := '1';
      end case;
   end loop;
end Drive_Wired_Or;
```

Figure 9.2 Subprogram body to convert a BIT_VECTOR into a wired-or strength value.

is said to dominate over 'L', whereas in the latter case '1' and '0' are said to be of the same strength. You can now see the potential problem that exists when working with the sole resolution function from the std_logic_1164 package. A modeler must always remember what strength to assign to a signal based on the type of bus being modeled. But most VHDL projects are quite demanding, and a modeler should not have to deal with the additional burden of remembering the various strength pattern combinations required to model a specific bus. So to relieve your VHDL colleagues of this burden I recommend that you write the following three functions: Drive_Tristate, Drive_Wired_Or, and Drive_Wired_And. Furthermore, they should be overloaded to support whatever nonresolved data type you are using inside your VHDL model. It should be emphasized that these subprograms are not resolution functions. Their sole task is to convert, if necessary, its input into an L (weak zero) or into an H (weak one), depending on the specific bussing scheme being modeled. This way the modeler only has to call the appropriate subprogram and assign the desired '0' or '1' value. Figure 9.2 shows the body for the function Drive_Wired_Or.

Several applications of this function are as follows:

```
Wired_Or_Std_Logic_Vctr_Sgnl <= Drive_Wired_Or (Bit_Vector_Var);
Wired_Or_Std_Logic_Vctr_Sgnl <= Drive_Wired_Or (BIT_
                                       VECTOR'("00110101"));
```

The BIT_VECTOR qualification may be required if overloaded Drive_Wired_Or subprograms were written to support BIT_VECTOR and std_ulogic_vector typed input parameters. The string literal "00110101" may be interpreted as either of these two data types. The given qualification informs the compiler to view this string literal as a BIT_VECTOR. This way the compiler will know which of the two overloaded subprograms should be called.

Part 4

Experimenting with VHDL

Chapter 10

Overview

10.1 Background

Several years ago I began writing a collection of VHDL experiments to isolate and explore:

- VHDL syntax and semantics.
- VHDL language subtleties.
- Conceptual modeling errors.
- New coding strategies.

These experiments may be viewed collectively as a logic analyzer that probes the VHDL language constructs and their concurrent interactions. Additionally, these experiments may also be thought of as a series of stress tests to evaluate the compliance of a VHDL product to the official IEEE 1076 standard. In this capacity, these experiments may be perceived as a subset of an all-encompassing VHDL validation suite.

There is no gradient pattern to these tests. Their identification scheme (Test_N) does not necessarily reflect their level of conceptual difficulty. Rather, their chronological order is a function of the sequential flow of VHDL topics that I have encountered in the real world of VHDL applications and projects.

The overall objectives of these experiments are as follows:

- To challenge and reinforce your understanding of VHDL.
- To provide a hands-on environment in which you can experience real-world VHDL coding scenarios.
- To enhance your VHDL problem-solving skills.
- *To have fun!*

10.2 Conducting the Experiments

You should first read the objectives of each test and then mentally run through its corresponding source code as if you would be a VHDL compiler and simulator. While traversing the experiment's source code you should at the same time make a conjecture about its behavior. The experiment should then be concluded by comparing your hypothesis with the expected observations that are listed with each test. If you have a VHDL tool suite available, then you may, instead, tangibly check out your conjecture by manually going through the exercise of compiling and then single stepping through each experiment's source code. Several experiments assume that your tool suite allows you to traverse the source code during the VHDL initialization phase. Be aware that failure to compile might be a key element of the test.

As you are conducting these experiments, keep in mind that they were all motivated by real-world VHDL coding scenarios. These experiments abstract many of the key issues and potential problems that you will come across in your day-to-day VHDL activities.

Chapter 11
The Experiments

In all, I have designed and tested well over 300 experiments. This chapter contains a sample of my favorite ones. As you get into the rhythm of these experiments, you will realize how easy it is to use a preexisting test as a template to generate newer ones. Each subsection below is devoted to a self-contained experiment and is concluded with the corresponding VHDL source code listing.

It should be kept in mind that the main thrust of each experiment is to focus on an individual VHDL construct or concept. The efficiency techniques shown in Chap. 8 might not have been applied.

11.1 Test_3 (wait until...)

Objective. To explore the VHDL statement:

```
wait until Sample = '1';
```

What will happen if Sample is already '1' when this wait statement is sequentially reached? Will the hosting construct (process or procedure) continue onto the next sequential statement or will it be suspended?

Expected observations. The assert statement in the Receiver process should not be executed, since the wait construct is waiting first for an event on the signal Sample. Only then will the condition (Sample = '1') be checked. Hence, the Receiver process will be suspended until Sample is updated from a non-'1' value to a '1'. The unconditional wait statement at the process' bottom should then terminate this process' further execution.

278 Experimenting with VHDL

Additional techniques. If you are modeling a latchlike device and do not wish to wait for an event when the condition is already satisfied, then the following code segment may be used:

```
if Sample /= '1' then

    wait on Sample;  -- Assumes that Sample is of type BIT.

end if;
```

Source code for experiment

```
-- File Name : test_3.vhd
--
-- Author    : Joseph Pick
--
entity Test_3 is
end Test_3;

architecture Behave_1 of Test_3 is
    signal Sample : BIT;
begin

    Sender:
    process
    begin
        Sample <= '1';
        wait for 20 ns;
        assert FALSE
            report "Terminated in Sender"
            severity NOTE;
        wait;
    end process;

    Receiver:
    process
    begin
        wait for 10 ns;
        wait until Sample = '1';
        assert FALSE
            report "Terminated in Receiver"
            severity NOTE;
        wait;
    end process;

end Behave_1;
```

11.2 Test_5 (wait on...until...)

Objective. To determine if the signal Enable is in the sensitivity list of the following VHDL statement:

```
wait on A, B until Enable = '1';
```

Expected observations. The assert statement in the Receiver process should not be executed, even though the signal Enable received an event and its updated value satisfies the given wait condition. This wait construct should first wait for an event on either signal A or signal B. Only then should the condition (Enable = '1') be checked.

Caveats. Those signals occurring in the until clause of a wait on...until...statement are not members of the corresponding sensitivity list. Compare this situation to the one highlighted in Test_3. In that experiment the sensitivity list is defined by those signals occurring in the until clause, because a wait on...construct was not explicitly specified. As these examples show, you must always be aware of what is in the sensitivity list when working with the different variants of the wait construct. Otherwise, your VHDL model will not behave as expected.

Source code for experiment

```
-- File Name : test_5.vhd
--
-- Author    : Joseph Pick

entity Test_5 is
end Test_5;

architecture Behave_1 of Test_5 is
signal  A : BIT := '0';
signal  B : BIT := '0';
signal  Enable : BIT := '0';
begin

  Sender:
  process
  begin
    Enable <= '1';
    wait for 20 ns;
    assert FALSE
      report "Terminated in Sender"
      severity NOTE;
    wait;
  end process;

  Receiver:
  process
  begin
    wait on A,B until Enable = '1';
    assert FALSE
      report "Terminated in Receiver"
      severity NOTE;
    wait;
  end process;

end Behave_1;
```

11.3 Test_7 (Simulation cycle)

Objective. To explore the VHDL simulation cycle via the statement:

```
wait on A, B until Enable = '1';
```

Signals A and Enable will simultaneously receive an event during the same simulation cycle (delta time). The event on A will then trigger the examination of the wait condition. This deliberately conceived scenario now highlights the following fundamental questions: Will the updating of Enable to its new value '1' occur before or after the inspection of this wait condition? Is there a well-defined, deterministic order that precisely stipulates the exact sequence in which signals are updated and processes are awakened?

Expected observations. Signals A and Enable should both be updated during the first half of the simulation cycle that occurs at 0 ns plus 1 delta time unit. The process waiting for an event on signal A should then be awakened during the second half of this same simulation cycle. But, by that time, Enable should already have the value '1' because it was updated to this value during the first half of this same simulation cycle. Hence the Receiver process should continue and execute its encapsulated assert statement.

Source code for experiment

```
-- File Name : test_7.vhd
--
-- Author    : Joseph Pick

entity Test_7 is
end Test_7;

architecture Behave_1 of Test_7 is
   signal A : BIT := '0';
   signal B : BIT := '0';
   signal Enable : BIT := '0';
begin

   Sender:
   process
   begin
     A <= '1';
     Enable <= '1';
     wait for 20 ns;
     assert FALSE
       report "Terminated in Sender"
       severity NOTE;
     wait;
   end process;

   Receiver:
   process
   begin
     wait on A,B until Enable = '1';
     assert FALSE
       report "Terminated in Receiver"
       severity NOTE;
     wait;
   end process;

end Behave_1;
```

11.4 Test_11a (wait until compound expression)

Objective. To investigate the ramifications of using the statement:

```
wait until ((Clock = '1') or (Clear = '0'));
```

Expected observations. The assertion statement in the Receiver process should be executed since the event on Clock should trigger the evaluation of the wait condition. So, even though Clock no longer has the value '1', the other or'ed condition (Clear = '0') is TRUE. Therefore the whole conditional expression is now consequently TRUE. Hence, the process should continue and execute the assert construct.

Caveats. You should always be aware of the ramifications of using a compound conditional expression when working with the VHDL construct, wait until.

Additional techniques. If necessary a finer granularity of control may be achieved via:

```
wait until ((Clock = '1' and Clock'EVENT) or
            (Clear = '0' and Clear'EVENT));
```

Source code for experiment

```
-- File Name : test_11a.vhd
--
-- Author    : Joseph Pick

entity Test_11a is
end Test_11a;

architecture Behave_1 of Test_11a is
  signal Clock : BIT := '1';
  signal Clear : BIT := '0';
begin

  Sender:
  process
  begin
    wait for 10 ns;
    Clock <= '0';
    wait for 20 ns;
    assert FALSE
      report "Terminated in Sender"
      severity NOTE;
    wait;
  end process;

  Receiver:
  process
  begin
    wait until ((Clock = '1') or (Clear = '0'));
    assert FALSE
      report "Terminated in Receiver"
      severity NOTE;
    wait;
  end process;

end Behave_1;
```

11.5 Test_13 (Infinite oscillation in delta time domain)

Objective. To illustrate a model that oscillates infinitely in the delta time domain.

Expected observations. The delta delay scheduling of signals A and B along with the wait conditions existing in processes Inc_A and Inc_B are such that simulation time should only advance in the delta time domain. Absolute simulation time should remain at 0 ns even though VHDL simulation time is sequencing through 0 ns plus N delta time units, where N goes from 1 out to infinity. Consequently, the if test in both processes should fail since the function Now returns only the current absolute simulation time, which is still 0 ns. Consequently, neither of the assert statements in process Inc_A or Inc_B should be executed. Instead, the simulation flow should oscillate infinitely between the processes Inc_A and Inc_B until the simulation is forcibly terminated by the user (Control-C) or by the VHDL simulation engine (current simulation cycle count exceeds a system- or user-specified default value).

Source code for experiment

```
-- File Name : test_13.vhd
--
-- Author    : Joseph Pick
--
entity Test_13 is
end Test_13;

architecture Behave_1 of Test_13 is

  signal A : NATURAL := 1;
  signal B : NATURAL := 1;

begin

  Inc_A:
  process
  begin
    A <= A + 1;
    wait on B;
    if (Now > 0 ns) then
      assert FALSE
        report "Absolute simulation time has progressed : Inc_A"
        severity NOTE;
    end if;
  end process;

  Inc_B:
  process
  begin
    wait on A;
    if (Now > 0 ns) then
      assert FALSE
        report "Absolute simulation time has progressed : Inc_B"
        severity NOTE;
    end if;
    B <= B + 1;
  end process;

end Behave_1;
```

11.6 Test_13a (Deadlock)

Objective. To illustrate a model that creates a deadlock.

Expected observations. In this experiment each process is waiting for an event on a signal that only the other process will induce. Hence, neither of the assert statements should be reached. It is very instructive to observe the behavior of your simulator under such a deadlock scenario. Every VHDL simulator has some sort of a time-ordered queue into which scheduled activities are placed. In our current experiment the simulator should execute the initialization phase and then terminate since there is essentially nothing for it to do. Under these circumstances most simulators will generate a message stating that simulation has been halted since all the scheduling queues are empty.

Source code for experiment

```
-- File_Name : test_13a.vhd
--
-- Author    : Joseph Pick
--
entity Test_13a is
end Test_13a;

architecture Behave_1 of Test_13a is

   signal A : NATURAL := 1;
   signal B : NATURAL := 1;

begin

   Inc_A:
   process
   begin
     wait on B;
     A <= A + 1;
     assert FALSE
        report "Simulation time has progressed : Inc_A"
        severity NOTE;
   end process Inc_A;

   Inc_B:
   process
   begin
     wait on A;
     B <= B + 1;
     assert FALSE
        report "Simulation time has progressed : Inc_B"
        severity NOTE;
   end process Inc_B;

end Behave_1;
```

11.7 Test_15 (Subtype constraint checking)

Objective. To explore how VHDL manages subtype objects.

Expected observations. A simulation run time error should occur because the assigned value to Var_LX01_ZZ is outside the bound constraints specified for the subtype LX01. A compilation error should not occur because the VHDL compiler is never aware of the function's returned value. But the VHDL compiler should generate additional range checking code to ensure that the function's returned value is within the range defined by the targeted signal's subtype LX01.

Caveats. Since the VHDL simulator must execute additional range-checking code, the run time efficiency of a model is reduced when working with subtype objects. Hence, the usage of subtypes should be used only when truly deemed necessary.

Additional information. Most VHDL tool suites give its users the option to turn off this range-checking feature. Though your simulation will run faster, this should only be done when you have a high level of confidence in your VHDL model. Otherwise, unexpected and erratic behavior will occur during your simulation that will be very difficult to track down, isolate, and resolve.

Source code for experiment

```
-- File Name : test_15.vhd
--
-- Author    : Joseph Pick

entity Test_15 is
end Test_15;

architecture Behave_1 of Test_15 is
begin

  Subtype_User:
    process
       type Logic4 is ('X','0','1','Z');
       subtype LX01 is LOGIC4 range 'X' to '1';
       type Logic4_Table is array (LOGIC4,LOGIC4) of LOGIC4;

       constant Table : Logic4_Table := (('X','0','X','X'),
                                         ('0','0','0','0'),
                                         ('X','0','1','X'),
                                         ('X','0','X','Z'));

       function Sample (L,R : Logic4) return Logic4 is
       begin
          return Table(L,R);
       end Sample;

       variable Var_LX01_XZ : LX01;
       variable Var_LX01_OX : LX01;
       variable Var_LX01_Z1 : LX01;
       variable Var_LX01_ZZ : LX01;

    begin
       Var_LX01_XZ := Sample('X','Z');
       Var_LX01_OX := Sample('0','X');
       Var_LX01_Z1 := Sample('Z','1');
                         -- During simulation the following line
                         -- should result in a run-time error.
       Var_LX01_ZZ := Sample('Z','Z');
       wait;
    end process;

end Behave_1;
```

11.8 Test_16 (Blocks and 'EVENT vs. 'STABLE)

Objective. To determine if an edge-triggering device may be modeled with a block construct having 'EVENT in its GUARD expression. Additionally, the behavior of guarded signal assignments are also investigated.

Expected observations. The block statement Blck_Test_1 having Clock'EVENT should behave as a latch and not as an edge-triggering device. On the other hand, when (not Clock'STABLE) is used, an edge-triggering device should be observed. Note also that the nonguarded signal assignments within a block construct should not be influenced by the block's GUARD expression. Instead, they should behave like any other concurrent signal assignment.

Caveats. Though equivalent in value to (not Clock'STABLE), Clock' EVENT cannot be used in the GUARD expression of a block that is modeling an edge-triggered device. Remember from the Excursion (Sec. 2.15) that a block's GUARD is only updated when an event occurs on any of its signal constituents. Since Clock'EVENT is not a signal, changes to its value will not cause the GUARD to be subsequently modified as well. If you are not aware of these key distinguishing details, then your VHDL model will not behave as desired.

Additional techniques. If Clock would have been initialized to '1', then the expected latchlike behavior of the block labeled Blck_Test_1 would not have been honored during the initialization phase. The central point of this experiment is to highlight that 'EVENT cannot be used to model an edge-triggering device. That is why 'EVENT was used in the GUARD expression. But to be precise a latch should be modeled without this attribute so that an assignment might conditionally be scheduled to occur because of the effects of the initialization phase. This modeled behavior would mirror the functionality of real hardware. Real-world signals will ripple into a latch even at startup if the latch enable is active. The correct GUARD expression for a latch, which is level sensitive to '1', should simply be:

```
Block_Latch:

block (Clock = '1');
```

Source code for experiment

```
-- File Name : test_16.vhd
--
-- Author    : Joseph Pick

entity Test_16 is
end Test_16;

architecture Behave_1 of Test_16 is
    signal  Source : NATURAL := 0;
    signal  Destination_1 : NATURAL := 0;
    signal  Destination_2 : NATURAL := 0;
    signal  Destination_3 : NATURAL := 0;
    signal  Destination_4 : NATURAL := 0;
    signal  Clock : BIT := '0';
begin

Blck_Test_1:
block (Clock = '1' and Clock'EVENT)
begin
    Destination_1 <= guarded Source;
    Destination_2 <= Source;
end block Blck_Test_1;

Blck_Test_2:
block (Clock = '1' and (not Clock'STABLE))
begin
    Destination_3 <= guarded Source;
    Destination_4 <= Source;
end block Blck_Test_2;

Monitor:
process
    variable Source_Var : NATURAL;
    variable Dest_1_Var : NATURAL;
    variable Dest_2_Var : NATURAL;
    variable Dest_3_Var : NATURAL;
    variable Dest_4_Var : NATURAL;
begin
    Source_Var := Source;
    Dest_1_Var := Destination_1;
    Dest_2_Var := Destination_2;
    Dest_3_Var := Destination_3;
    Dest_4_Var := Destination_4;
    wait on Destination_1, Destination_2,
            Destination_3, Destination_4;
end process Monitor;

Tick_Tock:
process
begin
    wait for 10 ns;
    Clock <= not Clock;
end process Tick_Tock;

Source_Wave: Source <= 1 after 8 ns,  2 after 15 ns,
                      3 after 16 ns,  4 after 17 ns,
                      5 after 18 ns,  6 after 19 ns;

end Behave_1;
```

11.9 Test_18b (Signal driver in inertial model)

Objective. To investigate how VHDL handles:

- Concurrent signal assignments.
- Projected signal drivers.

Expected observations. The projected driver for signal C should be updated as listed below. This projected driver may best be understood by analyzing the behavior of the process that is equivalent to the given, concurrent signal assignment. Note also that there should not be a preemption of the projected driver because the same value '0' is being scheduled on each inertial update. This non-preemption status should be contrasted with the preemption that takes place in Test_18c.

```
Time         Current status of C's projected driver
0 ns         {('0', 40 ns)}
5 ns         {('0', 40 ns), ('0', 45 ns)}
12 ns        {('0', 40 ns), ('0', 45 ns), ('0', 52 ns)}
```

Signal C's projected driver should then cause the following transactions to be observed:

```
Time of transaction                         Value of C
     40 ns                                     '0'
     45 ns                                     '0'
     52 ns                                     '0'
```

Caveats. Note that C'TRANSACTION fired off at 40 ns because of the signal scheduling that occurred during the initialization phase. This triggering of C'TRANSACTION might not be a desirable occurrence in your VHDL model. Hence, in general, you must always be aware of the ramifications of the initialization phase on those processes that are waiting for a signal transaction.

Additional techniques. To avoid initialization phase-induced 'TRANSACTION problems you should replace the concurrent signal assignment with a process that is analogous to the following code segment:

```
process
begin
   wait on A, B;  -- Initialization phase will stop here.
   C <= A and B after 40 ns;
end process;
```

Source code for experiment

```
-- File Name : test_18b.vhd
--
-- Author    : Joseph Pick

entity Test_18b is
end Test_18b;
```

```
architecture Behave_1 of Test_18b is
    signal A : BIT := '0';
    signal B : BIT := '0';
    signal C : BIT := '0';
begin

    Gen_Wave:
    process
    begin
        A <= '1' after 5 ns, '0' after 12 ns;
        wait;
    end process Gen_Wave;

    Analysis_C:
    process
        variable Var_C : BIT := '0';
    begin
        wait on C'TRANSACTION;
        Var_C := C;
    end process Analysis_C;

    Update_C: C <= A and B after 40 ns;
end Behave 1;
```

11.10 Test_18c (Glitches in inertial model)

Objective. To investigate how VHDL handles:

- Concurrent signal assignments.
- Projected signal drivers.
- Signal glitches in an inertial timing environment.

Expected observations. In the given, concurrent signal assignment the value of the expression (A or B) must persist for at least 40 ns or else it should not contribute to the updating of signal C. As per the VHDL inertial timing model, values that are maintained for less than the specified delay time should be interpreted as a glitch and should not be detected by the targeted signal of the assignment. In this particular example the specified delay is 40 ns and hence every projected update to C must persist for this time interval. Since the '0' and '1' updates only last for 5 and 7 ns, respectively, they should subsequently be deleted from C's driver. The VHDL inertial modeling algorithm implements this deletion by noting that the new value '1' added to the projected driver at 5 ns is not the same as the previously added value '0'. Therefore, as per the algorithm, the previous value must be deleted from the projected driver, thus ensuring that the 5 ns glitch '0' will not be assigned to the targeted signal C. Similarly, the 7 ns glitch should also not be propagated over to the targeted signal. The chronological status of C's projected driver is given below:

Time	Current status of C's projected driver
0 ns	{('0', 40 ns)}
5 ns	{~~('0', 40 ns)~~, ('1', 45 ns)}
12 ns	{~~('1', 45 ns)~~, ('0', 52 ns)}

Signal C's projected driver should then cause the following transaction to be observed:

```
Time of transaction    Value of C
        52 ns             '0'
```

Caveats. In an inertial timing assignment, the destination signal will not always take on the scheduled values. Preemption may occur in the signal's projected driver.

Source code for experiment

```
-- File Name : test_18c.vhd
--
-- Author    : Joseph Pick

entity Test_18c is
end Test_18c;

architecture Behave_1 of Test_18c is
    signal A : BIT := '0';
    signal B : BIT := '0';
    signal C : BIT := '0';
begin
Gen_Wave:
process
begin
    A <= '1' after 5 ns, '0' after 12 ns;
    wait;
end process Gen_Wave;

Analysis_C:
process
    variable Var_C : BIT := '0';
begin
    wait on C'TRANSACTION;
    Var_C := C;
end process Analysis_C;

Update_C: C <= A or B after 40 ns;

end Behave_1;
```

11.11 Test_18d (Glitches in transport model)

Objective. To investigate how VHDL handles

- Concurrent signal assignments.
- Projected signal drivers.
- Signal glitches in a transport timing environment.

Expected observations. Signal C should be updated whenever an event occurs on signal A or signal B. In this transport timing model C's projected driver should not be preempted even though the new added value '1' is different from the value '0' that already exists inside of this projected driver. Contrast this algorithm with the one that was used in

Test_ 18c. The projected driver for signal C should be updated and appended as follows:

```
Time        Current status of C's projected driver
0 ns        {('0', 40 ns)}
5 ns        {('0', 40 ns), ('1', 45 ns)}
12 ns       {('0', 40 ns), ('1', 45 ns), ('0', 52 ns)}
```

Signal C's projected driver should then cause the following transactions to be observed:

```
Time of transaction              Value of C
    40 ns                           '0'
    45 ns                           '1'
    52 ns                           '0'
```

Caveats. In a transport timing assignment, the destination signal may respond to glitches on the sourcing signal.

Source code for experiment

```
-- File Name : test_18d.vhd
--
-- Author    : Joseph Pick

entity Test_18d is
end Test_18d;

architecture Behave_1 of Test_18d is
    signal A : BIT := '0';
    signal B : BIT := '0';
    signal C : BIT := '0';
begin

Gen_Wave:
process
begin
    A <= '1' after 5 ns, '0' after 12 ns;
    wait;
end process Gen_Wave;

Analysis_C:
process
    variable Var_C : BIT := '0';
begin
    wait on C'TRANSACTION;
    Var_C := C;
end process Analysis_C;

Update_C: C <= transport A or B after 40 ns;

end Behave_1;
```

11.12 Test_113 (Transport driver preemption)

Objective. In the VHDL literature preemption of a projected driver in a transport delay environment is often demonstrated via a variation of the following sequential code:

```
Destination <= transport Source_2 after 26 ns;
Destination <= transport Source_1 after 22 ns;
```

Observe that the second assignment to Destination is to occur at an earlier time than the first assignment. Consequently, the first assignment will be preempted from Destination's projected driver. Destination will never be assigned the Source_2 value. Though correct, such an example of a transport-related preemption is very artificial and unrealistic. A VHDL modeler is not going to consciously make assignments to the same signal in two consecutive sequential statements such that the second assignment is scheduled to occur earlier in time than the first. The purpose of this experiment is to present a modeling scenario that mirrors the previous driver preemption but has a more realistic scope.

Expected observations. As the simulation progresses the following modifications to signal A's projected driver should be observed:

```
    Time              Current status of A's projected driver
    10 ns                     {('0', 26 ns)}
    12 ns                     {('0', 26 ns), ('1', 22 ns)}
```

Note well that the two scheduled assignments to signal A produce a situation that is similar to the earlier consecutive assignments to Destination. In both cases the second assignment is scheduled to occur before the first one. This timing inconsistency is resolved in VHDL by the deletion of any scheduled assignment from the driver that is to occur at a simulation time that is greater than or equal to the most recent addition to the projected driver. In this particular example the preempted driver should produce the following single transaction on A:

```
    Time of transaction                    Value of A
          22 ns                           '1' (NOT AN EVENT)
```

Caveats. Drivers may be preempted even in a transport timing model.

Additional information. The VHDL rule that the time values in a projected driver must be in an ascending order also carries over to inertial delay models. If the reserved word *transport* would be deleted from Test_113, then the projected driver would still be preempted, and the simulation flow of the resulting inertial delay model would be identical to that of the transport delay model that was just investigated.

By the way, did you notice the new modeling technique that was subtly infused into Test_113? For the first time in this book the delay time given in a signal assignment after clause was a function's returned value. Let your imagination run wild with this capability and you will realize that this implementation of high-to-low and low-to-high propagation timings just scrapes the surface. This VHDL feature opens the door to many other innovative algorithmic-based timing delays.

Source code for experiment

```
-- File Name : test_113.vhd
--
-- Author    : Joseph Pick

entity Test_113 is
end Test_113;

architecture Behave_1 of Test_113 is
    signal A : BIT := '1';
    signal B : BIT := '1';
    signal C : BIT := '1';

    function Delay(Value : BIT; Phl : TIME; Plh : TIME)
                            return TIME is
    begin
       if Value = '1' then
          return Plh;
       else
          return Phl;
       end if;
    end Delay;
begin
    Schedule_A:
    process
    begin
       wait on B, C;
       A <= transport B and C after Delay(B and C, 16 ns, 10 ns);
    end process Schedule_A;

    Analysis_A:
    process
      variable Var_A : BIT := '0';
    begin
      wait on A'TRANSACTION;
      Var_A := A;
    end process Analysis_A;

    B <= '0' after 10 ns, '1' after 12 ns;

end Behave_1;
```

11.13 Test_18e (More glitches in inertial model)

Objective. To further investigate how VHDL handles:

- Concurrent signal assignments.
- Projected signal drivers.
- Signal glitches in an inertial timing environment.

In this experiment signal A will have the value '1' as a glitch, and signal B will be updated with this same value '1' during this glitch. This test will determine in what way, if any, the signal A's glitch may contribute toward the updating of the destination signal C.

Expected observations. Signal C should be updated 40 ns after the occurrence of '1' on the signal A even though the duration of this '1' valued pulse is viewed as a glitch relative to the specified 40 ns delay time. The key point in this experiment is that the '1' value of signal B should cause the whole expression (A or B) to maintain this value for a duration greater than or equal to the specified projection time of 40 ns. The propagation of the right-hand side over to the left-hand side should depend solely on the net compound value irrespective of any glitches on the individual constituents of the compound expression. This is the way that real hardware behaves. So it should come as no surprise that VHDL's updating algorithms were designed to fully support this real-world functionality. The projected driver for signal C should be updated and modified as follows:

```
Time            Current status of C's projected driver
0 ns            {('0', 40 ns)}
5 ns            {~~('0', 40 ns)~~, ('1', 45 ns)}
8 ns            {('1', 45 ns), ('1', 48 ns)}
12 ns           {('1', 45 ns), ('1', 48 ns), ('1', 52 ns)}
```

Signal C's driver should then cause the following transactions to be observed:

```
Time of transaction                         Value of C
        45 ns                                   '1'
        48 ns                                   '1'
        52 ns                                   '1'
```

Observe that 45 ns (40 ns + 5 ns) will be 40 ns after the occurrence of '1' on signal A that happened at simulation time 5 ns.

Source code for experiment

```
-- File Name : test_18e.vhd
--
-- Author    : Joseph Pick

entity Test_18e is
end Test_18e;

architecture Behave_1 of Test_18e is
    signal A : BIT := '0';
    signal B : BIT := '0';
    signal C : BIT := '0';
begin

  Gen_Wave:
  process
  begin
    A <= '1' after 5 ns, '0' after 12 ns;
    B <= '1' after 8 ns;
    wait;
  end process Gen_Wave;
```

```
            Analysis_C:
            process
              variable Var_C : BIT := '0';
            begin
              wait on C'TRANSACTION;
              Var_C := C;
            end process Analysis_C;

            Update_C: C <= A or B after 40 ns;

        end Behave_1;
```

11.14 Test_28 (Multiple array inputs)

Objective. To illustrate why, in general, the VHDL alias construct should be used when a subprogram has more than one input array parameter.

Expected observations. A simulation run time error should occur while executing the function's for loop due to the range mismatch of the input actual parameters Vector_0_8(0 to 2) and Vector_3_11(8 to 10).

Possible solutions. One solution to this problem is to place the burden onto the user of this function to ensure that the inputs will always have matching ranges. This may be done by assigning the arrays having the mismatched ranges to user-defined array variables having matching range declarations. These temporary variables would then be used as the input parameters during the calling of the function. Clearly, such an approach is not very user friendly. Hence, the preferred solution is to apply VHDL aliasing techniques, as shown in the next experiment, Test_28a.

Source code for experiment

```
        -- File Name : test_28.vhd
        --
        -- Author    : Joseph Pick

        entity Test_28 is
        end Test_28;

        architecture Behave_1 of Test_28 is

            type Logic4 is ('X','0','1','Z');
            type Logic4_Vector is array (NATURAL range <>) of Logic4;
            type Logic4_Table is array (Logic4, Logic4) of Logic4;

            constant Or_Table : Logic4_Table := (('X', 'X', '1', 'X'),
                                                 ('X', '0', '1', 'Z'),
                                                 ('1', '1', '1', '1'),
                                                 ('X', 'Z', '1', 'Z'));

            function "or" (L_V, R_V : Logic4_Vector)
                                    return Logic4_Vector is
              variable Result : Logic4_Vector(L_V'LENGTH downto 1);
            begin
```

```
        assert L_V'LENGTH = R_V'LENGTH;
            report "Length mismatch of inputs"
                severity ERROR;
        for I in L_V'RANGE loop
            Result(I) := Or_Table(L_V(I), R_V(I));
        end loop;
        return result;
    end "or";

begin

    Or_Range_Test:
    process
        variable Vector_0_8   : Logic4_Vector(0 to 8);
        variable Vector_3_11  : Logic4_Vector(3 to 11);
        variable Vector_15_23 : Logic4_Vector(15 TO 23);
    begin
        Vector_0_8  := "000011111";
        Vector_3_11 := "111100000";
        Vector_15_23(18 to 20) := Vector_0_8(0 to 2) or
                                  Vector_3_11(8 to 10);
        wait for 50 ns;
    end process Or_Range_Test;

end Behave_1;
```

11.15 Test_28a (Alias and multiple array inputs)

Objective. To illustrate the use of VHDL aliasing techniques in a subprogram body having multiple input array parameters.

Expected observations. The model should successfully implement the overloaded "or" function.

Additional information. To alias an object is more than just a relabeling operation. The object's VHDL characteristics such as class and mode are also inherited. Herein lies the true power of the aliasing technique, since these inherited properties may now be explicitly or implicitly referenced. In this particular case the alias L V Alias inherits the *in* mode and *constant* class features of the input parameter L_V.

Source code for experiment

```
-- File Name : test_28a.vhd
--
-- Author    : Joseph Pick

entity Test_28a is
end Test_28a;

architecture Behave_1 of Test_28a is

    type Logic4 is ('X','0','1','Z');
    type Logic4_Vector is array (NATURAL range <>) of Logic4;
    type Logic4_Table is array (Logic4, Logic4) of Logic4;

    constant Or_Table : Logic4_Table := (('X', 'X', '1', 'X'),
                                         ('X', '0', '1', 'Z'),
                                         ('1', '1', '1', '1'),
                                         ('X', 'Z', '1', 'Z'));
```

```vhdl
      function "or" (L_V, R_V : Logic4_Vector)
                            return Logic4_Vector is
        alias L_V_Alias : Logic4_Vector(L_V'LENGTH - 1 downto 0 )
                          is L_V;
        alias R_V_Alias : Logic4_Vector(R_V'LENGTH - 1 downto 0)
                          is R_V;
        variable Result : Logic4_Vector(L_V'LENGTH downto 1);
      begin
        assert L_V'LENGTH = R_V'LENGTH
               report "Length mismatch of inputs"
               severity ERROR;
        for I in L_V_Alias'RANGE loop
           Result(I) := Or_Table(L_V_Alias(I), R_V_Alias(I));
        end loop;
        return Result;
      end "or";

begin

  Or_Range_Test:
  process
    variable Vector_0_8   : Logic4_Vector(0 to 8);
    variable Vector_3_11  : Logic4_Vector(3 to 11);
    variable Vector_15_23 : Logic4_Vector(15 TO 23);
  begin
    Vector_0_8  := "000011111";
    Vector_3_11 := "111100000";
    -- NOTE also that OR may optionally be referenced even though
    -- the function's name was originally declared using all
    -- lower cases letters in double quotes ("or").
    Vector_15_23(18 to 20) := Vector_0_8(0 to 2) OR
                              Vector_3_11(8 to 10);
    wait for 50 ns;
  end process Or_Range_Test;

end Behave_1;
```

11.16 Test_27d (Single array inputs)

Objective. To illustrate why, in general, the VHDL alias construct should be used when a subprogram has even just one input array parameter.

Expected observations. A simulation run time error should occur while executing the procedure's for loop due to the negative (non-NATURAL) indices of the input actual parameter.

Possible solutions. One solution to this problem is to place the burden on the user of this procedure to remember to ensure that the input will always have nonnegative indices. This may be done by assigning any troublesome, concatenation of arrays to a single, user-defined array variable having NATURAL indices. This temporary variable would then be used as the input parameter during the calling of the procedure. Clearly, such an approach is not very user friendly. Hence, the preferred solution is to apply VHDL aliasing techniques, as was shown in the previous experiment Test_28a.

Additional techniques. Note the usage of temporary variables to capture the VHDL attributes 'LEFT and 'RIGHT. Such an approach may be required if your simulator does not have the capability to examine any of the predefined VHDL attributes.

Additional information. As previously mentioned in this book, the VHDL'93 update corrects this situation via a more usable definition for the left bound and range of two concatenated arrays. When your tool suite is updated to support VHDL'93, you should repeat this test to confirm that the alias method is no longer required.

Source code for experiment

```
-- File Name : test_27d.vhd
--
-- Author    : Joseph Pick

entity Test_27d is
end Test_27d;

architecture Behave_1 of Test_27d is
type Nat_Array is array (NATURAL range <>) of NATURAL;

Procedure Proc_In (Nat_In : in Nat_Array;
                   Sum    : out NATURAL) is
    variable Current_Value : NATURAL;
    variable Current_Array : Nat_Array(Nat_In'RANGE);
    variable Sum_Var       : NATURAL;
    variable Nat_In_Left   : INTEGER;
    variable Nat_In_Right  : INTEGER;
begin
    -- The following two lines are required since
    -- neither 'LEFT nor 'RIGHT could be examined.
    Nat_In_Left  := Nat_In'LEFT;
    Nat_In_Right := Nat_In'Right;

    Sum_Var := 0;
    for I in Nat_In'RANGE loop
        -- Should get runtime error here when loop index
        -- has a value of -1.
        -- This is a stress test of the simulator suite.
        -- An error should be noted.
        Current_Array(I) := Nat_In(I);
        Current_Value    := Nat_In(I);
        Sum_Var          := Sum_Var + Current_Value;
    end loop;
    Sum := Sum_Var;
end Proc_In;

begin

  Ampersand_Test:
  process
    variable Vector_8_1 : Nat_Array(8 downto 1) :=
                          (8,7,6,5,4,3,2,1);
    variable Vector_25_17 : Nat_Array(25 downto 17) :=
                            (25,24,23,22,21,20,19,18,17);
    variable Sum : NATURAL;
  begin
    -- Procedure called with array having negative indices.
    Proc_In (Vector_8_1(3 downto 1) &
             Vector_25_17(22 downto 18), Sum);
    wait for 50 ns;
  end process Ampersand_Test;

end Behave_1;
```

11.17 Test_63 (Procedure location option)

Objective. To exhibit an optimum location for procedures, whereby key objects are automatically visible and hence do not have to be passed in as parameters.

Expected observations. Experiment should compile and simulate without any difficulties.

Additional information. The method shown in this test should only be used if a specific sequence of code occurs repeatedly in just one process. Otherwise, the common code must be encapsulated inside a package's procedure, and those external objects referenced inside the procedure body will have to be passed in as parameters.

Source code for experiment

```vhdl
-- File Name : test_63.vhd
--
-- Author    : Joseph Pick

entity Test_63 is
end Test_63;

architecture Behave_1 of Test_63 is
    signal Sig_Nat           : NATURAL := 0;
    signal Sig_Nat_Procedure : NATURAL := 0;

begin
    Proc_Scope:
    process
        variable Count        : Natural := 0;
        variable Count_Temp_1 : Natural := 0;
        variable Count_Temp_2 : Natural := 0;
        variable Exit_Loop    : BOOLEAN := FALSE;
        procedure Set_55 is
            begin
                If Sig_Nat = 1 then
                    wait until Sig_Nat = 2;
                    if Count_Temp_2 /= 55 then
                        Sig_Nat_Procedure <= 55;
                        Count := 55;
                        Exit_Loop := TRUE;
                    end If;
                end if;
            end Set_55;
    begin
        Loop_1:
        loop
            Count_Temp_2 := Count_Temp_1 + 1;
            Set_55;
            If Exit_Loop = TRUE then
                Count_Temp_1 := Count;
                wait on Sig_Nat_Procedure'TRANSACTION;
                Count_Temp_2 := Sig_Nat_Procedure;
                exit Loop_1;
            end if;
            wait for 15 ns;
        end loop loop_1;
        Count := Count + 1;
        wait for 20 ns;
    end process Proc_Scope;
```

```
        Gen_Sig_Nat:
        process
        begin
            wait for 10 ns;
            Sig_Nat <= Sig_Nat + 1;
        end process Gen_Sig_Nat;

    end Behave_1;
```

11.18 Test_110b (Procedure constraint)

Objective. To highlight a constraint that procedures have when their respective specification or body does not occur in the process declaration region as in Test_63.

Expected observations. Compilation should fail because the targets of signal assignments in procedures not declared within processes must be formals of either the procedure or one of the encapsulating chain of procedures in which it is hosted. In this specific example the signal assignment to C_Sig should yield a compilation error. Note well that a compilation error should not occur when the signals A_Sig and B_Sig are referenced. VHDL allows procedures to reference signals that are not passed into it so long as these signals are not written to within the procedure's body.

Source code for experiment

```
            -- File Name : test_110b.vhd
            --
            -- Author    : Joseph Pick

            entity Test_110b is
            end Test_110b;

            architecture Behave_1 of Test_110b is
                signal A_Sig : NATURAL := 0;
                signal B_Sig : NATURAL := 0;
                signal C_Sig : NATURAL := 0;

                procedure Concur_Proc is
                    variable Count_CP : NATURAL := 0;
                begin
                    Count_CP := Count_CP + 1;
                    if A_Sig = 55 then
                        Count_CP := Count_CP + 1;
                        wait until B_Sig = 65;
                        C_Sig <= 75 after 10 ns;
                    end if;
                    Count_CP := Count_CP + 1;
                end Concur_Proc;

            begin

                Gen_Signals:
                process
                    variable Count_GS : Natural := 0;
                begin
                    wait for 10 ns;
                    Count_GS := Count_GS + 1;
                    A_Sig <= 55;
```

```
                wait for 10 ns;
                Count_GS := Count_GS + 1;
                B_Sig <= 65;
                wait on C_Sig;
                Count_GS := C_Sig;
            end process Gen_Signals;

            Concur_Proc_Test: Concur_Proc;

        end Behave_1;
```

11.19 Test_64 ('TRANSACTION in procedures)

Objective. To show that a procedure may reference the 'TRANSACTION attribute of a signal whenever that signal is not a formal parameter. The key element of this experiment is the word *formal*.

Expected observations. Experiment should compile and simulate successfully since the signal Sig_Nat is not a formal parameter. It should also be noted that the only reason why Sig_Nat may be assigned to, in the first place, is that its host procedure is declared in the declarative part of a process.

Source code for experiment

```
            -- File Name : test_64.vhd
            --
            -- Author    : Joseph Pick

            entity Test_64 is
            end Test_64;

            architecture Behave_1 of Test_64 is
                signal Sig_Nat : NATURAL := 0;
                signal Sig_Nat_Procedure : NATURAL := 0;
            begin
                Proc_Scope:
                process
                    variable Count          : Natural := 0;
                    variable Count_Temp_1   : Natural := 0;
                    variable Count_Temp_2   : Natural := 0;
                    variable Exit_Loop      : BOOLEAN := FALSE;

                    procedure Set_55 is
                    begin
                        If Sig_Nat = 1 then
                            wait on Sig_Nat'TRANSACTION;
                            if Sig_Nat = 2 then
                                if Count_Temp_2 /= 55 then
                                    Sig_Nat_Procedure <= 55;
                                Count := 55;
                                Exit_Loop := TRUE;
                                end if;
                            end if;
                        end if;
                    end Set_55;

                begin
                    Loop_1:
                    loop
                        Count_Temp_2 := Count_Temp_1 + 1;
                        Set_55;
```

```
                If Exit_Loop = TRUE then
                   Count_Temp_1 := Count;
                   wait on Sig_Nat_Procedure'TRANSACTION;
                   Count_Temp_2 := Sig_Nat_Procedure;
                   exit Loop_1;
                end if;
                wait for 15 ns;
             end loop loop_1;
             Count := Count + 1;
             wait for 20 ns;
          end process Proc_Scope;

          Gen_Sig_Nat:
          process
          begin
             wait for 10 ns;
             Sig_Nat <= Sig_Nat + 1;
          end process Gen_Sig_Nat;

       end Behave_1;
```

11.20 Test_64a ('TRANSACTION of formal parameters)

Objective. To emphasize that the key element of the previous experiment Test_64 is the word *formal*.

Expected observations. Experiment should not compile since the signal Sig_Param is a formal parameter of the procedure and Sig_Param'TRANSACTION is referenced within the body of this procedure.

Source code for experiment

```
          -- File Name : test_64a.vhd
          --
          -- Author    : Joseph Pick

          entity Test_64a is
          end Test_64a;

          architecture Behave_1 of Test_64a is

             signal Sig_Nat : NATURAL := 0;

             procedure Set_55 (signal Sig_Param : in NATURAL;
                                      Count_Param : out NATURAL) is
             begin
                wait on Sig_Param'TRANSACTION;
                Count_Param := 55;
             end Set_55;

          begin

             Gen_Signal:
             process
                variable Count_PS : Natural := 0;
             begin
                Count_PS := Count_PS + 1;
                Set_55 (Sig_Nat, Count_PS);
                wait for 20 ns;
             end process Gen_Signal;

             Sig_Nat <= 22 after 20 ns;

          end Behave_1;
```

11.21 Test_65 (Solution for Test_64a)

Objective. To test a possible workaround to the language limitation that was demonstrated in Test_64a.

Expected observations. Experiment should compile and simulate without any difficulties.

Additional information. An alternative workaround is to call the procedure with the signal's attribute serving as an actual parameter. This technique is shown as a comment in the following source code:

Source code for experiment

```
-- File Name : test_65.vhd
--
-- Author    : Joseph Pick

entity Test_65 is
end Test_65;

architecture Behave_1 of Test_65 is

    signal Sig_Nat_Transaction : BIT;
    signal Sig_Nat : NATURAL := 1;

    procedure Wait_On_Sig (signal Sig_Nat_Transaction : in BIT) is
        variable Count_Procedure : NATURAL := 0;
    begin
        If Sig_Nat = 1 then
            Count_Procedure := Count_Procedure + 1;
            wait on Sig_Nat_Transaction;
            Count_Procedure := Count_Procedure + 2;
        else
            Count_Procedure := Count_Procedure + 55;
            wait for 50 ns;
        end if;
        return;
    end Wait_On_Sig;

begin

    Gen_Sig_Nat:
    process
    begin
        wait for 10 ns;
        Sig_Nat <= Sig_Nat + 1;
    end process Gen_Sig_Nat;

    -- **** Work-around requires temporary signal that is in essence
    -- **** the 'TRANSACTION signal that we want the procedure
    -- **** to be sensitive to.
    Cncrrnt_Trnsctn:  Sig_Nat_Transaction <= Sig_Nat'TRANSACTION;

    Call_Proc:
    process
      variable Count_Process : NATURAL := 0;
    begin
        -- Could alternatively have written
        --     Wait_On_Sig(Sig_Nat'Transaction);
        Wait_On_Sig(Sig_Nat_Transaction);
        Count_Process := Count_Process + 1;
    end process Call_Proc;

end Behave_1;
```

11.22 Test_67a (Nonefficient command decoding)

Objective. To illustrate a VHDL coding scenario that may be applied when modeling the decoding of instructions or commands.

Expected observations. Experiment should compile and simulate without any difficulties.

Caveat. The instruction decoding technique of this test is not efficient because of the numerous bit string comparisons that must be made as one traverses through the various when arms of the case statement.

Source code for experiment

```vhdl
-- File Name : test_67a.vhd
--
-- Author    : Joseph Pick

entity Test_67a is
end Test_67a;

architecture Behave_1 of Test_67a is
    signal Command_Bits : Bit_Vector(31 downto 26);

begin

    Proc_Execute:
    process
        variable Count_A : Natural := 5;
        variable Count_B : Natural := 2;
        variable Result  : Natural := 0;

    begin
        wait on Command_Bits;

        case Command_Bits is
            when "000000" =>
                Result := Count_A + Count_B;
            when "000001" =>
                Result := Count_A - Count_B;
            when "000010" =>
                Result := Count_A * Count_B;
            when others   =>
                assert FALSE
                    report "Illegal instruction identified"
                    severity ERROR;
        end case;

    end process Proc_Execute;

    Gen_Command:
    Command_Bits <= "000010" after 10 ns, "000011" after 20 ns;

end Behave_1;
```

11.23 Test_67b (Efficient command decoding)

Objective. To illustrate an efficient VHDL coding scenario that may be applied when modeling the decoding of instructions or commands.

Expected observations. Experiment should compile and simulate without any difficulties.

Caveat. However, the technique illustrated in this experiment requires the availability of an efficient algorithm to convert a BIT_VECTOR type into a NATURAL number. An efficient subprogram implementing this conversion has already been presented and discussed in Chap. 8, Model Efficiency.

Source code for experiment

```
-- File Name : test_67b.vhd
--
-- Author    : Joseph Pick

entity Test_67b is
end Test_67b;

architecture Behave_1 of Test_67b is

   type Instruction_Enum is (ADD, SUBTRACT, MULTIPLY);
   signal Command_Nat : NATURAL := 0;

begin

   Proc_Execute:
   process
      variable Count_A : Natural := 5;
      variable Count_B : Natural := 2;
      variable Result  : Natural := 0;
      variable Command_Enum : Instruction_Enum;
   begin
      -- In practise, the command will be a bit string that
      -- must be converted into a NATURAL number.
      wait on Command_Nat;

      -- Confirm that the converted NATURAL number will
      -- correctly, map onto an enumeration literal.
      if (Command_Nat >= 0) and
         (Command_Nat <= 2) then

         -- Convert this NATURAL number into the
         -- corresponding enumeration literal.
         Command_Enum := Instruction_Enum'VAL(Command_Nat);

      else

         assert FALSE
            report "Invalid command code sequence"
            report ERROR;

      end if;

      case Command_Enum is
         when ADD =>
            Result := Count_A + Count_B;
         when SUBTRACT =>
            Result := Count_A - Count_B;
         when MULTIPLY =>
            Result := Count_A * Count_B;
      end case;
   end process Proc_Execute;

   Gen_Command:
   Command_Nat <= 1 after 10 ns, 0 after 20 ns, 2 after 30 ns,
                  0 after 40 ns, 1 after 50 ns;

end Behave_1;
```

11.24 Test_68 (File declaration: Case 1)

Objective. To illustrate how a process may reference a unique output file name, even when its host entity or architecture pair is multiply instantiated.

Expected observations. Experiment should successfully write into the files, test_file_1 and test_file_2.

Source code for experiment

```
-- File Name : test_68.vhd
--
-- Author    : Joseph Pick

entity File_Test is
    generic(Filename : STRING);
end File_Test;

use STD.TEXTIO.all;
architecture Behavior of File_Test is

 begin

    Process_Print:
    process
      file Outfile     : TEXT is out Filename;
      -- In VHDL'93 the above line MUST be replaced by
      -- file Outfile : TEXT open WRITE_MODE is Outfile;
      variable Outline : LINE;
      variable Count   : INTEGER := 0;
    begin
        write (Outline,Count);
        writeline (Outfile,Outline);

        Count := Count + 1;
        wait for 10 ns;
    end process Process_Print;
 end Behavior;

entity Test_68 is
end Test_68;

architecture Structure_1 of Test_68 is

  component File_Test
  end component;

begin
  arch_1: File_Test;
  arch_2: File_Test;
end Structure_1;

configuration Universe_68 of Test_68 is
   for Structure_1
      for arch_1: File_Test
         use entity WORK.File_Test(Behavior)
            generic map ("test_file_1");
      end for;
      for arch_2: File_Test
         use entity WORK.File_Test(Behavior)
            generic map ("test_file_2");
      end for;
   end for;

end Universe_68;
```

11.25 Test_69 (File declaration: Case 2)

Objective. To illustrate how different processes within the same architecture body may reference unique output file names, even when its host entity or architecture pair is multiply instantiated.

Expected observations. Experiment should successfully write into the files, test_file_1_Proc_1, test_file_1_Proc_2, test_file_2_Proc_1, and test_file_2_Proc_2.

Source code for experiment

```vhdl
-- File Name : test_69.vhd
--
-- Author    : Joseph Pick

entity File_Test_2 is
    generic(Filename : STRING);
end File_Test_2;


use STD.TEXTIO.all;
architecture Behavior of File_Test_2 is
begin

   Proc_1:
   process
      file Outfile     : TEXT is out Filename & "Proc_1";
      -- In VHDL'93 the above line MUST be replaced by
      -- file Outfile : TEXT open WRITE_MODE is
      --                          Filename & "Proc_1";
      variable Outline : LINE;
      variable Count   : INTEGER := 0;
   begin
      write (Outline,Count);
      writeline (Outfile,Outline);

      Count := Count + 1;
      wait for 10 ns;
   end process Proc_1;

   Proc_2:
   process
      file Outfile     : TEXT is out Filename & "Proc_2";
      -- In VHDL'93 the above line MUST be replaced by
      -- file Outfile : TEXT open WRITE_MODE is
      --                          Filename & "Proc_2;

      variable Outline : LINE;
      variable Count   : INTEGER := 0;
   begin
      write (Outline,Count);
      writeline (Outfile,Outline);

      Count := Count + 1;
      wait for 10 ns;
   end process Proc_2;
end Behavior;

entity Test_69 is
end Test_69;

architecture Structure_1 of Test_69 is

   component File_Test_2
   end component;
```

```
begin
   arch_1: File_Test_2;
   arch_2: File_Test_2;
end Structure_1;

configuration Universe_69 of Test_69 is
   for Structure_1
      for arch_1: File_Test_2
         use entity WORK.File_Test_2(Behavior)
            generic map ("test_file_1");
      end for;
      for arch_2: File_Test_2
         use entity WORK.File_Test_2(Behavior)
            generic map ("test_file_2");
      end for;
   end for;
end Universe_69;
```

11.26 Test_74 ('TRANSACTION and resolved signals)

Objective. To demonstrate that a multisourced signal is updated whenever any of its multiple sources has a transaction.

Expected observations. The assignment of 'Z' to the multisourced signal called Signal_Tristate_RS_S should cause a transaction to be associated with this signal. Note well that the effective value of Signal_Tristate_RS_S should still have its previous value of '1', even though an assignment of 'Z' and not '1' was made.

Source code for experiment

```
-- File Name : test_74.vhd
--
-- Author    : Joseph Pick

entity Test_74 is
end Test_74;

architecture Behave_1 of Test_74 is

   type Logic4 is ('X', '0', '1', 'Z');
   type Logic4_Vector is array (NATURAL range <>) of Logic4;
   type Logic4_Table is array (Logic4, Logic4) of Logic4;

   constant Table : Logic4_Table := (('X', 'X', 'X', 'X'),
                                     ('X', '0', 'X', '0'),
                                     ('X', 'X', '1', '1'),
                                     ('X', '0', '1', 'Z'));

   function Tristate_RF (V : Logic4_Vector) return Logic4 is
            variable Result : Logic4      := 'Z';

   begin
      for I in V'RANGE loop
         Result := Table (Result, V(I));
         exit when Result = 'X';
      end loop;
      return Result;
   end Tristate_RF;
      subtype Tristate_RS is Tristate_RF Logic4;
```

```vhdl
        signal Signal_Tristate_RS_S : Tristate_RS := 'Z';
        signal Count : NATURAL := 0;
begin
    Tristate_Source_1:
    process
    begin
        wait for 10 ns;
        Signal_Tristate_RS_S <= '1';
        wait for 30 ns;
    end process Tristate_Source_1;

    Tristate_Source_2:
    process
    begin
        wait for 20 ns;
        Signal_Tristate_RS_S <= 'Z';
    end process Tristate_Source_2;

    Test_Proc:
    process
    begin
        wait on Signal_Tristate_RS_S'TRANSACTION;
        --      ****      RESULT       ****
        -- As expected, wait condition passed when Z was placed
        -- on one of the signal's multi-sources.
        Count <= Count + 1;
    end process Test_Proc;

end Behave_1;
```

11.27 Test_105 (Function slices)

Objective. To investigate the referencing of function slices.

Expected observations. Though the slicing of a function's returned value is valid in VHDL, this model should, nonetheless, still incur a simulation run time error due to the usage of an inappropriate index. A detailed description of the expected errors is documented in the following source code.

Recommendation. The user of a function must have knowledge of the range boundaries of the array that is being returned. Therefore, the function's author is professionally bound to somehow communicate information regarding the returned array's range definition. The only avenue available to do so is in the form of a comment just prior to the function's specification statement as follows:

```vhdl
        -- Function returns an array having the range
        -- Input'LENGTH - 1 downto 0.
```

Source code for experiment

```vhdl
        -- File Name : test_105.vhd
        --
        -- Author    : Joseph Pick

        entity Test_105 is
        end Test_105;
```

```vhdl
architecture Behave_1 of Test_105 is
   type Logic4 is ('X','0','1','Z');
   type Logic4_Vector is array (NATURAL range <>) of Logic4;
   type Logic4_Table is array (Logic4, Logic4) of Logic4;

   constant Sample_Tbl : Logic4_Table := (('X', 'X', '1', 'X'),
                                          ('X', '0', '1', 'Z'),
                                          ('1', '1', '1', '1'),
                                          ('X', 'Z', '1', 'Z'));

   function Manipulate_Vectors (L_V, R_V : Logic4_Vector)
                                return Logic4_Vector is
      alias L_V_Alias : Logic4_Vector(L_V'LENGTH downto 1)
                        is L_V;
      alias R_V_Alias : Logic4_Vector(R_V'LENGTH downto 1)
                        is R_V;
      variable Result : Logic4_Vector(L_V'LENGTH downto 1);
   begin
      for I in Result'RANGE loop
         Result(I) := Sample_Tbl(L_V_Alias(I), R_V_Alias(I));
      end loop;
      return Result;
   end Manipulate_Vectors;

begin

Funct_Slice_Test:
process
   variable Vector_A : Logic4_Vector(0 to 7);
   variable Vector_B : Logic4_Vector(0 to 7);
   variable Vector_C : Logic4_Vector(0 to 3);
begin
   Vector_A := "00000000";
   Vector_B := "11110000";
   Vector_C := Manipulate_Vectors(Vector_A, Vector_B)
                                  (8 downto 5);

   -- Following line should result in a run-time error
   -- due to the fact that the function slice has an
   -- ascending range whereas the returned Result is declared
   -- to have a descending one.
   Vector_C := Manipulate_Vectors(Vector_A, Vector_B)(0 to 3);

   -- Following line results in a run-time error due to the fact
   -- that element Result(0) does not exist.
   Vector_C := Manipulate_Vectors(Vector_A, Vector_B)
                                  (3 downto 0);

   wait for 50 ns;
end process Funct_Slice_Test;

end Behave_1;
```

11.28 Test_120 (Circuit oscillation)

Objective. To investigate the initialization difficulties of an RS flip-flop modeled at the gate level. Another goal is to illustrate that the initial value assigned to a signal may be overridden by the effective default of the formal out port with which it is associated (connected).

In essence, the device under test is the RS flip-flop configuration shown in Fig. 11.1, with Temp_Q_Bar assigned an initial value of '1'. All the other signals in this diagram have '0' as their respective defaults.

Suggestion. Before viewing the VHDL source code, you should first manually trace the data flow through the given RS flip-flop. Assume

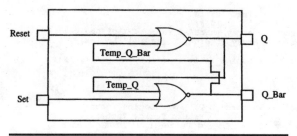

Figure 11.1 Nor gate implementation of an RS flip-flop.

that the gates have a propagation delay of 1 delta time unit. Your manual inspections should convince you that there really is an infinite oscillation problem.

Expected observations. Temp_Q_Bar's default value of '1' should be overwritten by the port Cout's default value (= BIT'LEFT). Moreover, the circuit should oscillate forever though absolute simulation time should never increase. Instead, delta time should increment indefinitely until the simulation is forcibly aborted by the user (Control-C) or by the simulation engine (current delta time count surpasses a system- or user-specified default parameter).

Caveat. This experiment yields two immediate caveats. The first is operational, whereas the second one has important ramifications on how models with multiple instantiations and feedback loops may have to be handled. The first caveat is merely a reminder from the Excursion that the default value assigned to a signal of an actual association list will be overwritten by the effective initial value of its corresponding formal (inout) out port. The second caveat, on the other hand, addresses a much more dangerous dilemma. As exposed in this experiment, multiple instantiations of the same component will all inherit the same default value of the corresponding formal out port. Consequently, be aware that a feedback circuit having multiple instantiations of the same component might oscillate indefinitely, as was the case in this experiment. (Test_121 and Test_118 will show you how to solve this oscillation problem.)

Additional information. Architecture Behave_1 of the entity RS_Flip_Flop also reviews a standard technique that was already discussed in the Compilation Caveats chapter (Chap. 6). It is not valid to use an *out* port from the parent entity as a port map actual that is associated with (connected to) a component's local *in* port parameter. For example, it would have been incorrect to use the port map association, Bin => Q_Bar. To overcome this constraint, you must, instead, declare intermediate signals, use them in the port map associations, and then concurrently assign them to the corresponding out ports of the parent entity. This workaround technique is vividly illustrated by the way in which Temp_Q_Bar, Temp_Q, Q_Bar, and Q are used in this model.

Source code for experiment

```
-- File Name : test_120.vhd
--
-- Author    : Joseph Pick

entity Nor_Gate_Bit is
      port (Ain: in BIT; Bin: in BIT; Cout: out BIT);
end Nor_Gate_Bit;

architecture Behave_1 of Nor_Gate_Bit is

begin

   Gate:
   process
   begin
       Cout <= Ain nor Bin;
       wait on Ain, Bin;
   end process;

end Behave_1;

entity RS_Flip_Flop_Bit is
      port (Set : In BIT;  Reset : in BIT;
            Q   : out BIT; Q_Bar : out BIT);
end RS_Flip_Flop_Bit;

architecture Behave_1 of RS_Flip_Flop_Bit is

   component Nor_Gate_Bit
             port (Ain: in BIT; Bin: in BIT; Cout: out BIT);
   end component;

   signal Temp_Q     : BIT := '0';
   signal Temp_Q_Bar : BIT := '1';

begin
   Out_Q     : Nor_Gate_Bit
               port map (Ain  => Reset,
                         Bin  => Temp_Q_Bar,
                         Cout => Temp_Q);

   Out_Q_Bar : Nor_Gate_Bit
               port map (Ain  => Set,
                         Bin  => Temp_Q,
                         Cout => Temp_Q_Bar);

   Q     <= Temp_Q;
   Q_Bar <= Temp_Q_Bar;

   Intermediate_Monitor:
   process
      variable Temp_Q_Var     : BIT := '0';
      variable Temp_Q_Bar_Var : BIT := '0';
   begin
      Temp_Q_Var     := Temp_Q;
      Temp_Q_Bar_Var := Temp_Q_Bar;
      wait on Temp_Q, Temp_Q_Bar;
   end process Intermediate_Monitor;
```

```vhdl
    end Behave_1;
entity Test_120 is
end Test_120;
architecture Behave_1 of Test_120 is
    component RS_Flip_Flop_Bit
        port (Set  : in BIT;  Reset : in BIT;
              Q    : out BIT; Q_Bar : out BIT);
    end component;

    signal Set   : BIT := '0';
    signal Reset : BIT := '0';

    signal Q     : BIT := '0';
    signal Q_Bar : BIT := '0';
begin
    RS_FF_1: RS_Flip_Flop_Bit
               port map (Set => Set, Reset => Reset,
                         Q   => Q,   Q_Bar => Q_Bar);

    RS_Stimulus:
    process
        variable Stimulus_Count : NATURAL := 0;
    begin
        Set <= '1' after 5 ns, '0' after 10 ns;
        Stimulus_Count := Stimulus_Count + 1;
        wait for 40 ns;
        Reset <= '1' after 15 ns, '0' after 20 ns;
        Stimulus_Count := Stimulus_Count + 1;
        wait for 40 ns;
    end process RS_Stimulus;

    RS_Monitor:
    process
        variable Q_Var     : BIT := '0';
        variable Q_Bar_Var : BIT := '0';
    begin
        Q_Var     := Q;
        Q_Bar_Var := Q_Bar;
        wait on Q, Q_Bar;
    end process RS_Monitor;

end Behave_1;

configuration Config_Test_120 of Test_120 is
    for Behave_1
        for RS_FF_1: RS_Flip_Flop_Bit
            use entity WORK.RS_Flip_Flop_Bit(Behave_1);
            for Behave_1
                for all : Nor_Gate_Bit
                    use entity WORK.Nor_Gate_Bit(Behave_1);
                end for;
            end for;
        end for;
    end for;
end Config_Test_120;
```

11.29 Test_121 (Solution 1 for Test_120)

Objective. To illustrate a solution to the oscillation problem exhibited by Test_120.

Expected observations. The circuit should not oscillate because appropriate default values have now been allocated to the multiply instantiated out port Cout via the generic parameter, DEFAULT_OUT. Note also that the initial values of Temp_Q and Temp_Q_Bar should be overwritten again, but this time by the respective values assigned to the generic parameter DEFAULT_OUT.

Recommendation. The assignment of a formal generic parameter as a port default should, in general, be a valuable tool in your repertoire of modeling tricks and techniques. However, in all honesty, I am certain that you do not want to selectively assign values to DEFAULT_OUT when your VHDL model contains 50,000 gates. It just is not practical to do so.

Additional information. Instead of using generic parameters, this circuit could also have been forced to stabilize by initializing the top level signals, Set and Reset, to complementary values. But here again it is not practical to assign values to Set and Reset when there are 50,000 gates in your model.

Incidentally, the real crux of this infinite oscillation problem is that the updating of the nor gates is done in the delta time domain instead of using a nonzero absolute amount of time. If this updating had been written as

```
Cout <= Ain nor Bin after NON_ZERO_DELAY;
```

then all our problems would have been solved. Yes, there would still be some initial oscillation but, in this case, absolute simulation time will advance to the point (5 ns as per these models) where the signal Set will be assigned the value '1'. As this update to Set ripples through the circuit, it will cause all the other signals to settle down and stabilize. In many ways this is analogous to what happens in the real world. When you turn on a circuit, initially there is chaos and perhaps even oscillation until the global reset signal is activated.

Source code for experiment

```
-- File Name : test_121.vhd
--
-- Author    : Joseph Pick
entity Nor_Gate_Bit is
      generic (DEFAULT_OUT : Bit);
      port (Ain: in BIT;
            Bin: in BIT;
            Cout: out BIT := DEFAULT_OUT);
end Nor_Gate_Bit;

architecture Behave_1 of Nor_Gate_Bit is

begin

   Gate:
   process
```

```vhdl
    begin
        Cout <= Ain nor Bin;
        wait on Ain, Bin;
    end process;
end Behave_1;

entity RS_Flip_Flop_Bit is
        port (Set : in BIT;  Reset : in BIT;
              Q   : out BIT; Q_Bar : out BIT);
end RS_Flip_Flop_Bit;

architecture Behave_1 of RS_Flip_Flop_Bit is

    component Nor_Gate_Bit
            generic (DEFAULT_OUT : Bit);
            port (Ain: in BIT; Bin: in BIT; Cout: out BIT);
    end component;

    signal Temp_Q     : BIT := '0';
    signal Temp_Q_Bar : BIT := '1';

begin
    Out_Q      : Nor_Gate_Bit
                 generic map (DEFAULT_OUT => '0')
                 port map (Ain => Reset,
                           Bin => Temp_Q_Bar,
                           Cout => Temp_Q);

    Out_Q_Bar : Nor_Gate_Bit
                 generic map (DEFAULT_OUT => '1')
                 port map (Ain => Set,
                           Bin => Temp_Q,
                           Cout => Temp_Q_Bar);

    Q     <= Temp_Q;
    Q_Bar <= Temp_Q_Bar;

    Intermediate_Monitor:
    process
        variable Temp_Q_Var     : BIT := '0';
        variable Temp_Q_Bar_Var : BIT := '0';
    begin
        Temp_Q_Var     := Temp_Q;
        Temp_Q_Bar_Var := Temp_Q_Bar;
        wait on Temp_Q, Temp_Q_Bar;
    end process Intermediate_Monitor;

end Behave_1;

entity Test_121 is
end Test_121;

architecture Behave_1 of Test_121 is

    component RS_Flip_Flop_Bit
        port (Set : in BIT; Reset : in BIT;
              Q   : out BIT; Q_Bar : out BIT);
    end component;

    signal Set   : BIT := '0';
    signal Reset : BIT := '0';

    signal Q     : BIT := '0';
    signal Q_Bar : BIT := '0';

begin
    RS_FF_1: RS_Flip_Flop_Bit
             port map (Set => Set, Reset => Reset,
                       Q   => Q,   Q_Bar => Q_Bar);

    RS_Stimulus:
    process
        variable Stimulus_Count : NATURAL := 0;
```

```
      begin
         Set <= '1' after 5 ns, '0' after 10 ns;
         Stimulus_Count := Stimulus_Count + 1;
         wait for 40 ns;
         Reset <= '1' after 15 ns, '0' after 20 ns;
         Stimulus_Count := Stimulus_Count + 1;
         wait for 40 ns;
      end process RS_Stimulus;

      RS_Monitor:
      process
         variable Q_Var     : BIT := '0';
         variable Q_Bar_Var : BIT := '0';
      begin
         Q_Var     := Q;
         Q_Bar_Var := Q_Bar;
         wait on Q, Q_Bar;
      end process RS_Monitor;

   end Behave_1;

   configuration Config_Test_121 of Test_121 is
      for Behave_1
         for RS_FF_1: RS_Flip_Flop_Bit
            use entity WORK.RS_Flip_Flop_Bit(Behave_1);
            for Behave_1
               for all : Nor_Gate_Bit
                  use entity WORK.Nor_Gate_Bit(Behave_1);
               end for;
            end for;
         end for;
      end for;
   end Config_Test_121;
```

11.30 Test_118 (Solution 2 for Test_120)

Objective. To illustrate another solution to the oscillation problem exhibited by Test_120.

Expected observations. The Cout port default value of 'X' (= Logic4'LEFT) should now be the effective, initial value of any actual element associated with this formal out port. As per the constant Nor_Table 'X' nor'ed with the '0' default value of both Set and Reset should result in an 'X'. Consequently, the circuit should stabilize within one delta time after the initialization phase.

Recommendation. To stabilize complex feedback circuits, it is more preferable to use a non-BIT_VECTOR approach analogous to the solution demonstrated in Test_118. As previously stated the solution described in Test_121 might be operationally difficult to implement for large intricate feedback networks.

Source code for experiment

```vhdl
-- File Name : test_118.vhd
--
-- Author    : Joseph Pick
package Logic4_Pkg_Sampler is
    type Logic4 is ('X', '0', '1', 'Z');
    type Logic4_Vector is array (NATURAL range <>) of Logic4;
    function "nor" (L, R : Logic4) return Logic4;
end Logic4_Pkg_Sampler;

package body Logic4_Pkg_Sampler is
    type Logic4_Table is array (Logic4, Logic4) of Logic4;

        constant Nor_Table : Logic4_Table := (('X', 'X', '0', 'X'),
                                              ('X', '1', '0', 'X'),
                                              ('0', '0', '0', '0'),
                                              ('X', 'X', '0', 'X'));

    function "nor" (L, R : Logic4) return Logic4 is
    begin
        return Nor_Table(L, R);
    end "nor";
end Logic4_Pkg_Sampler;

use WORK.Logic4_Pkg_Sampler.all;
entity Nor_Gate is
        -- Logic4'LEFT (= 'X') will be assigned by the
        -- simulation engine to the out port, Cout.
        port (Ain: in Logic4; Bin: in Logic4; Cout: out Logic4);
end Nor_Gate;

architecture Behave_1 of Nor_Gate is
begin
    Gate:
    process
    begin
        Cout <= Ain nor Bin;  -- Overloaded "nor" must be used.
        wait on Ain, Bin;
    end process;
end Behave_1;

use WORK.Logic4_Pkg_Sampler.all;
entity RS_Flip_Flop is
        port (Set : in Logic4;  Reset : in Logic4;
              Q   : out Logic4; Q_Bar : out Logic4);
end RS_Flip_Flop;

architecture Behave_1 of RS_Flip_Flop is

    component Nor_Gate
            port (Ain: in Logic4;
                  Bin: in Logic4;
                  Cout: out Logic4);
    end component;

    signal Temp_Q     : Logic4 := '0';
    signal Temp_Q_Bar : Logic4 := '1';

begin
    Out_Q     : Nor_Gate
                port map (Ain => Reset,
                          Bin => Temp_Q_Bar,
                          Cout => Temp_Q);

    Out_Q_Bar : Nor_Gate
                port map (Ain => Set,
                          Bin => Temp_Q,
                          Cout => Temp_Q_Bar);

    Q     <= Temp_Q;
    Q_Bar <= Temp_Q_Bar;
```

```vhdl
    Intermediate_Monitor:
    process
       variable Temp_Q_Var     : Logic4 := 'X';
       variable Temp_Q_Bar_Var : Logic4 := 'X';
    begin
       Temp_Q_Var     := Temp_Q;
       Temp_Q_Bar_Var := Temp_Q_Bar;
       wait on Temp_Q, Temp_Q_Bar;

       end process Intermediate_Monitor;

end Behave_1;

entity Test_118 is
end Test_118;

use WORK.Logic4_Pkg_Sampler.all;
architecture Behave_1 of Test_118 is
    component RS_Flip_Flop
        port (Set : in Logic4; Reset : in Logic4;
              Q   : out Logic4; Q_Bar : out Logic4);
    end component;

    signal Set   : Logic4 := '0';
    signal Reset : Logic4 := '0';

    signal Q     : Logic4 := 'X';
    signal Q_Bar : Logic4 := 'X';

begin
    RS_FF_1: RS_Flip_Flop
             port map (Set => Set, Reset => Reset,
                       Q   => Q,   Q_Bar => Q_Bar);

    RS_Stimulus:
    process
       variable Stimulus_Count : NATURAL := 0;
    begin
       Set <= '1' after 5 ns, '0' after 10 ns;
       Stimulus_Count := Stimulus_Count + 1;
       wait for 40 ns;
       Reset <= '1' after 15 ns, '0' after 20 ns;
       Stimulus_Count := Stimulus_Count + 1;
       wait for 40 ns;
    end process RS_Stimulus;

    RS_Monitor:
    process
       variable Q_Var     : Logic4 := 'X';
       variable Q_Bar_Var : Logic4 := 'X';
    begin
       Q_Var     := Q;
       Q_Bar_Var := Q_Bar;
       wait on Q, Q_Bar;
    end process RS_Monitor;

end Behave_1;

configuration Config_Test_118 of Test_118 is
    for Behave_1
        for RS_FF_1: RS_Flip_Flop
            use entity WORK.RS_Flip_Flop(Behave_1);
            for Behave_1
                for all : Nor_Gate
                    use entity WORK.Nor_Gate(Behave_1);
                end for;
            end for;
        end for;
    end for;
end Config_Test_118;
```

11.31 Test_89 (TIME to NATURAL)

Objective. To illustrate the various options available to convert from type TIME to type NATURAL.

Expected observations. The derived NATURAL values should agree with those documented in the source code below.

Recommendation. Of the two presented options, you should select the conversion technique that best suits your modeling needs.

Source code for experiment

```
-- File Name : test_89.vhd
--
-- Author    : Joseph Pick

entity Test_89 is
end Test_89;

architecture Behave_1 of Test_89 is

begin

  Time_Convert:
  process
    variable Time_Nat_NS : NATURAL;
    variable Time_Nat_Div: NATURAL;
  begin
    wait for 20 ns;                         -- Results
    Time_Nat_NS := TIME'POS(Now);           -- 20 * (10 ** 6)
    Time_Nat_Div := Now / 1 ns;             -- 20
  end process Time_Convert;

end Behave_1;
```

11.32 Test_1 (Reference for Test_1a, 1b, 1d, and 1e)

Objective. To become familiar with a valid coding scenario, against which later experimental results will be contrasted. The theme in this set of experiments is the manner in which VHDL creates drivers for array signals in a concurrent modeling environment.

Expected observations. The experiment should successfully compile and simulate. The key point here is that there should not be a need for a resolution function. The individual signals, Sample(4) and Sample(1), are not being multiply driven by two concurrent constructs. By the way, just for comparison, let's suppose that both processes would have assigned a value to the same subelement Sample(4). Then this signal Sample(4) would have been multiply driven by two processes. Sample(4) would then have to be associated with a resolution function.

Source code for experiment

```
-- File Name : test_1.vhd
--
-- Author    : Joseph Pick

entity Test_1 is
end Test_1;

architecture Behave_1 of Test_1 is

  signal Sample : BIT_VECTOR (4 downto 0) := (others => '1');

begin

  Load_N:
  process
  begin
    Sample(4) <= '1';
    wait for 2 ns;
    assert FALSE
      report "Test went OK"
      severity NOTE;
    wait;
  end process;

  Load_M:
  process
  begin
    Sample(1) <= '1';
    wait for 12 ns;
    assert FALSE
      report "Test went OK"
      severity NOTE;
    wait;
  end process;

end Behave_1;
```

11.33 Test_1a (Unexpected multisources: Case 1)

Objective. To explore how VHDL creates drivers for arrays when the specific index into the array is dynamically derived during a simulation. Note well that once the for loops are unraveled, the resulting code segments are exactly the same as those in Test_1, which had no multiple sourcing conflicts.

Expected observations. The experiment should successfully compile. But (!!) when the model is entered into the simulation environment, the VHDL elaboration phase should abort because of the existence of unresolved signals. The error message should report that every element of the signal array Sample is multisourced even though they have not been associated with a user-defined resolution function.

The root of the problem highlighted by this experiment is that the specific value of Index cannot be determined during either the compilation or the elaboration phase. The compiler therefore has no choice but to conjecture that every element of the array Sample could, during a simulation run, potentially be the target of a signal assignment. Under such an assumption, a driver should be created in both processes for the longest static prefix (a VHDL LRM term) of the signal Sample(Index). In this case the whole array identifier Sample is the longest static prefix and hence, drivers should be established in both processes for each and every element of the array. However, these individual signal elements are not associated with any resolution function, and as a result they should subsequently be identified as unresolved signals during the elaboration phase as indicated earlier. Incidentally, I have seen VHDL tool suites that caught this multisourcing error during compilation. So do not be surprised if this experiment does not even make it to the elaboration phase.

Caveat. Though the questionable looping setup of this experiment (for Index in 4 to 4) is unlikely to be repeated in a real VHDL model, analogous situations may unintentionally be implemented (see Test_341).

Source code for experiment

```
-- File Name : test_1a.vhd
--
-- Author    : Joseph Pick

entity Test_1a is
end Test_1a;

architecture Behave_1 of Test_1a is

  signal Sample : BIT_VECTOR (4 downto 0);
begin

  Load_N:
  process
  begin
    for Index in 4 to 4 loop
      Sample(Index) <= '1';
      wait for 2 ns;
      assert FALSE
             report "Test went OK"
             severity NOTE;
    end loop;
    wait;
  end process;

  Load_M:
  process
  begin
    for Index in 1 to 1 loop
      Sample(Index) <= '1';
      wait for 12 ns;
      assert FALSE
             report "Test went OK"
             severity NOTE;
    end loop;
    wait;
  end process;

end Behave_1;
```

11.34 Test_1b (Generics and array signal drivers)

Objective. To explore how VHDL creates drivers for arrays when the specific index into the array is a generic parameter and hence known during the elaboration phase. Note well that once the generic values are elaborated (assigned), the resulting code segments are exactly the same as those in Test_1, which had no multiple sourcing conflicts.

Expected observations. The experiment should successfully compile and simulate since the respective values for INDEX_N and INDEX_M should be viewed by the VHDL tool suite as constants that can be determined during the elaboration phase. Hence, drivers should only be created for those array subelements specifically pointed to by INDEX_N (= 4) and INDEX_M (= 1). Drivers should not be created for the whole array as was the case in Test_1a. In fact, this coding scenario postpones the creation of the respective drivers until the elaboration phase when the generic values are known. Incidentally, if INDEX_N and INDEX_M would inadvertently be assigned the same value, then the array subelement Sample(INDEX_N) would be driven by two processes. A multisourcing error would subsequently be identified and reported during the elaboration phase since Sample(INDEX_N) is not associated with any resolution function.

Source code for experiment

```
-- File Name : test_1b.vhd
--
-- Author    : Joseph Pick

entity Component_Test_1b is
    generic (INDEX_N : NATURAL; INDEX_M : NATURAL);
end Component_Test_1b;

architecture Behave_1 of Component_Test_1b is

    signal Sample : BIT_VECTOR (4 downto 0);

begin

  Load_N:
  process
  begin
      Sample(INDEX_N) <= '1';
      wait for 2 ns;
      assert FALSE
              report "Test went OK"
              severity NOTE;
      wait;
  end process;

  Load_M:
  process
  begin
      Sample(INDEX_M) <= '1';
      wait for 12 ns;
```

```
            assert FALSE
                   report "Test went OK"
                   severity NOTE;
            wait;
        end process;

    end Behave_1;


    entity Test_1b is
    end Test_1b;

    architecture Behave_1 of Test_1b is
        component Component_Test_1b
            generic (INDEX_N : NATURAL; INDEX_M : NATURAL);
        end component;
    begin
        Instance_1 : Component_Test_1b
            generic map (INDEX_N => 4, INDEX_M => 1);
    end Behave_1;
```

11.35 Test_1d (Unexpected multisources: Case 2)

Objective. To explore how VHDL creates drivers for arrays when the specific index into the array is dependent on a generic parameter and hence, globally static (known during the elaboration phase). Note well that once the generic values are elaborated and the subsequent element, NATURAL_ARRAY(INDEX_N) is accessed, then the resulting code segments are exactly the same as those in Test_1.

Expected observations. Unfortunately, discrepancies exist among the various VHDL tool suites in their implementation of this experiment. VHDL'87 is somewhat vague and ambiguous regarding this issue, and so different vendors took different interpretations of the LRM. Some VHDL tool suites will create a driver for only the element Sample(4) in the Load_N process, while others (because of the usage of NATURAL_ARRAY) will create one for each and every element of the signal Sample. The latter approach will result in multiple sourcing problems for Sample(1).

Additional information. Test_1e will further isolate and explore the key dilemma of this experiment. Incidentally, VHDL'93 has a clearer posture regarding this coding scenario. Test_1d should not experience any difficulties in a VHDL'93 compliant environment.

Source code for experiment

```
    -- File Name : test_1d.vhd
    --
    -- Author    : Joseph Pick

    entity Component_Test_1d is
        generic (INDEX_N : NATURAL; INDEX_M : NATURAL);
    end Component_Test_1d;
```

```
architecture Behave_1 of Component_Test_1d is

   signal Sample : BIT_VECTOR (4 downto 0);
   type Natural_Array_Type is array (4 downto 0) of NATURAL;
   constant NATURAL_ARRAY : Natural_Array_Type := (0,1,2,3,4);

begin

   Load_N:
   process
   begin
      -- NOTE that the generic actual is 0 and hence
      -- the following is equivalent to Sample(4) <= '1'.
      Sample(NATURAL_ARRAY(INDEX_N)) <= '1';
      wait for 2 ns;
      assert FALSE
             report "Test went OK"
             severity NOTE;
      wait;
   end process;

   Load_M:
   process
   begin
      -- NOTE that the generic actual is 1 and hence
      -- the following is equivalent to Sample(1) <= '1'.
      Sample(INDEX_M) <= '1';
      wait for 12 ns;
      assert FALSE
             report "Test went OK"
             severity NOTE;
      wait;
   end process;

end Behave_1;


entity Test_1d is
end Test_1d;

architecture Behave_1 of Test_1d is
   component Component_Test_1d
      generic (INDEX_N : NATURAL; INDEX_M : NATURAL);
   end component;
begin
   Instance_1 : Component_Test_1d
      -- These actuals should yield Load_N/Sample(4) <= '1'
      -- and Load_M/Sample(1) <= '1'.
      -- NOTE that there really is no multi-sourcing here.
      generic map (INDEX_N => 0, INDEX_M => 1);
end Behave_1;
```

11.36 Test_1e (Unexpected multisources: Case 3)

Objective. To expose that the indexing of an array type signal via an array element is the real culprit in Test_1d. In this experiment only constant values are used when indexing into the arrays Sample and NATURAL_ARRAY. Note well that once the element NATURAL_ARRAY(0) is accessed, then the resulting code segments are exactly the same as those in Test_1, which had no multiple sourcing conflicts.

Expected observations. Unfortunately, discrepancies exist among the various VHDL tool suites in their implementation of this experiment. Some compilers will create a driver for only the element Sample(4) in the Load_N process. While other compilers will not realize the intuitive and obvious fact that a constant literal indexing into a constant array will, in effect, produce a constant value. Instead, such compilers will automatically create a driver for each and every element of the signal Sample simply because of the presence of the array NATURAL_ARRAY. For these compilers it did not matter at all that this array in question is a constant.

Additional information. The different behavior of tool suites on this issue is due to their various interpretations of the subtle wording in the VHDL LRM regarding the term static. Intuitively, one would like Test_1e to compile and simulate without any difficulties. VHDL'93 is more explicit (or more relaxed ??) on this matter, and compliance with this update will mean that the desired, intuitive response must be honored. Test_1e should not experience any difficulties in a VHDL'93 compliant environment.

Source code for experiment

```
-- File Name : test_1e.vhd
--
-- Author    : Joseph Pick

entity Test_1e is
end Test_1e;

architecture Behave_1 of Test_1e is

   signal Sample : BIT_VECTOR (4 downto 0);
   type Natural_Array_Type is array (4 downto 0) of NATURAL;
   constant NATURAL_ARRAY : Natural_Array_Type := (0,1,2,3,4);

begin

   Load_N:
   process
   begin
      Sample(NATURAL_ARRAY(0)) <= '1';
      wait for 2 ns;
      assert FALSE
             report "Test went OK"
             severity NOTE;
      wait;
   end process;

   Load_M:
   process
   begin
      Sample(1) <= '1';
      wait for 12 ns;
      assert FALSE
             report "Test went OK"
             severity NOTE;
      wait;
   end process;

end Behave_1;
```

11.37 Test_124_0 (Port map actuals)

Objective. To highlight the significance of the requirement that in VHDL'87 the actuals of a port association must be a signal.

Expected observations. The experiment should not compile because the port map actual associated with the port Data_In cannot be an enumeration literal, such as '0'.

Additional information. The VHDL'93 LRM update allows users to write port map associations, such as Data_In <= '0'.

Source code for experiment

```
-- File Name : test_124_0.vhd
--
-- Author    : Joseph Pick

entity Test_124_0_Component is
     port    (Data_In : in BIT;  Data_Out : out BIT);
end Test_124_0_Component;

architecture Behave_1 of Test_124_0_Component is
begin
  Receive:
  process
     variable Boolean_Var : BOOLEAN := FALSE;
  begin
     wait on Data_In;
     Data_Out <= not Data_In;
  end process Receive;
end Behave_1;

entity Test_124_0 is
end Test_124_0;

architecture Behave_1 of Test_124_0 is
   component Test_124_0_Component
            port    (Data_In : in BIT; Data_Out : out BIT);
   end component;

   signal    Signal_A      : BIT := '1';

begin
   Device_1 : Test_124_0_Component
              port map (Data_In  => '0',
                        Data_Out => Signal_A);
end Behave_1;

configuration Config_Test_124_0 of Test_124_0 is
   for Behave_1
      for Device_1 : Test_124_0_Component
         use entity WORK.Test_124_0_Component(Behave_1);
      end for;
   end for;
end Config_Test_124_0;
```

11.38 Test_124 (Solution for Test_124_0)

Objective. To illustrate a solution to the problem that was exposed in Test_124_0. Keep in mind that this solution is only required by a VHDL'87 compliant tool suite. As stated above, Test_124_0 should compile successfully in VHDL'93.

Expected observations. The experiment should compile and simulate successfully.

Source code for experiment

```vhdl
-- File Name : test_124.vhd
--
-- Author    : Joseph Pick

entity Test_124_Component is
        port    (Data_In : in BIT;  Data_Out : out BIT);
end Test_124_Component;

architecture Behave_1 of Test_124_Component is
begin
   Receive:
   process
      variable Boolean_Var : BOOLEAN := FALSE;
   begin
      wait on Data_In;
      Data_Out <= not Data_In;
   end process Receive;
end Behave_1;

entity Test_124 is
end Test_124;

architecture Behave_1 of Test_124 is
   component Test_124_Component
             port    (Data_In : in BIT; Data_Out : out BIT);
   end component;

   signal   Ground_Signal : BIT := '0';
   signal   Signal_A      : BIT := '1';
begin
   Device_1 : Test_124_Component
              port map (Data_In => Ground_Signal,
                        Data_Out => Signal_A);
end Behave_1;

configuration Config_Test_124 of Test_124 is
   for Behave_1
      for Device_1 : Test_124_Component
         use entity WORK.Test_124_Component(Behave_1);
      end for;
   end for;
end Config_Test_124;
```

11.39 Test_158 (Scope and visibility)

Objective. To illustrate a visibility and scoping problem that is similar to a scenario that was unintentionally created by an automatic VHDL code generator.

Expected observations. The experiment should not compile because the scope of the STANDARD package's physical identifier ns should be overridden by the declared signal having the same name. Consequently, the statement wait for 10 ns should no longer be valid since the desired physical unit ns is no longer visible at this point. Instead, ns should now refer to the name of a signal and in this context, the wait construct should be void of any valid interpretation. A unit of VHDL time measurement is expected to follow the literal 10 in the wait for construct. It is totally inappropriate for this value 10 to be followed by the name of an object.

Recommendation. Designers of automatic code generators should ensure that their products have a built-in list of words that cannot be generated as object names. In addition to VHDL reserved words this list should include ns, ps, us, or any other name appearing in the packages, STANDARD, TEXTIO, and STD_Logic_1164. And speaking of TEXTIO, it is recommended that when modeling pins or signals with names such as Read or Write that you either not use these names in your VHDL models or that you use the VHDL'93 back slash notation, as in \Read\. Otherwise, the read and write procedures will no longer be available to your model since these identifiers will not be visible, as highlighted by this experiment.

Source code for experiment

```
-- File Name : test_158.vhd
--
-- Author    : Joseph Pick

entity Test_158 is
end Test_158;

architecture Behave_1 of Test_158 is
    signal Ns : NATURAL := 55;
begin

    process
    begin
        Ns <= 77;
        -- The following line should not compile because
        -- "ns" is now viewed as a signal name and not
        -- as a physical time unit.
        wait for 10 ns;
    end process;

end Behave_1;
```

11.40 Test_159 (Solution for Test_158)

Objective. To illustrate the usage of extended names to solve the visibility and scoping problem exposed in Test_158.

Expected observations. The experiment should compile and simulate successfully.

Recommendation. Designers of automatic VHDL code generators should build products that do not require further adjustments to the generated output. Such modifications are not only extremely tedious but, if done automatically via a postprocessor, they are also very risky and susceptible to a wide range of additional errors. The whole problem should be completely bypassed via a built-in list of words that cannot be generated as VHDL object names.

Source code for experiment

```
-- File Name : test_159.vhd
--
-- Author    : Joseph Pick

entity Test_159 is
end Test_159;

architecture Behave_1 of Test_159 is
    signal Ns : NATURAL := 55;
begin

    process
    begin
        Ns <= 77;
        wait for 10 STD.STANDARD.ns;
    end process;

end Behave_1;
```

11.41 Test_190 (Scope and visibility [cont])

Objective. To further explore VHDL's scoping and visibility rules. Specifically, this experiment creates a coding scenario whereby the identical function specification is contained within two packages, both of which are made visible via a use clause. This experiment aims to determine how this ambiguous situation is treated by VHDL. Should the function declaration in the second package supersede the one in the first or should it be the other way around?

Expected observations. VHDL acknowledges the ambiguity of the coding scenario created by this experiment. The names of these identifiers are therefore nullified by the VHDL scoping and visibility rules. This nullification of the function's name means that, as per the compiler's perspective, the name Return_Natural does not even exist at all. An error message should be generated by the compiler indicating, strangely enough, that the referenced name Return_Natural has not been declared.

Additional information. Return_Natural's visibility can be achieved via an extended name such as WORK.Pkg_B.Return_Natural. This extended name informs the compiler to choose the overloaded function from the package Pkg_B that is in the current working library. Incidentally, if a function named Return_Natural had been declared in the processes' declarative part, then there would not have been any visibility problem at all. The compiler would then unquestionably identify a call to Return_Natural as a call to the function declared in this process declarative part. In actuality, the real source of the ambiguity encountered in this experiment is that the visibility of both Pkg_A and Pkg_B are hierarchically equivalent. In VHDL terminology they are both visible in the same declaration region.

A similar ambiguous situation can occur when identical component declarations are encapsulated in two different files. Suppose that the component And_Gate has the same profile declaration in both the packages Gate_Components and Gate_Parts. Suppose further that these packages were compiled into libraries having the logical names TTL_Lib and CMOS_Lib, respectively. If both these packages would be made visible to the same model, then the instantiation of the component And_Gate would require an extended name. Otherwise, a compilation error analogous to the one illustrated by Test_190 would be reported. Here is how the instantiation would look:

```
U1 : TTL_Lib.Gate_Components.And_Gate
        port map (Sig_A, Sig_B, Sig_C);

U2 : CMOS_Lib.Gate_Parts.And_Gate
        port map (Sig_D, Sig_E, Sig_F);
```

Source file for experiment

```
-- File Name : test_190.vhd
--
-- Author    : Joseph Pick

package Pkg_A is

    function Return_Natural (Value_In : Integer) return NATURAL;

end Pkg_A;

package body Pkg_A is

    function Return_Natural (Value_In : Integer) return NATURAL is
    begin
        if Value_In > 55 then
            return 500;
        else
            return 55;
        end if;
    end Return_Natural;
```

```
end Pkg_A;

package Pkg_B is

    function Return_Natural (Value_In : Integer) return NATURAL;

end Pkg_B;

package body Pkg_B is

    function Return_Natural (Value_In : Integer) return NATURAL is
    begin
        if Value_In > 22 then
            return 200;
        else
            return 22;
        end if;
    end Return_Natural;

end Pkg_B;

entity Test_190 is
end Test_190;

use WORK.Pkg_A.all;
use WORK.Pkg_B.all;
architecture Behave_1 of Test_190 is
begin
   process
      variable Value_Returned : NATURAL;
      variable Value_In       : NATURAL := 10;
   begin
      -- ***** RESULT *****
      --       The following line does not compile.
      Value_Returned := Return_Natural(Value_In);
      wait for 10 ns;
      Value_In := Value_In + 10;
   end process;
end Behave_1;
```

11.42 Test_162 (& in case selector)

Objective. To illustrate an unexpected error in the case selector expression.

Expected observations. This experiment should not compile because of the concatenation operator in the case selector expression. On the surface all looks well, but here is what actually goes wrong. The concatenation operator & is really an overloaded function. In the current experiment the concatenation of two BIT-type objects implies that of all the overloaded & subprograms the one that should be called is the one that has two BIT-type inputs and returns a BIT_VECTOR value. Herein lies the problem. The returned value BIT_VECTOR is an unconstrained array and so, as far as the compiler is concerned the previous expression In_B & In_C is an unconstrained array. But (wink!) you and I both know that this expression is only two bits wide. Nevertheless, VHDL has a different interpretation, which really would not be so bad except that any array in a case selector expression must be a constrained array. Now you can see why a VHDL compliant compiler should reject the concatenation In_B & In_C in a case selector expression. The expression In_B & In_C is not a constrained array data type.

Caveat. Unfortunately, some of the VHDL tool suites currently on the market accept the coding scenario of this experiment. Consequently, if your compiler does not find any fault with the previous test, then be aware that your model may not be portable to another vendor's VHDL environment.

Recommendation. The way to solve this problem, in general, is to use a variation of the following code segment:

```
subtype Two_Bits is BIT_VECTOR(0 to 1);

.......................

case Two_Bits'(In_B & In_C) is
```

By qualifying the concatenated expression you are, in effect, requesting the compiler to interpret the expression as a constrained array. Since this request is reasonable and not out of line (INTEGER'(In_B & In_C) would be ceremoniously rejected), the compiler is now acquiesced regarding this matter.

Additional information. Similarly, if In_B and In_C are BIT_VECTOR(0 to 1) objects, it is incorrect to write:

```
case In_B or In_C is -- ERROR
```

Here again, the aforementioned qualification technique may be successfully applied to solve this unconstrained array problem.

But if In_D and In_E are each a single BIT, then the overloaded or function will return a BIT instead of a BIT_VECTOR. Since a BIT is a scalar the constrained array rule does not even apply. Hence for these objects it is valid to write:

```
case In_D or In_E is -- OK
```

Source code for experiment

```
-- File Name : test_162.vhd
--
-- Author    : Joseph Pick

package Test_162 is
    function Sample_Func (In_B, In_C : BIT) return INTEGER;
end Test_162;

package body Test_162 is
```

```
            function Sample_Func (In_B, In_C: BIT) return INTEGER is
                variable Result : INTEGER;
            begin
                    case In_B & In_C is
                        when "00" =>
                                Result := 50;
                        when "01" =>
                                Result := 51;
                        when "10" =>
                                Result := 52;
                        when "11" =>
                                Result := 53;
                    end case;

                    return Result;

            end Sample_Func;

        end Test_162;
```

11.43 Test_172 (Reading and writing via the same buffer line)

Objective. To expose a problem that occurs when the same buffer line (pointer) is used for both reading and writing from an ASCII file.

Expected observations. Suppose that the input file DATAIN contains the ASCII string 123456789ABCDEF. Since only the first six characters are read from the buffer line Linebuf_In_Out, any data written to this buffer line should be appended to its remaining unread characters. Consequently, the output file DATAOUT should contain 789ABCDEF123456.

Recommendation. The buffer line used for writing to a file should not be the same as the one used to read from a file.

Source code for experiment

```
        -- File Name : test_172.vhd
        -- Author    : Joseph Pick
        --
        entity Test_172 is
        end Test_172;

        use STD.TEXTIO.all;
        architecture Behave_1 of Test_172 is

            file Datain  : TEXT is in "DATAIN";
            file Dataout : TEXT is out "DATAOUT";
            -- In VHDL'93 the above two lines MUST be replaced
            -- by the following two file declarations.
            -- file Datain  : TEXT open READ_MODE is "DATAIN";
            -- file Dataout : TEXT open WRITE_MODE is DATAOUT";

        begin

            process
                variable Linebuf_In_Out : LINE;
                variable SomeText : STRING(1 to 6);
            begin
```

```
                wait for 10 ns;
                if Endfile(Datain) = FALSE then
                    Readline (Datain, Linebuf_In_Out);
                    Read (Linebuf_In_Out, SomeText);

                    Write(Linebuf_In_Out, SomeText);
                    Writeline (Dataout, Linebuf_In_Out);
                else
                    assert FALSE
                        report "End of file reached"
                        severity NOTE;
                end if;

            end process;

        end Behave_1;
```

11.44 Test_178 (Overriding of generic defaults)

Objective. To investigate the hierarchical overriding of default (initial) values assigned to generic and port parameters.

Expected observations. The observed values should agree with those that are documented in the following source code. These listed values should occur during the first pass through the model during the initialization phase. The main point is that the default value of 77 assigned to the local (component declared) generic parameter DEFAULT (Fact #1) should override the default value of 66 that was assigned to the formal (entity declared) generic parameter DEFAULT (Fact #2). This overriding may be confirmed by noting that the formal parameter DEFAULT now has the value 77 instead of its originally specified default value of 66 (Fact #3). Consequently, the formal out port Component_Port_Out should have the initial value of 77 instead of 66 (Fact #4). Also since Component_Port_Out is a formal out port, its initial value of 77 should override the initial value assigned to the signal that is port mapped onto its corresponding local port parameter. Hence, Signal_Out's initial value of 44 should be overridden by 77 (Fact #5). This overriding of 44 by 77 may be verified by examining the value of Signal_Out during the initialization phase (Fact #6). Since Signal_Out is also connected to the local in port Component_Port_In_B (Fact #7) the initial value of the corresponding formal port parameter should be overridden by the initial value of Signal_Out. This overriding action may be verified by noting that during the initialization phase the formal parameter Component_Port_In_B has the value 77 (Fact #9) instead of its user-specified initial value of 5555 (Fact #8). This experiment also has another example of the overriding of a formal in port's initial value by the initial value of a signal that is port mapped to its corresponding local parameter. This overriding action may be verified via the subsequent lines that are identified as Facts #10 through #13. An interesting point of Facts #10 through #13 is that they illustrate that the initial value of the formal in port Component_Port_In_A should be overridden twice by the VHDL tool suite. Once because of the hierarchical overriding of the formal generic parameter DEFAULT (66 should become 77), and the second overriding (77 should become 88) is due to the port mapping between its corresponding local parameter and the signal Signal_In.

Caveat. You should always be aware of the hierarchical overriding of generic defaults. Otherwise, your anticipated value for a generic parameter might not occur. An analogous caveat is also applicable to the initial values that you assign to signals that are used as actuals in a port mapping construct.

Source code for experiment

```
-- File Name : test_178.vhd
--
-- Author    : Joseph Pick
--
entity Component_Test_178 is
    generic (DEFAULT : NATURAL := 66); -- Fact #2: 77
    port (Component_Port_In_A : in NATURAL := DEFAULT; -- Fact #12: 77
          Component_Port_In_B : in NATURAL := 5555;    -- Fact #8: 5555
          Component_Port_Out  : out NATURAL := DEFAULT); -- Fact #4: 77
end Component_Test_178;

architecture Behave_1 of Component_Test_178 is
begin
    Component_Proc_Monitor:
    process
        variable Port_In_A_Value : NATURAL := 55;
        variable Port_In_B_Value : NATURAL := 55;
        variable Default_Value   : NATURAL := 55;
    begin
        Port_In_A_Value := Component_Port_In_A; -- Fact #13: 88

        Port_In_B_Value := Component_Port_In_B; -- Fact #9: 77

        Default_Value := DEFAULT;  -- Fact #3: 77

        wait for 25 ns;
    end process Component_Proc_Monitor;
end Behave_1;


entity Test_178 is
end Test_178;

architecture Behave_1 of Test_178 is
    component Component_Test_178
        generic (DEFAULT : NATURAL := 77); -- Fact #1
        port (Component_Port_In_A : in NATURAL := DEFAULT;
              Component_Port_In_B : in NATURAL := DEFAULT;
              Component_Port_Out  : out NATURAL := DEFAULT);
    end component;

    signal Signal_In  : NATURAL := 88;  -- Fact #10: 88
    signal Signal_Out : NATURAL := 44;

begin
    Instance_A: Component_Test_178
        port map (Component_Port_In_A => Signal_In,   -- Fact #11: 88
                  Component_Port_In_B => Signal_Out,  -- Fact #7: 77
                  Component_Port_Out  => Signal_Out); -- Fact #5: 77
```

```
    Top_Level_Proc_Monitor:
    process
        variable Signal_Out_Var : NATURAL := 101;
    begin
        Signal_Out_Var := Signal_Out;  -- Fact #6: 77
        wait for 25 ns;
    end process Top_Level_Proc_Monitor;

end Behave_1;

configuration Config_Test_178 of Test_178 is
    for Behave_1
        for Instance_A : Component_Test_178
            use entity WORK.Component_Test_178(Behave_1);
        end for;
    end for;
end Config_Test_178;
```

11.45 Test_179 (Overriding of generic defaults [cont])

Objective. To continue investigating the hierarchical overriding of default values assigned to generic parameters.

Expected observations. The actual value 33 (Fact #1) assigned (via a generic map construct) to the local generic parameter DEFAULT should override the default values assigned to both this local generic parameter (Fact #2) and the formal generic parameter (Fact #3) to which it corresponds to. This overriding of values may be confirmed by examining the final effective value of the generic parameter DEFAULT (Fact #4) during a simulation run.

Additional information. Ideally, it would be nice to be able to assign an actual to a generic parameter as late as possible in the design cycle. The best place to do so would be in a configuration declaration. However, VHDL'87 is very restrictive on this issue. The coding scenario of Test_179 cannot include the following generic mapping:

```
use entity WORK.Component_Test_179(Behave_1)  -- No Semi-colon!
    generic map (DEFAULT => 99);
```

This annoying VHDL'87 restriction has been removed in VHDL'93.

Source code for experiment

```
    -- File Name : test_179.vhd
    --
    -- Author    : Joseph Pick
    --
    entity Component_Test_179 is
        generic (DEFAULT : NATURAL := 66);   -- Fact #3
        port (Component_Port_In  : in NATURAL := DEFAULT;
              Component_Port_Out : out NATURAL := DEFAULT);
    end Component_Test_179;
```

```vhdl
architecture Behave_1 of Component_Test_179 is
begin
   process
      variable Port_In_Value : NATURAL := 55;
      variable Default_Value : NATURAL := 55;
   begin
      Port_In_Value := Component_Port_In;
      Default_Value := DEFAULT;      -- Fact #4: 33
      wait for 25 ns;
   end process;
end Behave_1;

entity Test_179 is
end Test_179;

architecture Behave_1 of Test_179 is
   component Component_Test_179
      generic (DEFAULT : NATURAL := 77);  -- Fact #2
      port (Component_Port_In : in NATURAL := DEFAULT;
            Component_Port_Out : out NATURAL := DEFAULT);
   end component;

   signal Signal_In  : NATURAL := 88;
   signal Signal_Out : NATURAL := 44;

begin
   Instance_A: Component_Test_179
          generic map (DEFAULT => 33)  -- Fact #1
          port map (Component_Port_In => Signal_In,
                    Component_Port_Out => Signal_Out);

   Monitor:
   process
      variable Signal_Out_Var : NATURAL := 101;
   begin
      Signal_Out_Var := Signal_Out; -- 33 So overrode 44.
      wait for 25 ns;
   end process Monitor;

end Behave_1;

configuration Config_Test_179 of Test_179 is
   for Behave_1
      for Instance_A : Component_Test_179
         use entity WORK.Component_Test_179(Behave_1);
      end for;
   end for;
end Config_Test_179;
```

11.46 Test_59d (Overriding of out port initial values)

Objective. To investigate how the initial values of out mode ports are overridden when these out ports are hierarchically connected. In our particular experiment the innermost out port's initialization to '1' occurs on the line identified as Fact #1 in the experiment's source code listed subsequently. To easily capture the initial value of the outermost out port a design entity is modeled at this same outermost hierarchical level, and its in mode port is connected to this outermost out mode port. The architecture body associated with this monitoring entity will, during the initialization phase, read this in port and place it into a variable. In effect, this read value will be the initial value of the outermost out port that is hierarchically connected to the innermost out port. The objective of this experiment is to determine what influence, if any, the initial value of the innermost out port has on the initial value of the outermost out port that it is hierarchically connected to.

Expected observations. The initial value assigned to the innermost out port should propagate outward and override the initial values of all out ports that it is hierarchically connected to. This VHDL rule is illustrated in Fig. 11.2, in which rectangles and circles represent entities and architectures, respectively. The arrow in this diagram emphasizes that the overriding of out port initial values propagate outward. The initial value of '1' that is assigned to the innermost out port should propagate outward and override any out port that this out port is hierarchically connected to. This overriding feature should be observed on the line identified as Fact #2. Examination of the in mode port Local_LevelC_2 should yield the value '1', thus confirming that the innermost out port's initial value did indeed propagate up the hierarchical chain of connections.

Additional information. It must always be remembered that an instance of a component declaration is considered to be a concurrent construct. Furthermore, when a local out port is connected to a formal out port via an instantiation's port map association, then the local port is consid-

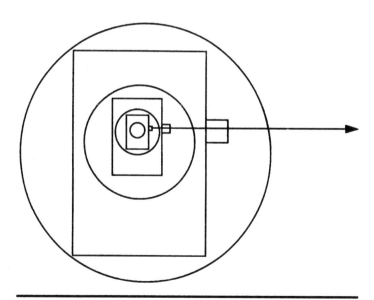

Figure 11.2 Initial values assigned to out ports propagate outward and override the initial values assigned to ports at a higher hierarchical level.

ered as a signal source for this out port. Consequently, it would be incorrect to assign to this out port via another concurrent construct if this out port is not associated with a resolution function. These facts are highlighted by the lines identified as Additional Info #1 and #2. If these two lines would not be commented out then a multisourcing error would occur since, in both cases, the respective out ports would be multiply driven even though they are not resolved signals.

Incidentally, pay close attention to how the following configuration declaration provides a helpful bird's-eye view of the hierarchical decomposition of this experiment.

Source code for experiment

```
-- File Name : test_59d.vhd
--
-- Author   : Joseph Pick
--
package MVL_4 is
  type Logic4 is ('X', '0', '1', 'Z');
end MVL_4;

use WORK.MVL_4.all;
entity Component_LevelA_59d is
      port (Local_LevelA : out Logic4 := '1');   -- *** Fact #1 ***
end Component_LevelA_59d;

architecture Behave_1 of Component_LevelA_59d is
  begin
     process
     begin
        wait for 10 ns;
        Local_LevelA <= '1';

     end process;
  end Behave_1;

use WORK.MVL_4.all;
entity Component_LevelB_59d is
      port (Local_LevelB : out Logic4 := 'Z');
end Component_LevelB_59d;

architecture Behave_1 of Component_LevelB_59d is
  component Component_LevelA_59d
      port (Local_LevelA : out Logic4);
  end component;
  begin

     Instant_LevelA : Component_LevelA_59d
              port map (Local_LevelA => Local_LevelB);
     process
        variable Count_B : NATURAL;
     begin
        wait for 10 ns;
        Count_B := Count_B + 1;
        -- Local_LevelB <= '1';   -- *** Additional Info #1 ***
     end process;
  end Behave_1;

use WORK.MVL_4.all;
entity Component_LevelC_1_59d is
      port (Local_LevelC_1 : out Logic4 := '0');
end Component_LevelC_1_59d;

architecture Behave_1 of Component_LevelC_1_59d is
  component Component_LevelB_59d
      port (Local_LevelB : out Logic4);
  end component;
  begin
```

```vhdl
            Instant_LevelB : Component_LevelB_59d
                    port map (Local_LevelB => Local_LevelC_1);
            process
              variable Count_C_1 : NATURAL;
            begin
              wait for 10 ns;
              Count_C_1 := Count_C_1 + 1;
              -- Local_LevelC_1 <= '1';    -- *** Additional Info #2 ***
            end process;
        end Behave_1;

        use WORK.MVL_4.all;
        entity Component_LevelC_2_59d is
            port (Local_LevelC_2 : in Logic4 := 'X');
        end Component_LevelC_2_59d;

        architecture Behave_1 of Component_LevelC_2_59d is
          begin
            Capture_Values:
            process
              variable Local_LevelC_2_Var : Logic4;
            begin
              Local_LevelC_2_Var := Local_LevelC_2;  -- *** Fact #2 ***
              wait on Local_LevelC_2'TRANSACTION;
            end process;
        end Behave_1;

use WORK.MVL_4.all;
entity Test_59d is
end Test_59d;

architecture Behave_1 of Test_59d is

  component Component_LevelC_1_59d
      port (Local_LevelC_1 : out Logic4);
  end component;

  component Component_LevelC_2_59d
      port (Local_LevelC_2 : in Logic4);
  end component;

  signal Connect_LevelC_1_2 : Logic4;
begin

  Instant_LevelC_1 : Component_LevelC_1_59d
              port map (Local_LevelC_1 => Connect_LevelC_1_2);

  Instant_LevelC_2 : Component_LevelC_2_59d
              port map (Local_LevelC_2 => Connect_LevelC_1_2);

end Behave_1;

configuration Config_59d of Test_59d is

   for Behave_1

       for Instant_LevelC_1 : Component_LevelC_1_59d
           use entity WORK.Component_LevelC_1_59d(Behave_1);

           for Behave_1
               for Instant_LevelB : Component_LevelB_59d
                   use entity WORK.Component_LevelB_59d(Behave_1);
                   for Behave_1
                       for Instant_LevelA : Component_LevelA_59d
                           useentityWORK.Component_LevelA_59d(Behave_1);
                       end for;
                   end for;
               end for;
           end for;
       end for;

       for Instant_LevelC_2 : Component_LevelC_2_59d
           use entity WORK.Component_LevelC_2_59d(Behave_1);
       end for;
   end for;

end Config_59d;
```

11.47 Test_182 (Behavior of 'DELAYED(25 ns))

Objective. This test aims to determine how the signal S'DELAYED(25 ns) behaves when S is updated in the delta time domain.

Expected observations. The signal S should take on the following consecutive values: 22 at 10 ns, 33 at 10 ns plus 1 delta, 44 at 10 ns plus 2 delta, and 55 at 10 ns plus 3 delta. The progressive updating of S'DELAYED(25 ns) may be manually verified by keeping in mind that it will always be equal to the hypothetical signal S_Delayed_25_Value, which should be updated according to the activities of the following process:

```
Proc_S_Delayed_25
process (S)
begin
    S_Delayed_25_Value <= transport S after 25 ns;
end process Proc_S_Delayed_25;
```

This process yields all the key features regarding the signal S'DELAYED(25 ns). First S_Delayed_25_Value is updated according to the transport-delay model, in that pulse widths less than 25 ns will still propagate to it. And second, all delta time updates to S will be absorbed because the after clause contains a nonzero absolute amount of time. This second piece of information, coupled with the fact that the times in the projected driver must be in an ascending order, should preempt the following ordered pairs from S_DELAYED_25_Value's projected driver: (22, 35 ns), (33, 35 ns), and (44, 35 ns). Only the last ordered pair (55, 35 ns) appended to the projected driver should remain in it. The upshot of all this analysis is that S_Delayed_25_Value, and hence S'DELAYED(25 ns), should be updated only once to the value 55, and this update should occur at 35 ns. Consequently, the process Monitor_1 of Test_182 should fire only once at exactly 35 ns (10 ns + 25 ns) instead of firing multiple times at 35 ns plus K delta time units, where K = 1,2,3.

Source code for experiment

```
-- File Name : test_182.vhd
--
-- Author    : Joseph Pick
--
entity Test_182 is
end Test_182;

architecture Behave_1 of Test_182 is

    signal S : NATURAL := 88;

begin

    Update_S:
    process
    begin
```

```
          S <= 22 after 10 ns;
          wait for 10 ns;
          S <= 33;
          wait for 0 ns;
          S <= 44;
          wait for 0 ns;
          S <= 55;
          wait;
      end process Update_S;

      Monitor_1:
      process
          variable S_Var_M1 : NATURAL := 101;
          variable Count   : NATURAL := 0;
      begin
          wait on S'DELAYED(25 ns);
          S_Var_M1 := S;
          Count    := Count + 1;
      end process Monitor_1;

  end Behave_1;
```

11.48 Test_184 (Constant assignment via a function call)

Objective. To investigate the possibility of assigning the returned value of a function call to a constant object during the elaboration phase.

Expected observations. Unfortunately, there is no uniformity in how the various VHDL vendors handle the coding scenario of this experiment. The VHDL'87 LRM is slightly vague on whether or not it should allow a function call to be made during the elaboration phase. Consequently, some tool suites will accept the constant declaration of this experiment, whereas others will not allow it to pass compilation.

Additional information. VHDL'93 has firmed up the language's definition on this issue. In VHDL'93 constants can be assigned values via a function call during the elaboration phase. Hence, every VHDL'93 compliant tool suite should successfully compile Test_184.

Source code for experiment

```
-- File Name : test_184.vhd
--
-- Author    : Joseph Pick
--
entity Test_184 is
end Test_184;

use STD.TEXTIO.all;
architecture Behave_1 of Test_184 is

begin

   process
      function Get_Data (File_Name : STRING; N : NATURAL)
                                        return BIT_VECTOR is
         file Local_File_Name : TEXT is in File_Name;
         -- In VHDL'93 the above line MUST be replaced
```

```
            -- by the following file declaration.
            -- file Datain : TEXT open READ_MODE is File_Name;
            variable Read_Data : BIT_VECTOR (1 to N);
            variable Buffer_Line : LINE;
         begin
            readline(Local_File_Name, Buffer_Line);
            Read (Buffer_Line, Read_Data);
            return Read_Data;
         end Get_Data;

         constant DATA : BIT_VECTOR(0 to 3) := Get_Data("stimulus_184",4);

         variable Captured_Data : BIT_VECTOR(0 to 3) := "0000";

      begin
            Captured_Data := Data;
            wait for 25 ns;
         end process;

      end Behave_1;
```

11.49 Test_185 (Reading both array size and data from same file)

Objective. To investigate the possibility of using the same file from which to successively read both the size and the contents of an array and to, respectively, assign them to constants during the elaboration phase.

Expected observations. Assuming that this model compiles (see Test_184), it will still not provide the desired result. The first constant declaration for SIZE should be executed successfully. However once the function Get_Size is completed the file stimulus_185 should be closed. But the function call to Get_Data, in the subsequent constant declaration for DATA, should cause this file to be reopened anew. Hence, the function Get_Data should once again begin reading from the file's top instead of from the second line which is where the desired data begins. This read operation done in Get_Data should now erroneously access an INTEGER data-type value (intended for SIZE) instead of an object of type BIT_VECTOR (intended for DATA), which is what is really desired. This data-typing discrepancy should be found during the elaboration phase and should cause the simulator to abruptly (perhaps even ungracefully) terminate its activities.

Additional information. The capability to use a function to read in both an array size and the corresponding array's data is very important. It is very desirable to create parameterizable models of memories as was attempted in this experiment. In VHDL'87 a workaround to the problem illustrated in this experiment may be achieved by simply using two separate files: one for the SIZE value and another for the array's intended DATA values. But, clearly, the model's testing environment would be more manageable if these two pieces of information would not be scattered around in separate files. Fortunately, one of VHDL'93's enhancements will allow both the size and the array's data to be resident in the same file. VHDL'93 has introduced a new category of functions called impure functions. An impure function is a function that can also reference and update objects that were not passed to it as subprogram parameters. Function specifications preceded by the reserved word *impure* imply that the declared function is to behave as a VHDL'93 impure function. The objectives of Test_185 can be achieved in VHDL'93 by creating two impure functions analogous to Get_Size and Get_Data. But this time you should make only one file declaration and place it in the architecture declarative region. As impure functions Get_Size and Get_Data will both be able to access this same externally declared file. And since the file declaration is outside of these impure functions it will not be closed when these functions complete their respective executions. Consequently, the impure Get_Size will read the first line and the impure Get_Data will then correctly read the second line.

Source code for experiment

```
-- File Name : test_185.vhd
--
-- Author    : Joseph Pick
--
entity Test_185 is
end Test_185;

use STD.TEXTIO.all;
architecture Behave_1 of Test_185 is

begin
process

  function Get_Size (File_Name_GS : String) return NATURAL is
    file Local_File_Name_GS : TEXT is in File_Name_GS;
    -- In VHDL'93 the above line MUST be replaced
    -- by the following file declaration.
    -- file Datain : TEXT open READ_MODE is File_Name_GS;

    variable Buffer_Line_GS : LINE;
    variable Read_Size : NATURAL;
  begin
    readline (Local_File_Name_GS,Buffer_Line_GS);
    read (Buffer_Line_GS, Read_Size);
    return Read_Size;
  end Get_Size;

  constant SIZE : NATURAL := Get_Size("stimulus_185");

  function Get_Data (File_Name_GD : STRING) return BIT_VECTOR is
    file Local_File_Name_GD : TEXT is in File_Name_GD;
    -- In VHDL'93 the above line MUST be replaced
    -- by the following file declaration.
    -- file Datain : TEXT open READ_MODE is File_Name_GD;

    variable Buffer_Line_GD : LINE;
    variable Read_Data : BIT_VECTOR (1 to SIZE) := (others => '0');
  begin
    readline(Local_File_Name_GD, Buffer_Line_GD);
    Read (Buffer_Line_GD, Read_Data);
    return Read_Data;
  end Get_Data;
```

```
        constant DATA : BIT_VECTOR := Get_Data("stimulus_185");
        variable Captured_Data : BIT_VECTOR(1 to SIZE) := (others => '0');
    begin
        Captured_Data := Data;
        wait for 25 ns;
    end process;

end Behave_1;
```

11.50 Test_187 (Usage of others in subprograms)

Objective. To investigate the usage of the reserved word *others* in subprograms.

Expected observations. This experiment should not compile because the reserved word *others* may not appear in the aggregate of an unconstrained array type. The aggregate should be interpreted by your compiler as an unconstrained array, since the formal parameter C_Out, to which the aggregate is assigned, is declared to be an unconstrained array.

Source code for experiment

```
-- File Name : test_187.vhd
--
-- Author    : Joseph Pick

package Excerpt is
    type Logic4 is ('X','0','1','Z');
    type Logic4_Vector is array (NATURAL range <>) of Logic4;

    procedure Sample_Proc (A_In : in Logic4;
                           B_In : in Logic4_Vector;
                           C_Out : out Logic4_Vector);
end Excerpt;

package body Excerpt is

    procedure Sample_Proc (A_In : in Logic4;
                           B_In : in Logic4_Vector;
                           C_Out : out Logic4_Vector) is
    begin

        if A_In = '1' then
            C_Out := B_In;
        else
            C_Out := (others => 'X');
        end if;

    end Sample_Proc;

end Excerpt;
```

11.51 Test_188 (Solution for Test_187)

Objective. To illustrate a solution to the problem exposed in Test_187.

Expected observations. Applying an appropriate type qualification to the aggregate should result in the successful compilation of this experiment.

Additional information. Since subprograms are often written with unconstrained parameters the type qualification method of this experiment is a very valuable tool to add to your repertoire of tricks and techniques.

Source code for experiment

```
-- File Name : test_188.vhd
--
-- Author    : Joseph Pick

package Excerpt is
   type Logic4 is ('X','0','1','Z');
   type Logic4_Vector is array (NATURAL range <>) of Logic4;

   procedure Sample_Proc (A_In : in Logic4;
                          B_In : in Logic4_Vector;
                          C_Out : out Logic4_Vector);
end Excerpt;

package body Excerpt is

   procedure Sample_Proc (A_In : in Logic4;
                          B_In : in Logic4_Vector;
                          C_Out : out Logic4_Vector) is
      subtype Temp_Typ is Logic4_Vector(1 to C_Out'LENGTH);
   begin

      if A_In = '1' then
         C_Out := B_In;
      else
         -- ***** RESULT *****
         --       Successfully compiled
         C_Out := Temp_Typ'(others => 'X');
      end if;

   end Sample_Proc;

end Excerpt;
```

11.52 Test_194 (Constancy of for loop bounds)

Objective. This experiment will explore the effects of modifying the bounds of a for loop within the for loop itself.

Expected observations. The for loop bounds should be treated as fixed values. Any changes made inside the body of the for loop should have no effect on these bounds.

Source code for experiment

```
-- File Name : test_194.vhd
--
-- Author    : Joseph Pick
--

entity Test_194 is
end Test_194;

architecture Behave_1 of Test_194 is
begin

   process
      variable Lower : Natural := 5;
      variable Upper : Natural := 10;
      variable Count : Natural := 0;
   begin
      for I in Lower to Upper loop
         Count := Count + I;
         Lower := Lower + 1;
         Upper := Upper - 1;
      end loop;
      wait for 50 ns;
   end process;

end Behave_1;
```

11.53 Test_195 (Dangers of the artificial usage of inout ports)

Objective. This experiment will explore what might go wrong when the inout mode is artificially used just so that an intended out port may appear on both sides of a signal assignment.

Expected observations. The intended periodic waveform on the signal Clock should not occur. The inout port Clock is connected to a multisourced signal. Hence, the resolved value of this multisourced signal should flow back into the design entity Component_Test_195 via the inout port Clock. Any other nonresolved port or signal connected to Clock should therefore see only this resolved value. The intended periodic waveform characteristics of Clock should be nullified. In our particular example the signal Clock should have a periodic waveform only until the simulation time of 200 ns. From that point on its value should stay constant at '1'.

Caveat. An artificial (unnecessary) usage of inout ports could potentially be disastrous since it might accidentally be connected to a multisourced signal having the same data type. Compilation will not be able to identify that anything is wrong since the types in the port map association are of the same type. Another point to keep in mind is that the artificial usage of inout ports will obstruct the readability of your model. An inout port intuitively implies bidirectionality. At the entity declaration level there is no reason to believe that a port was declared inout only for the sake of convenience so that the port's name may appear on both sides of a signal assignment.

Recommendation. The best solution to this problem in VHDL'87 is to declare the port as being of mode out and then to use an intermediate variable as was done for the waveform generator in the Excursion part of this book. However, VHDL'93 offers an even better solution. Keep the port as mode out but instead of declaring your own intermediate variable, rely on the new VHDL attribute 'DRIVING_VALUE and write Clock <= not Clock'DRIVING_VALUE.

Additional information. Note that it is possible to have a resolved signal that is of type BIT. A resolved signal does not necessarily have to belong to a well-known resolved data type such as Tristate_RS or std_logic.

Source code for experiment

```
-- File Name : test_195.vhd
--
-- Author    : Joseph Pick
--
entity Component_Test_195 is
    port (Clock : inout BIT);
end Component_Test_195;

architecture Behave_1 of Component_Test_195 is
begin
    Gen_Waveform:
    process
    begin
        wait for 50 ns;
        Clock <= not Clock;
    end process;
end Behave_1;

entity Test_195 is
end Test_195;

architecture Behave_1 of Test_195 is
    component Component_Test_195
        port (Clock : inout BIT);
    end component;

    function Or_Pull_Down (V : BIT_VECTOR) return BIT is
        variable Result : BIT := '0';
    begin
        for I in V'RANGE loop
            if V(I) = '1' then
                Result := '1';
            end if;
        end loop;
        return Result;
    end Or_Pull_Down;
    signal Clock_Signal_A : Or_Pull_Down BIT;
    signal Clock_Signal_B : BIT;
begin

    Instance_A: Component_Test_195
                port map (Clock => Clock_Signal_A);
```

```
              Instance_B: Component_Test_195
                         port map (Clock => Clock_Signal_B);
              Waveform: Clock_Signal_A <= '0' after 75 ns,
                                          '1' after 125 ns;

              Monitor_Clock_Signal_A:
              process
                 variable Clock_Var_A : BIT;
              begin
                 wait on Clock_Signal_A'TRANSACTION;
                 Clock_Var_A := Clock_Signal_A;
              end process Monitor_Clock_Signal_A;

              Monitor_Clock_Signal_B:
              process
                 variable Clock_Var_B : BIT;
              begin
                 wait on Clock_Signal_B'TRANSACTION;
                 Clock_Var_B := Clock_Signal_B;
              end process Monitor_Clock_Signal_B;

          end Behave_1;
```

11.54 Test_35_0 (The signal multiplexing problem)

Objective. To investigate the concurrent signal multiplexing problem.

Expected observations. You should not be able to simulate this experiment because there are multisources on the unresolved signal X. Depending on whose VHDL tool suite you are working with, this multisourcing error will either be identified during the compilation or the elaboration phase.

Additional information. The simple scenario of this experiment may be readily solved by combining the two processes into one and then doing the signal multiplexing in a sequential if...then...else coding environment. However (and this is a big "HOWEVER") such an amalgamation might not be possible when modeling more complex and comprehensive devices. So it is very important that you master the following sequence of experiments that deal exclusively with this topic.

Source code for experiment

```
              -- File Name : test_35_0.vhd
              --
              -- Author    : Joseph Pick

              entity Test_35_0 is
              end Test_35_0;

              architecture Behave_1 of Test_35_0 is
                 signal X     : BIT := '0';
                 signal Set   : BIT := '0';
                 signal Reset : BIT := '0';
```

```
begin
   Proc_1:
   process(Set)
   begin
      if Set = '1' then
         X <= '1';
      end if;
   end process Proc_1;

   Proc_2:
   process(Reset)
   begin
      if Reset = '1' then
         X <= '0';
      end if;
   end process Proc_2;

   Set <= '1' after 10 ns, '0' after 20 ns;
   Reset <= '1' after 30 ns, '0' after 40 ns;

end Behave_1;
```

11.55 Test_35 (Standard solution to scalar signal multiplexing)

Objective. To become familiar with the standard solution to the scalar version of the concurrent signal multiplexing problem. The names of the intermediate signals used in this solution were deliberately chosen to improve the readability of this model. For instance, the signal X_Proc_1 consists of the two parts X and Proc_1. The X part is the name of the signal that is being multiplexed. The Proc_1 part reflects the name of the unique process in which the intermediate signal X_Proc_1 is being assigned to.

Expected observations. Signal X should be updated to '1' at time (10 ns + 2 deltas) and then to '0' at (30 ns + 2 deltas). There should not be any multisourcing problem since X, X_Proc_1, and X_Proc_2 are single-source signals.

Additional information. You must always remember which signal attributes are signals and which are not. Recall Test_16 in which the nonsignaling attribute 'EVENT could not be used to model an edge-triggered device. Instead 'STABLE had to be used. An analogous situation occurs in this experiment. Though X_Proc_1'ACTIVE is functionally equivalent to the expression (not X_Proc_1'QUIET), it still cannot be an element of a sensitivity list since it is not a signal. X_Proc_1'QUIET, on the other hand, is a signal and so can contribute toward the sensitivity list implied by the solution given in this Test_35. And so the following conditional concurrent signal assignment will not work as desired, since updates to the nonsignal X_Proc_1'ACTIVE (or X_Proc_2'ACTIVE) will not reawaken this concurrent statement:

```
X <= X_Proc_1 when X_Proc_1'ACTIVE else  -- *** WILL ***
     X_Proc_2 when X_Proc_2'ACTIVE else  -- *** NOT  ***
     X;                                  -- *** WORK ***
```

Another significant point is that the attribute 'QUIET was used instead of 'STABLE. The reasoning behind this choice is that, possibly aside from their very first assignments, both X_Proc_1 and X_Proc_2 will henceforth only have transactions in this model. Since the attribute 'STABLE zeroes in on events, while 'QUIET zeroes in on transactions, the former would not trigger the awakening of Test_35's conditional concurrent signal assignment, whereas the latter would.

The given solution can be made more efficient by taking advantage of the fact that the assigned values X_Proc_1 and X_Proc_2 will always be '1' and '0', respectively. Hence the conditional concurrent assignment to X should, in this case, be rewritten as follows:

```
X <= '1' when not X_Proc_1'QUIET else
     '0' when not X_Proc_2'QUIET else
     X;
```

Source code for experiment

```
-- File Name : test_35.vhd
--
-- Author    : Joseph Pick

entity Test_35 is
end Test_35;

architecture Behave_1 of Test_35 is
    signal X_Proc_1 : BIT := '0';
    signal X_Proc_2 : BIT := '0';
    signal X        : BIT := '0';
    signal Set      : BIT := '0';
    signal Reset    : BIT := '0';
begin

Proc_1:
process(Set)
begin
  if Set = '1' then
     X_Proc_1 <= '1';
  end if;
end process Proc_1;

Proc_2:
process(Reset)
begin
  if Reset = '1' then
     X_Proc_2 <= '0';
  end if;
end process Proc_2;

X <= X_Proc_1 when not X_Proc_1'QUIET else
     X_Proc_2 when not X_Proc_2'QUIET else
     X;

Set   <= '1' after 10 ns, '0' after 20 ns;
Reset <= '1' after 30 ns, '0' after 40 ns;

end Behave_1;
```

11.56 Test_35a (Side effects of Test_35)

Objective. To demonstrate that the Test_35 solution to the concurrent signal multiplexing problem may have an undesirable side effect. This experiment may be best understood by examining the process that is equivalent to the following conditional concurrent signal assignment:

```
X <= X_Proc_1 when not X_Proc_1'QUIET else
     X_Proc_2 when not X_Proc_2'QUIET else
     X;
```

Expected observations. An event should be observed on the signal X'TRANSACTION 1 delta delay after the initialization phase. The assignment of X to itself was scheduled because of the else clause occurring in the conditional concurrent signal assignment.

Caveat. When using the attribute 'TRANSACTION, always be aware that this signal might take on an unexpected (and unwanted) event because of a signal assignment that was scheduled during the initialization phase.

Additional information. The VHDL'93 update will allow the following syntax to be used:

```
X <= X_Proc_1 when not X_Proc_1'QUIET else
     X_Proc_2 when not X_Proc_2'QUIET else
     unaffected;
```

The new VHDL reserved word *unaffected* will, in essence, not schedule an assignment to X when all the conditional statements are false. Hence, the error outlined in this experiment may be easily avoided in VHDL'93. Another option that you have available in VHDL'93 is to avoid the problem completely by simply omitting the else clause.

Source code for experiment

```
-- File Name : test_35a.vhd
--
-- Author    : Joseph Pick

entity Test_35a is
end Test_35a;

architecture Behave_1 of Test_35a is
  signal X_Proc_1   : BIT := '0';
  signal X_Proc_2   : BIT := '0';
  signal X          : BIT := '0';
  signal Set        : BIT := '0';
  signal Reset      : BIT := '0';
begin

  Proc_1:
  process(Set)
```

```
      begin
        if Set = '1' then
            X_Proc_1 <= '1';
        end if;
      end process Proc_1;

      Proc_2:
      process(Reset)
      begin
        if Reset = '1' then
            X_Proc_2 <= '0';
        end if;
      end process Proc_2;

      X <= X_Proc_1 when not X_Proc_1'QUIET else
           X_Proc_2 when not X_Proc_2'QUIET else
           X;

      Set   <= '1' after 10 ns, '0' after 20 ns,
               '1' after 50 ns, '0' after 60 ns;
      Reset <= '1' after 30 ns, '0' after 40 ns,
               '1' after 70 ns, '0' after 80 ns;

      Trans_Proc:
      process
          variable Count : NATURAL := 0;
      begin
          wait on X'TRANSACTION;
          Count := Count + 1;
      end process Trans_Proc;

    end Behave_1;
```

11.57 Test_39 (Solution for Test_35a)

Objective. To demonstrate a solution to the problem highlighted in Test_35a.

Expected observations. Signal X should be updated to '1' at time (10 ns + 2 deltas) and then to '0' at (30 ns + 2 deltas). Hence, an assignment to X should not be scheduled during the initialization phase as was the case in Test_35a.

Source code for experiment

```
      -- File Name : test_39.vhd
      --
      -- Author    : Joseph Pick

      entity Test_39 is
      end Test_39;

      architecture Behave_1 of Test_39 is

        signal X_Proc_1 : BIT := '0';
        signal X_Proc_2 : BIT := '0';
        signal X        : BIT := '0';
        signal Set      : BIT := '0';
        signal Reset    : BIT := '0';
      begin

        Proc_1:
        process(Set)
```

```
      begin
         if Set = '1' then
            X_Proc_1 <= '1';
         end if;
      end process Proc_1;

      Proc_2:
      process(Reset)
      begin
         if Reset = '1' then
            X_Proc_2 <= '0';
         end if;
      end process Proc_2;

      Time_Mux:
      process(X_Proc_1'TRANSACTION, X_Proc_2'TRANSACTION)
      begin
         if X_Proc_1'ACTIVE then
            X <= X_Proc_1;
         end if;
         if X_Proc_2'ACTIVE then
            X <= X_Proc_2;
         end if;
      end process Time_Mux;

      Set   <= '1' after 10 ns, '0' after 20 ns;
      Reset <= '1' after 30 ns, '0' after 40 ns;

   end Behave_1;
```

11.58 Test_35b (Array signal multiplexing: error)

Objective. To illustrate, via a counterexample, that the Test_35 solution to the concurrent signal multiplexing problem is, in general, not appropriate for arrays. The type LX01 is used only to more rapidly identify the problem. The notation Set_X_Proc_1_2 implies that the element X_Proc_1(2) is to be assigned a value. Since X_Proc_1 is one of the intermediate signals contributing toward the value of the multiplexed signal X, it follows that the notation Set_X_Proc_1_2 really intends to assign a value to X(2).

Expected observations. The assignment of array X_Proc_2 to X should unintentionally replace the '1' value in the subelement X(2) with the undesirable value '0'. Consequently, the array X, as a whole, should be updated to an undesirable and incorrect value.

Source code for experiment

```
      -- File Name : test_35b.vhd
      --
      -- Author    : Joseph Pick

      entity Test_35b is
      end Test_35b;

      architecture Behave_1 of Test_35b is

         type LX01 is ('X', '0', '1');
         type LX01_Vector is array (NATURAL range <>) of LX01;
```

```vhdl
    signal X                 : LX01_Vector(3 downto 0) :=
                                              (others => 'X');
    signal X_Proc_1          : LX01_Vector(3 downto 0) :=
                                              (others => '0');
    signal X_Proc_2          : LX01_Vector(3 downto 0) :=
                                              (others => '0');
    signal Set_X_Proc_1_2    : BOOLEAN := FALSE;
    signal Set_X_Proc_2_3    : BOOLEAN := FALSE;
    signal Count_1           : NATURAL := 0;
    signal Count_2           : NATURAL := 0;
begin

  Proc_1:
  process
  begin
    wait until Set_X_Proc_1_2 = TRUE;
    X_Proc_1(2) <= '1';
    wait for 200 ns;
    Count_1 <= Count_1 + 1;
  end process Proc_1;

  Proc_2:
  process
  begin
    wait until Set_X_Proc_2_3 = TRUE;
    X_Proc_2(3) <= '1';
    wait for 200 ns;
    Count_2 <= Count_2 + 1;
  end process Proc_2;

  X <= X_Proc_1 when not X_Proc_1'QUIET else
       X_Proc_2 when not X_Proc_2'QUIET else
       X;
  Set_X_Proc_1_2 <= TRUE after 10 ns, FALSE after 400 ns;
  Set_X_Proc_2_3 <= TRUE after 50 ns, FALSE after 500 ns;

end Behave_1;
```

11.59 Test_35l (Brute force solution for Test_35b)

Objective. To illustrate a possible solution to the array version of the concurrent signal multiplexing problem.

Expected observations. The various subelements of the array X should be appropriately set and reset.

Additional information. Although this solution works it is neither satisfying nor aesthetically appealing. Test_35l is analogous to accessing all the subelements of an array Data by explicitly referencing each and every one of them via hard-coded indices as follows: Data(1), Data(2), Data(3), ...Data(256). This method is very impractical and, in fact, is somewhat barbaric. It would be much more convenient, productive, and memory efficient to use a for loop parameter instead to iteratively access the K'th element of the array using the notation Data(K). In addition to not being compact, Test_35l's solution also has a major weakness that one of my colleagues encountered. The following experiment Test_40 abstracts a real-world scenario that captures the breakdown of Test_35l's solution.

Source code for experiment

```
-- File Name : test_351.vhd
--
-- Author    : Joseph Pick
--
entity Test_351 is
end Test_351;

architecture Behave_1 of Test_351 is
  type LX01 is ('X', '0', '1');
  type LX01_Vector is array (NATURAL range <>) of LX01;
  signal X_Proc_1_2       : LX01;
  signal X_Proc_1_1       : LX01;
  signal X_Proc_2_2       : LX01;
  signal X                : LX01_Vector(3 downto 0) := (others => 'X');
  signal Set_X_Proc_1_2   : BOOLEAN := FALSE;
  signal Set_X_Proc_1_1   : BOOLEAN := FALSE;
  signal Set_X_Proc_2_2   : BOOLEAN := FALSE;
  signal Count_1          : NATURAL := 0;
  signal Count_2          : NATURAL := 0;

begin

  Proc_1:
  process
  begin
    wait until Set_X_Proc_1_2 = TRUE;
    X_Proc_1_2 <= '1';
    wait until Set_X_Proc_1_1 = TRUE;
    X_Proc_1_1 <= '1';
    wait for 200 ns;
    Count_1 <= Count_1 + 1;
  end process Proc_1;

  Proc_2:
  process
  begin
    wait until Set_X_Proc_2_2 = TRUE;
    X_Proc_2_2 <= '0';
    wait for 200 ns;
    Count_2 <= Count_2 + 1;
  end process Proc_2;

  Time_Mux_1:
  X(2) <= X_Proc_1_2 when not X_Proc_1_2'QUIET else
          X_Proc_2_2 when not X_Proc_2_2'QUIET else
          X(2);

  X(1) <= X_Proc_1_1 when not X_Proc_1_1'QUIET else
          X(1);

  Set_X_Proc_1_2 <= TRUE after 10 ns, FALSE after 400 ns;
  Set_X_Proc_2_2 <= TRUE after 20 ns, FALSE after 400 ns;
  Set_X_Proc_1_1 <= TRUE after 30 ns, FALSE after 400 ns;

  Monitor:
  process
    variable X_Var : LX01_Vector(3 downto 0) := (others => 'X');
  begin
    wait on X;
    X_Var := X;
  end process Monitor;
end Behave_1;
```

11.60 Test_40 (Counterexample for Test_35l's solution)

Objective. To demonstrate that the solution given in Test_35l might not always be appropriate.

Expected observations. As per the experiments Test_1a through Test_1e it readily follows that there should be a multisourcing error due to the signal assignments to the indexed array X(Count) and to X(2). We now know that the tool suite should single out the nonresolved signal X(2) as erroneously having multiple sources.

Recommendation. A solution as per Test_35l may be used whenever the array is indexed only by fixed subscripts. Nonetheless, Test_35l's solution, even if it works, is not general enough. Later experiments will explore alternative solutions that are much more flexible.

Source code for experiment

```
-- File Name : test_40.vhd
--
-- Author    : Joseph Pick
--
entity Test_40 is
end Test_40;
architecture Behave_1 of Test_40 is
   type Nat_Array is array (1 to 6) of NATURAL;
   signal X        : Nat_Array := (5,5,5,5,5,5);
   signal X_Proc_1_2 : NATURAL;
   signal X_Proc_2_2 : NATURAL;
   signal Count    : NATURAL := 1;
begin

   Proc_1:
   process(Count)
   begin
     if Count = 2 then
        X_Proc_1_2 <= 2;
     else
        X(Count) <= 66;    -- *** Cause of multi-sourcing problem
     end if;
   end process Proc_1;

   Proc_2:
   process(Count)
   begin
     if Count = 5 then
        X_Proc_2_2 <= 3;
     end If;
   end process Proc_2;

   Inc_Count:
   process
   begin
     Count <= Count + 1;
     wait for 20 ns;
   end process Inc_Count;

   X(2) <= X_Proc_1_2 when not X_Proc_1_2'QUIET else
           X_Proc_2_2 when not X_Proc_2_2'QUIET else
           X(2);
end Behave_1;
```

11.61 Test_35b_1 (Ideal solution for Test_35b)

Objective. To investigate the ideal solution for the array version of the concurrent signal multiplexing problem.

Expected observations. A compile error should be generated, since the prefix X_Proc_1(K) of the attribute 'ACTIVE is unfortunately not permitted by VHDL.

Recommendation. Actually, this is a recommendation to the VHDL language committee.

Future enhancements to the VHDL standard should allow constructs like X_Proc_1(K)'ACTIVE and X_Proc_1(K)'EVENT. Regrettably, the VHDL'93 update does not support this highly desirable capability.

Source code for experiment

```
-- File Name : test_35b_1.vhd
--
-- Author    : Joseph Pick

entity Test_35b_1 is
end Test_35b_1;

architecture Behave_1 of Test_35b_1 is

    type LX01 is ('X', '0', '1');
    type LX01_Vector is array (NATURAL range <>) of LX01;

    signal X              : LX01_Vector(3 downto 0) :=
                                        (others => 'X');
    signal X_Proc_1       : LX01_Vector(3 downto 0) :=
                                        (others => '0');
    signal X_Proc_2       : LX01_Vector(3 downto 0) :=
                                        (others => '0');
    signal Set_X_Proc_1_2 : BOOLEAN := FALSE;
    signal Set_X_Proc_2_3 : BOOLEAN := FALSE;
    signal Count_1        : NATURAL := 0;
    signal Count_2        : NATURAL := 0;

begin

    Proc_1:
    process
    begin
      wait until Set_X_Proc_1_2 = TRUE;
      X_Proc_1(2) <= '1';
      wait for 200 ns;
      Count_1 <= Count_1 + 1;
    end process Proc_1;

    Proc_2:
    process
    begin
      wait until Set_X_Proc_2_3 = TRUE;
      X_Proc_2(3) <= '1';
      wait for 200 ns;
      Count_2 <= Count_2 + 1;
    end process Proc_2;
```

```
Time_Mux:
process
   variable X_Update : LX01_Vector(3 downto 0) :=
                                     (others => 'X');
begin
   wait on X_Proc_1'TRANSACTION, X_Proc_2'TRANSACTION;
   X_Update := X;  -- Capture current/old value of X.

   if X_Proc_1'ACTIVE then
      for K in X_Proc_1'RANGE loop
         -- Following line does not compile because prefix of
         -- 'ACTIVE must be a static name.
         if X_Proc_1(K)'ACTIVE then
            X_Update(K) := X_Proc_1(K);
         end if;
      end loop;
   end if;

   if X_Proc_2'ACTIVE then
      for K in X_Proc_2'RANGE loop
         -- Following line does not compile because prefix of
         -- 'ACTIVE must be a static name.
         if X_Proc_2(K)'ACTIVE then
            X_Update(K) := X_Proc_2(K);
         end if;
      end loop;
   end if;

   -- Update only those elements of the array, X, that have
   -- received a transaction.
   X <= X_Update;
end process Time_Mux;

Set_X_Proc_1_2 <= TRUE after 10 ns, FALSE after 400 ns;
Set_X_Proc_2_3 <= TRUE after 50 ns, FALSE after 500 ns;

end Behave_1;
```

11.62 Test_35j (Shadow signals and 'LAST_VALUE)

Objective. To investigate the usage of shadow signals to implement a workaround technique which, in spirit, captures the essence of the ideal, though noncompilable solution shown in Test_35b_1. Here is the algorithm in a nutshell. Whenever a value is assigned to X_Proc_1(K) I will toggle the K'th element of its corresponding shadow signal T_X_Proc_1. It should come as no surprise that the T in the identifier T_X_Proc_1 stands for Toggle. The process implementing the actual signal multiplexing will first compare the current value of T_X_Proc_1 to its previous value. This information will then be used to determine which element of X_Proc_1 received a transaction. Suppose that X_Proc_1(5) received a transaction. Then this individual element X_Proc_1(5) will be assigned to its corresponding signal X(5). Observe that by making assignments to the individual elements of X we are bypassing the array assignment problems of the previous Test_35b. Incidentally, note well that I have manually determined the value of X_Proc_1(K)'ACTIVE, even though VHDL does not permit this expression to be explicitly accessed (see Test_35b_1).

Expected observations. Setting Last_Value_Var to T_X_Proc_1'LAST_VALUE, at time (30 ns + 1 delta) should cause signal X to be erroneously updated during the next delta with the incorrect value ('X', '1', '1', 'X'). The value ('X', '0', '1', 'X') should have been used instead. The root of the problem is not the overall philosophy of the solution but rather the misapplication of the attribute 'LAST_VALUE.

Caveat. The attribute S'LAST_VALUE works as expected when S is a scalar signal. It provides the previous value of S just before that signal had its most recent event. Unfortunately, VHDL'87 has a very strange definition for V'LAST_VALUE when its prefix V is an array (vector). The attribute 'LAST_VALUE attacks each element of the array bit-wise. Consequently, what you end up with is the composite of the previous value of each individual subelement just before that subelement had an event. This amalgamation is totally impractical. Hence the VHDL'87 version of the attribute 'LAST_VALUE should never be applied to arrays, since its array definition is void of any meaningful interpretation.

Additional information. The VHDL'93 LRM update has revised the definition of the attribute 'LAST_VALUE for composite type objects. This attribute will no longer be applied individually to each of the subelements of a composite type object. Instead, it will be equal to the previous value of the object before the object as a whole was identified as having received an event. Consequently, the VHDL'93 update will permit Test_35j to be used as a solution to the array version of the concurrent multiplexing signal problem.

Source code for experiment

```
-- File Name : test_35j.vhd
--
-- Author    : Joseph Pick

entity Test_35j is
end Test_35j;

architecture Behave_1 of Test_35j is
    type LX01 is ('X', '0', '1');
    type LX01_Vector is array (NATURAL range <>) of LX01;

    signal X              : LX01_Vector(3 downto 0) :=
                                         (others => 'X');
    signal X_Proc_1       : LX01_Vector(3 downto 0) :=
                                         (others => '0');
    signal T_X_Proc_1     : BIT_VECTOR(3 downto 0) :=
                                         (others => '0');
    signal X_Proc_2       : LX01_Vector(3 downto 0) :=
                                         (others => '0');
    signal T_X_Proc_2     : BIT_VECTOR(3 downto 0) :=
                                         (others => '0');
    signal Set_X_Proc_1_2 : BOOLEAN := FALSE;
    signal Set_X_Proc_2_2 : BOOLEAN := FALSE;
```

```vhdl
    signal Set_X_Proc_1_1 : BOOLEAN := FALSE;
    signal Count_1        : NATURAL := 0;
    signal Count_2        : NATURAL := 0;
begin

    Proc_1:
    process
    begin
      wait until Set_X_Proc_1_2 = TRUE;
      X_Proc_1(2) <= '1';
      T_X_Proc_1(2) <= not T_X_Proc_1(2);
      wait until Set_X_Proc_1_1 = TRUE;
      -- **** NOTE that the intent here is to assign 1 ONLY
      -- ****         to bit #1. BUT because 'LAST_VALUE
      -- ****         operates bitwise, bit #2 will ALSO be
      -- ****         set to 1, which was NOT desired.
      X_Proc_1(1) <= '1';
      T_X_Proc_1(1) <= not T_X_Proc_1(1);
      wait for 200 ns;
      Count_1 <= Count_1 + 1;
    end process Proc_1;

  Proc_2:
  process
  begin
    wait until Set_X_Proc_2_2 = TRUE;
    -- **** NOTE that the intent here is to reset the bit #2
    -- ****         that was set in the process, Proc_1.
    -- ****         Such a situation can easily occur in
    -- ****         the real world.
    X_Proc_2(2) <= '0';
    T_X_Proc_2(2) <= not T_X_Proc_2(2);
    wait for 200 ns;
    Count_2 <= Count_2 + 1;
  end process Proc_2;

  Time_Mux:
  process
    variable Last_Value_Var : BIT_VECTOR(3 downto 0);
    variable X_Update_Var   : LX01_Vector(3 downto 0) :=
                                        (others => 'X');

  begin
      wait on T_X_Proc_1, T_X_Proc_2;
      X_Update_Var := X; -- Capture current/old value
      if T_X_Proc_1'EVENT then
        -- Determine which element had an activity by
        -- comparing the 'LAST_VALUE (WRONG !!) attribute
        -- of the toggling shadow signal with its current value.
        Last_Value_Var := T_X_Proc_1'LAST_VALUE;
        for I in 3 downto 0 loop
           if Last_Value_Var(I) /= T_X_Proc_1(I) then
              X_Update_Var(I) := X_Proc_1(I);
           end if;
        end loop;
      end if;

      if T_X_Proc_2'EVENT then
        Last_Value_Var := T_X_Proc_2'LAST_VALUE;
        for I in 3 downto 0 loop
           if Last_Value_Var(I) /= T_X_Proc_2(I) then
              X_Update_Var(I) := X_Proc_2(I);
           end if;
        end loop;
      end if;

      X <= X_Update_Var;
  end process Time_Mux;

  Set_X_Proc_1_2 <= TRUE after 10 ns, FALSE after 400 ns;

     Set_X_Proc_2_2 <= TRUE after 20 ns, FALSE after 400 ns;
     Set_X_Proc_1_1 <= TRUE after 30 ns, FALSE after 400 ns;

     end Behave_1;
```

11.63 Test_35k (Array signal multiplexing: OK)

Objective. To investigate the usage of shadow signals to implement a workaround technique that, in spirit, captures the essence of the ideal, though noncompilable solution shown in Test_35b_1. This experiment illustrates the usage of variables to capture the previous value of the shadow signals since, as shown in Test_35j, 'LAST_VALUE is totally inappropriate for arrays.

Expected observations. The time multiplexed array, X, should be correctly updated so that at time (30 ns + 2 deltas) its value should be ('X', '0', '1', 'X').

Recommendation. When applying this multiplexing algorithm, make certain that the variables and their corresponding shadow signals both begin with the same initial value.

The next recommendation is for the VHDL language committee: Future VHDL language updates should allow constructs like X_Proc_1(K)'ACTIVE so that the run time overhead of managing shadow signals will not be necessary. There should not be any technical difficulties hampering this feature's inclusion into the language. Test_35k shows that I can, in essence, manually determine the value of X_Proc_1(K)'ACTIVE using my shadow signals. So if I can do it, then surely the simulation kernel must also be able to determine these values because, in the final analysis, this kernel has more control and internal hooks into the simulation than I do.

Additional information. The VHDL'93 definition of 'LAST_VALUE permits the previous experiment Test_35j to be used as a solution to the array version of the concurrent signal multiplexing problem. In a VHDL'93 compliant environment it is no longer necessary to introduce variables to capture the previous values of the shadow registers. The VHDL'93 version of 'LAST_VALUE may be used instead. Unfortunately, the use of shadow signals will still be required since VHDL'93 does not permit the usage of X_Proc_1(K)'ACTIVE.

Source code for experiment

```
-- File Name : test_35k.vhd
--
-- Author    : Joseph Pick

entity Test_35k is
end Test_35k;

architecture Behave_1 of Test_35k is

  type LX01 is ('X', '0', '1');
  type LX01_Vector is array (NATURAL range <>) of LX01;
```

```vhdl
    signal X                  : LX01_Vector(3 downto 0) :=
                                            (others => 'X');
    signal X_Proc_1           : LX01_Vector(3 downto 0) :=
                                            (others => '0');
    signal T_X_Proc_1         : BIT_VECTOR(3 downto 0) :=
                                            (others => '0');
    signal X_Proc_2           : LX01_Vector(3 downto 0) :=
                                            (others => '0');
    signal T_X_Proc_2         : BIT_VECTOR(3 downto 0) :=
                                            (others => '0');
    signal Set_X_Proc_1_2 : BOOLEAN := FALSE;
    signal Set_X_Proc_2_2 : BOOLEAN := FALSE;
    signal Set_X_Proc_1_1 : BOOLEAN := FALSE;
    signal Count_1 : NATURAL := 0;
    signal Count_2 : NATURAL := 0;

begin

Proc_1:
process
begin
  wait until Set_X_Proc_1_2 = TRUE;
  X_Proc_1(2) <= '1';
  T_X_Proc_1(2) <= not T_X_Proc_1(2);
  wait until Set_X_Proc_1_1 = TRUE;
  X_Proc_1(1) <= '1';
  T_X_Proc_1(1) <= not T_X_Proc_1(1);
  wait for 200 ns;
  Count_1 <= Count_1 + 1;
end process Proc_1;

Proc_2:
process
begin
  wait until Set_X_Proc_2_2 = TRUE;
  X_Proc_2(2) <= '0';
  T_X_Proc_2(2) <= not T_X_Proc_2(2);
  wait for 200 ns;
  Count_2 <= Count_2 + 1;
end process Proc_2;

Time_Mux:
process
    variable X_Update_Var : LX01_Vector(3 downto 0) :=
                                        (others => 'X');
    -- Variables required because 'LAST_VALUE in VHDL'87
    -- is defined bitwise and hence, is not appropriate for
    -- this concurrent signal multiplexing algorithm.
    variable Previous_Value_T_X_Proc_1 : BIT_VECTOR(3 downto 0)
                                        := (others => '0');
    variable Previous_Value_T_X_Proc_2 : BIT_VECTOR(3 downto 0)
                                        := (others => '0');
begin
  wait on T_X_Proc_1, T_X_Proc_2;
  X_Update_Var := X;  -- Capture current/old value

  if T_X_Proc_1'EVENT then
      for I in 3 downto 0 loop
          if Previous_Value_T_X_Proc_1(I) /= T_X_Proc_1(I) then
              X_Update_Var(I) := X_Proc_1(I);
          end if;
      end loop;

      -- Capture the current value to be used next time.
      Previous_Value_T_X_Proc_1 := T_X_Proc_1;
  end if;

  if T_X_Proc_2'EVENT then
      for I in 3 downto 0 loop
          if Previous_Value_T_X_Proc_2(I) /= T_X_Proc_2(I) then
              X_Update_Var(I) := X_Proc_2(I);
          end if;
      end loop;
```

```
        -- Capture the current value to be used next time.
        Previous_Value_T_X_Proc_2 := T_X_Proc_2;
    end if;

    X <= X_Update_Var;
end process Time_Mux;

    Set_X_Proc_1_2 <= TRUE after 10 ns, FALSE after 400 ns;
    Set_X_Proc_2_2 <= TRUE after 20 ns, FALSE after 400 ns;
    Set_X_Proc_1_1 <= TRUE after 30 ns, FALSE after 400 ns;

  end Behave_1;
```

Appendix A

Responsibilities of the In-House VHDL Guru(s)

- To develop an in-house VHDL course and then to subsequently train current and future VHDL team members.
  - If no time has been allocated by management for the design and implementation of an in-house VHDL training program, then the VHDL guru must identify and select the training consultant who will best prepare the company's VHDL team members. Ideally, this training should be completed as soon as possible and prior to the announcement of any VHDL modeling requirements. That way the company's engineers will already be in the rhythm of hardware modeling with the VHDL language even before the start of the VHDL project.
- To help management in their evaluation of commercially available VHDL-related products.
- To write and maintain the official companywide VHDL style guide.
- To present oneself as a general, all-purpose VHDL problem solver.
- To serve as a clearinghouse for VHDL coding techniques and language subtleties.
- To generate tests and experiments that identify and explore:
  - New coding design strategies.
  - Team members' conceptual errors.
  - VHDL language subtleties.
  - VHDL tool suite errors.

- To straddle multiple projects in order to ensure that common utilities and modeling parts are not redesigned and reimplemented.
- To design, code, and test those general, all-purpose utility VHDL subprograms and packages that may potentially be needed across the board throughout the company.
- To guide newly trained VHDL engineers with their initial modeling efforts.
- To participate in all design reviews and VHDL code walk-throughs.
- To assist VHDL team members in finding ways to minimize memory utilization and to improve the overall run time efficiency of their models.
- To maintain contact with newly published VHDL books, proceedings, articles, and techniques.
- To generate an atmosphere of cooperation and esprit de corps among the VHDL team members.
  - This responsibility is the most important one and emphasizes the fundamental fact that in addition to being highly knowledgeable of VHDL, the in-house guru(s) must also be a team player and a people person. Hardware designs are getting so complex that it is no longer possible for only one engineer to do it all. Teams and concurrent engineering methodologies are now a must for a company's survival. Teams require team players who are successful communicators. And always remember that, in the final analysis, VHDL is about communicating ideas and concepts.

Appendix

B

Chronology of a VHDL Modeling Project

1. Identify and train the VHDL designers and engineers.
2. Develop a VHDL style guide that everyone must adhere to.
3. Set up a subdirectory tree and naming convention for the various physical libraries and model subcomponents.
4. Decide on the data types to be used by the core model. (Note that the actual types might be contractually mandated.)
5. Design and code the entity declaration for the core model.
6. Conduct a preliminary design review.
7. Design and code the architecture body associated with the entity declaration completed in step 5. This architecture body may be a complex behavioral description or merely a structural decomposition consisting of only component declarations and instantiations.
8. If the behavior of this architecture is fairly complex, then write an intermediary test bench to verify the model's functionality.
9. Modify any code that needs to be corrected. Redesign and retest the model if necessary.
10. Conduct a preliminary code walk-through. If the architecture body is fairly simple, then step 6 may be done in conjunction with this step.
11. If the core architecture body is decomposed into components, then recursively apply steps 5 through 10 to the children entity/architecture pairs that constitute this core parent entity/architecture pair.

12. Design and test a suitable stimulus generator for your core VHDL model.
    - Based on the complexity of this core model the appropriate test vectors may be generated either internally or read from an input file.
13. Determine which signals should be monitored during a simulation run to best verify the correct behavior of your core model.
14. Design and test a design entity that will serve as a logic analyzer to monitor and record these preselected signals.
15. Encapsulate within a top level VHDL test bench the following components:
    - Core model.
    - Stimulus generator for the core model.
    - Logic analyzer to capture and store pertinent signals of the core model.
16. Use the VHDL configuration construct to hierarchically bind the test bench components.
17. Simulate this configuration to test your core VHDL model.
18. Conduct a critical design review for the overall VHDL model. Make the recommended modifications and retest the changes.
19. Write a user's manual to fully describe the
    - Required compilation dependencies.
    - Functionality of the core model and its constituent components.
    - Test bench philosophy. This information may be applied by the user to create additional test vectors and to analyze and interpret the data sensed and saved by the logic analyzer component.

Appendix

C

Code Walk-through Checkoff List

- Adherence to the companywide VHDL coding style guide.
- Consistency with the companywide VHDL design methodologies.
  - For instance, everyone's finite state machine should have the same look and feel.
- Commonality of parts.
  - Common devices should be modeled only once and then analyzed or compiled into a predefined library.
  - Repeated code segments should be placed into a subprogram.
- Intent or behavior of the VHDL model may be understood without the aid of any preexisting schematic diagrams.
- Available system designer's timing diagrams are supported by the VHDL model.
- Code is scalable.
  - New code may be easily and surgically incorporated into currently existing VHDL model.
  - Unwanted code may be easily and surgically removed from currently existing VHDL model.
- Each primary and secondary design unit is introduced by a documentation header outlining its purpose, author, and modification history.
  - All such headers should concur with the standard header template as specified in the companywide VHDL style guide.

- Referenced libraries and packages are all used.
- Port names reflect schematic pin and signal names.
- Each process and subprogram is preceded by a brief description.
- Source code is liberally supplemented with VALID comments that assist in the reader's understanding of the code's intent.
- Device is modeled at the appropriate level of complexity and datatype granularity, as defined by the requirements and scope of the assigned VHDL modeling task.
- Values are not inadvertently recomputed.
  - Recomputed computations may be repositioned to a common area so that the same result is computed only once. Additionally, depending on the specific coding scenario, recomputed computations may alternatively be stored in temporary variables or constants that may be repeatedly referenced.
- There are no infinite loops during initialization.
- There are no ill effects during the initialization phase when the simulation engine executes each explicit or implicit equivalent process until the first wait statement is reached.
- Signals and variables are initialized appropriately.
- Variables are reinitialized whenever its host process is using it to compute an algorithm.
- Out and inout ports are initialized to a (neutral) value that will prevent the circuit from oscillating indefinitely during initialization.
- Inout ports are really multisourced and not just a convenient workaround so that an intended out directional port may be read as well as written to.
  - If an out port must be accessible for both reading and writing, then intermediate variables or the VHDL'93 attribute 'DRIVING_VALUE should be used.
- Buffer ports are not used since they have many port-mapping restrictions that make them very impractical in a multilevel hierarchical design.
- Active polarity of device's ports and internal signals is honored by the VHDL model.
- Name of active low signals should be easily identified by the unique postfix letter that is specified in the companywide VHDL style guide.
  - This postfix letter should agree with the nomenclature assigned to the active low signal by the digital designer.

- For run time optimization, variables are used instead of signals whenever possible.
- There are no ill effects from delta delay signal updates.
- Sourcing signals have appropriate old or new value.
- Sourcing variables have appropriate old or new value.
  - This is especially important when using variables to model combinational circuits.
- Subprograms should not be written that duplicate the functionality of subprograms already existing and archived in an available library package.
- Synchronous or asynchronous behavior is correctly handled.
- VHDL optimization techniques, as specified in this book, are appropriately applied.
- Delta delay times are not counted in the VHDL model.
  - Consecutive (wait for 0 ns;) statements is an indication that the VHDL user is counting the number of delta delay time advancements. This approach results in nonscalable code, and hence, the model's overall design should be reevaluated and modified.
- State machine models are simple and to the point.
- When waiting for events on several signals, determination is always made as to which of these signals triggered the process reactivation.
- Waiting for an event on the attribute, 'TRANSACTION, is not erroneously triggered by signal updates scheduled during the VHDL initialization phase.
- There are no ill side effects from the methodology used to implement concurrent signal (time) multiplexing.
- An appropriate test bench has been designed to validate the VHDL model.
  - The complexity and the data-type granularity of the test bench is a function of the device under test.

Appendix D

VHDL'87 Reserved Words

abs	downto	linkage
access		loop
after	else	
alias	elsif	map
all	end	mod
and	entity	
architecture	exit	nand
array		new
assert	file	next
attribute	for	nor
	function	not
begin		null
block	generate	
body	generic	of
buffer	guarded	on
bus		open
	if	or
case	in	others
component	inout	out
configuration	is	
constant		package
	label	port
disconnect	library	procedure

373

process	signal	use
range	subtype	
record		variable
register	then	
rem	to	wait
report	transport	when
return	type	while
		with
select	units	
severity	until	xor

Appendix E

VHDL'93 Reserved Words

abs	elsif	loop
access	end	
after	entity	map
alias	exit	mod
al		
and	file	nand
architecture	for	new
array	function	next
assert		nor
attribute	generate	not
	generic	null
begin	group	
block	guarded	of
body		on
buffer	if	open
bus	impure	or
	in	others
case	inertial	out
componen	inout	
configuration	is	package
constant		port
	label	postponed
disconnect	library	procedure
downto	linkage	process
else	literal	pure

range	shared	units
record	sla	until
register	sll	use
reject	sra	
rem	srl	variable
report	subtype	
return		wait
rol	then	when
ror	to	while
	transport	
select	type	xnor
severity		xor
signal	unaffected	

References

1. *IEEE Standard VHDL Language Reference Manual,* Std 1076-1987, IEEE, New York, 1988.
2. *IEEE Standard VHDL Language Reference Manual,* Std 1076-1993, IEEE, New York, 1994.

Index

'ACTIVE, 190–191, 194, 218–219
'ASCENDING, 187
'DELAYED, 189, 191–193, 201–202, 340
'DRIVING_VALUE, 99, 190, 200
'EVENT, 20, 124–126, 190–191, 194, 198, 218, 235, 285
'HIGH, 182, 187
'IMAGE, 182, 184–185, 267
'LAST_ACTIVE, 190, 192, 194
'LAST_EVENT, 190, 192, 194, 197, 235
'LAST_VALUE, 190, 192, 196, 235, 359
'LEFT, 65, 141, 181, 183, 186
'LENGTH, 151, 187
'LOW, 182, 187
'POS, 183, 263–264
'PRED, 184
'QUIET, 189, 191, 194, 349–351
'RANGE, 149, 187
'REVERSE_RANGE, 187
'RIGHT, 182–183, 187
'STABLE, 90–91, 189, 191, 193–194, 201–202, 205, 285
'SUCC, 184
'TRANSACTION, 79, 190, 192, 196, 200, 287, 351
'VAL, 183, 263–264
'VALUE, 185

activity (*see* transaction)
after, 17, 38
aggregate:
    caveats of, 221–223
    examples of, 54, 62
alias, 150–151, 188, 237, 294–297
analogy:
    circuit board design, 112
    road map, 147–148
    schematic entry, 50–51, 56
and, 84
architecture, 5
architecture body, 5, 13, 29, 48

array:
    attributes of, 187
    caveats of, 225–226, 232–233
    example of, 53
    index of, 53
assert statement:
    behavior of, 151
    caveats of, 228–229, 261

base notation, 224
binding of models, 44–46, 110–111, 118–119
BIT, 14, 65
BIT_VECTOR, 52–53
block:
    caveat of, 285
    example of, 89
BOOLEAN, 20
buffer, disadvantages of, 98–99, 211–212

case statement:
    caveats of, 209–210
    example of, 69–71
closely related types, 144, 156–157, 267–270
comment notation, 11
compilation dependencies, 29, 64, 110, 146
component declaration, 50–52, 57, 161
component instantiation, 54–57
concatenation:
    behavior of, 84, 237
    caveats of, 209–210, 216–217, 237, 330
concurrency:
    caveats of, 41–42
    modeling of, 15–27
concurrent assert statement:
    behavior of, 202
    caveats of, 238
concurrent signal assignment, 32–33
concurrent multiplexing, 348–363

Index

conditional concurrent assignment, behavior of, 236
configuration declaration, 45, 109–113, 118–119, 131–136
conformance rules, caveats of, 214
constant:
　declaration of, 100
　efficient usage of, 251–252
current working library, 30, 109–110, 116, 121, 146

deadlock, 283
default binding, 44, 109
default value:
　'LEFT option for, 65–66
　assigned to port, 153
　of attributes, 191–192
　notation for, 14
　overriding of, 162, 310
delta time, 34–40, 87
design entity, 3, 58
discrete type, 69
driver, 176, 318–324
dynamic sensitivity list, 30

elaboration phase, 55, 96, 100, 115, 119, 162, 319, 321, 341–342, 348
entity, 5
entity declaration, 3, 11–16, 29, 58–59, 114–115
enumeration literal:
　caveats of, 141, 219
　example of, 67
　position of (see 'POS)
enumeration type:
　attributes of, 183
　user-defined, 67, 140–141
event, 18–19
exit, 64
extended name, 328–329

FILE_OPEN_KIND, 105
for statement, 63, 183
function, 105, 116–117, 291, 308, 341–343

generic map, 55, 95, 118, 130
generic parameter:
　behavior as constant, 97

generic parameter (*Cont.*):
　declaration, 51
　default value, 95–96
　for entity parameterization, 114–115
　overriding default value, 95, 333–335
global variables, nondeterministic behavior of, 40–42
GUARD (*see* guarded assignment)
guarded assignment, 90–93, 285

identifier, 14, 219
impure functions, 343
inertial, 175, 178–180, 287–289, 292–293
infinite oscillation, 282, 309, 313
initialization phase:
　activities of, 16, 21, 31
　caveats of, 124, 234, 239
　and concurrent signal assignments, 33
　and events, 19
inout, artificial usage of, 97, 346
INTEGER, 181, 250

label, 15, 54, 89
library clause, 116, 121
loop iterator, 63

min-max-typical, modeling of, 205
mode, 52

NATURAL, 53, 182
not, 17
now, 107

object classes, 14
old value vs. new value, 44, 80
others:
　caveats of, 221, 344
　usage of, 54, 153, 345
out port, inability to be read, 85, 97, 211–213

package, 104, 116, 138–140, 161, 229, 328–329
parameterized entity, 114–115
parentheses, requirement of, 226
passive construct, 239

Index

port, 3, 13, 52–53, 57, 65, 153, 336–338
port map:
   caveats of, 211, 217–218, 325
   examples of, 55, 118, 128, 130
postponed process, 240–241
procedure:
   declared within a process, 298, 300
   purpose of, 105
process:
   concurrency of, 15–16
   infinite looping of, 16, 106–107
   sequentiality of, 16
   with static sensitivity list, 30
   wait statement requirement, 20
   wait statement restriction, 31
projected driver, updating of, 177–180, 286–293

record, 223
reject, 180
report statement, 262
reset, modeling of, 198–199
resolution function, 142, 156, 307
resource library, 116, 122

sensitivity list, 18, 30
sequential signal assignment, 17, 32
setup and hold, 202–205
shift operators, 241–242
short circuit, 238, 260
signal:
   architectural visibility, 15, 25
   declaration of, 13
   initialization of, 14, 27
   port is, 65
signal assignment:
   concurrent, 32–33, 88
   conditional concurrent, 122
   deterministic nature of, 42–43
   done in background, 21
   multiple sources for, 138, 319–324, 338, 348–363
   notation for, 17
   scheduling of, 17, 35
   sequential, 17, 32, 87–88
   variable as assignment source, 65
simulation cycle:
   activities of, 22, 280
   and delta time, 36
simulation time, 35

slice direction, caveats of, 224, 238, 308
STANDARD, 104
state machine, modeling of, 67–81
static sensitivity list, 30
STD, 104, 159
std_logic_1164, 140, 165–166, 271–272
STRING, 104
subprogram:
   achieving elasticity of, 149, 155
   caveats of, 214–217, 227–228, 237, 294–297
   default class, 215
   default mode, 215
   overloading of, 106, 143, 159, 215–216
   purpose of, 105
subtype, inefficiency of, 283–284

test bench, 49
TEXTIO:
   applications of, 104, 184–185, 305–306
   caveats of, 210, 243–245, 332, 342
   limitations of, 157, 184
TIME, 52
transaction, 79, 307
transport, 175–178, 289–291
type mark:
   caveats of, 226
   definition of, 65
type qualification, 106, 210–211, 345

unaffected, 259
unconstrained array:
   caveats of, 220
   examples of, 53, 141–142, 186
use clause, 104, 116

variable:
   declaration of, 25
   initialization of, 26
   process visibility of, 25
variable assignment:
   importance of assignment sequence, 84
   notation, 26
   occurrence immediately, 26, 82–85
visibility:
   caveats of, 211, 327–329
   examples of, 64, 153–155

wait, 277
wait for, 34, 36, 81
wait on:
  caveats of, 231, 279
  examples of, 17, 30, 79
wait statement:
  implicit existence, 30, 76
  locational importance of, 27, 71
  process requirement of, 20

wait until:
  behavior of, 19
  caveats of, 231–232, 234, 277–278
waveform generation:
  caveat of, 242–243
  format of, 100, 192
WORK, 30, 111, 121, 162

xor, 84

ABOUT THE AUTHOR

Joseph Pick is a Staff Trainer at Synopsys Inc. and an instructor specializing in computer architectures and simulation languages at The Johns Hopkins University. He has presented numerous VHDL-related papers and tutorials at conferences throughout North America, Europe, and Asia. Since joining Synopsys Mr. Pick has applied his VHDL expertise to explore the interplay between VHDL coding styles and synthesis.